25 AUG 1970

£8 0 0

D0584977

Department of Agriculture & Fisheries
LIBRARY
56796
DUBLIN.

59.9 (42)

GROWTH AND DEVELOPMENT OF MAMMALS

GROWTH AND DEVELOPMENT OF MAMMALS

Proceedings of the Fourteenth Easter School in Agricultural Science, University of Nottingham, 1967

Edited by

G. A. Lodge
Reader in Animal Production

G. E. Lamming
Professor of Animal Physiology

University of Nottingham School of Agriculture
Sutton Bonnington, Loughborough

LONDON

BUTTERWORTHS

ENGLAND: BUTTERWORTH & CO. (PUBLISHERS) LTD.
LONDON: 88 Kingsway, W.C.2

AUSTRALIA: BUTTERWORTH & CO. (AUSTRALIA) LTD.
SYDNEY: 20 Loftus Street
MELBOURNE: 343 Little Collins Street
BRISBANE: 240 Queen Street

CANADA: BUTTERWORTH & CO. (CANADA) LTD.
TORONTO: 14 Curity Avenue, 16

NEW ZEALAND: BUTTERWORTH & CO. (NEW ZEALAND) LTD.
WELLINGTON: 49/51 Ballance Street
AUCKLAND: 35 High Street

SOUTH AFRICA: BUTTERWORTH & CO. (SOUTH AFRICA) LTD.
DURBAN: 33/35 Beach Grove

©

The several contributors named on pp. v to vii
1968

Suggested U.D.C. *No.* 591.134: 636.08

Made and printed in Great Britain by
William Clowes and Sons, Limited, London and Beccles

CONTENTS

a* v

CONTENTS

CONTENTS

VIII. PRACTICAL IMPLICATIONS OF FACTORS AFFECTING GROWTH

PREFACE

AN understanding of the factors which govern the growth and development of animals is fundamental to any attempt to influence the productivity of farm livestock, not only in meat characteristics but also in reproductive function and lactation. Arbitrary attempts to influence animal performance which ignore the underlying physiological principles involved have only a limited chance of success and allow no scope for deviation from the confines of the original work. As it becomes increasingly important to improve the efficiency of livestock production and to direct tissue development further away from what might be regarded as its 'natural' form, it also becomes important to establish the physiological limitations to growth and to the development of the body tissues.

In view of the great importance of animal tissues as a source of human food it is remarkable that relatively little is known about their development in the economically important species. Since the classic Cambridge work initiated by the late Sir John Hammond there has been a lack of concerted effort in this field. The object of this Symposium was to bring together many of those workers interested in the subject of animal growth and to review the state of existing knowledge, drawing where appropriate from research in the medical field where more advance has been made in recent years than with farm livestock.

The Symposium had the additional objective of allowing interchange of ideas between workers in the fundamental sciences related to growth and those whose interests lie mainly in its application to the control of growth in farm species. It was considered that each group might have new information of use to the other. That this objective was justified is indicated by the wealth of information presented in the papers and by the ideas developed in the discussions which followed them.

G.A.L.
G.E.L.

ACKNOWLEDGEMENTS

THE success of the Fourteenth Easter School is attributed to all those who presented papers and participated in the ensuing discussion. We are indebted to Professor Dainton, Vice-Chancellor of Nottingham University, who opened the Symposium, and to those who kindly acted as chairmen of the Sessions: Dr. F. C. Greenwood of the Imperial Cancer Research Fund, Professor F. W. Rogers Brambell of University College of North Wales, Dr. H. Palsson of the Agricultural Society of Iceland, Professor W. E. Howell of the University of Saskatchewan, Dr. J. C. Shaw of Mercke, Sharpe and Dohme (Europe) Inc., Dr. M. Ingram of the A.R.C. Meat Research Institute and Professor M. M. Cooper of the University of Newcastle upon Tyne.

We are particularly indebted to the following organizations for financial contributions towards the expenses of speakers from overseas:

> Australian Meat Research Committee.
> British Oil and Cake Mills Ltd.
> Colborn Vitafeeds Ltd.
> Eastern Counties Farmers Ltd.
> Fisons Pharmaceuticals Ltd.
> Imperial Chemical Industries Ltd.
> Levers Feeds Ltd.
> Pauls Foods Ltd.
> R. Silcock and Sons Ltd.

We thank the staff of the School for their assistance with the Symposium.

I. TISSUE GROWTH

GROWTH OF BONE, MUSCLE AND FAT DURING CHILDHOOD AND ADOLESCENCE

J. M. TANNER

Institute of Child Health, University of London

INFORMATION about the amount and location of bone, muscle and fat in the living body can be obtained by anthropometric, radiographic or physiological methods. The last includes underwater weighing to determine specific gravity and hence calculate the percentage of fat in the body; estimation of the amount of potassium in the body by measuring the radiation produced by the naturally occurring isotope ^{40}K, and hence calculating the amount of muscle; and a variety of isotopic or chemical dilution methods for measuring body water, fat and other constituents.

Physiological and chemical dilution methods give better estimates of the total amount of fat and perhaps of muscle than anthropometric and radiological methods, but they cannot indicate *whereabouts* in the body the muscle or fat is abundant and where little. This the radiographic technique, and to a much less extent, anthropometry, can do. The best use of these two approaches is therefore a complementary one, the physiological methods estimating the 'size', in the formal sense, of body tissue components, and the radiographic method measuring 'shape'.

The *physiological methods* all involve considerable assumptions. Some of these, for example, that the percentage composition of the non-fat part of the body is constant in all persons, are known to be untrue. The methods are nevertheless useful in at least indicating approximate values. A particular difficulty in studying the growth of body tissues by these methods is that the assumptions made for adults are very probably erroneous when applied to children. It is by no means certain, for example, that the equations relating the percentage of body fat to specific gravity in adult men apply even to adolescent boys, let alone pre-pubertal children. Up to the present these methods have given less information about the growth of tissues than the anthropometric and radiographic techniques, though there are a few useful reports (*see* Parizkova, 1961 and Heald,

3

Hunt, Schwartz, Cook, Elliot and Vajdan, 1963 for specific gravity; Forbes, 1964 for ^{40}K and Friis-Hansen, 1965 for body water; also reviews by Brozek, 1965 and 1966).

The most useful of the *anthropometric methods* is the measurement of subcutaneous fat by the skinfold technique. At a few places in the body the skin and the fat under it can be pulled up away from the underlying muscle and pinched into a double fold whose thickness is measured with a special constant-pressure caliper (Tanner and Whitehouse, 1955; Tanner, 1962; and for illustrations *see* Tanner, 1964). The best sites are over the triceps muscle and under the angle of the scapula but others are usually possible, including over the biceps, above the patella, and a variety of places in the thorax and abdomen. The other anthropometric technique of relevance is the measurement of circumferences, but these are of limited usefulness in that even in the limbs they include all three tissues. The thickness of skinfolds can be subtracted, for example in the upper arm, but one is still left with an inseparable mixture of bone and muscle. However, circumferences have some usefulness for fieldwork on large numbers of children, as in nutritional surveys, under circumstances where radiographs cannot be obtained. The *ultrasonic* technique may perhaps be included under anthropometric methods; present equipment leaves a great deal to be desired but can be used to measure thickness of fat at certain sites. Its extension to measure also muscle and bone has not yet been successful, at least in our hands.

SKINFOLD GROWTH CURVES

Figures 1 and *2* show the growth curves for the triceps and subscapular skinfolds in girls and boys. *Figure 1* represents the amount of the skinfold at each age; *Figure 2* its change from one age to another. The caliper measurements at each age are distributed in a non-Gaussian fashion but can be made much more nearly Normal by a log transformation (Edwards, Hammond, Healy, Tanner and Whitehouse, 1955).

Subcutaneous fat increases from birth up to a peak reached, on average, at 9 months. It then decreases (hence has a negative velocity in *Figure 2*) until around 6–8 years, when it begins to increase again slowly. Girls have a little more fat than boys at birth and the difference becomes more marked during the period of loss, which is quantitatively greater in boys. From 8 years onwards,

4

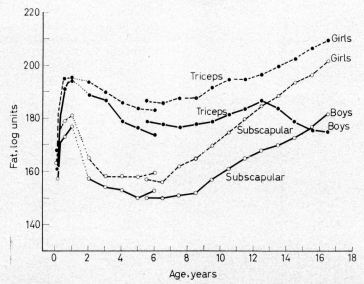

Figure 1. Distance curve of subcutaneous tissue measured by Harpenden skinfold calipers over triceps and under scapula. Logarithmic transformation units. Data 0–1 year, pure longitudinal, 74 boys and 65 girls (Brussels Child Study, Graffar, Asiel and Corbier, unpub.); 2–6 London Child Study Centre (Tanner, unpub.) with pure longitudinal core 4–6 of 59 boys and 57 girls and actual mean increments subtracted or added to get means at 2 and 3; 5–16 London County Council (Scott, 1961) cross-sectional, 1,000 to 1,600 of each sex at each year of age from 5 to 14; 500 at 15; 250 at 16 (From Tanner, 1962).

By courtesy of Blackwell Scientific Publications

however, the curves for girls and boys diverge more radically. Girls have a fairly continuous gain in subscapular skinfold throughout adolescence; in triceps skinfold the gain ceases for a year or so during the maximal velocity of growth in height and then resumes. In boys, on the other hand, an actual fat loss occurs in the triceps fold, coincident with maximal height velocity, and a considerably greater slowing up in gain in the subscapular fold. The events in the limbs are made much clearer by the radiographic data given below; external trunk fat clearly is less influenced by androgenic hormones than the fat on the limbs. It is usually assumed that the intra-abdominal fat follows the same growth curve as the extra-abdominal but at present we have no direct evidence for this.

Radiographic methods have been most useful in demonstrating how bone, muscle and fat develop during childhood and the remainder of this paper will be concerned with them. Most of the data quoted are from the Harpenden Growth Study, a longitudinal study of

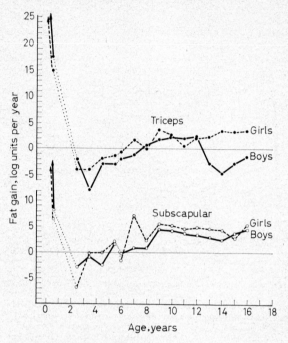

Figure 2. Velocity curve of subcutaneous tissue as measured by Harpenden Skinfold Calipers over triceps and under angle of scapula. Data as for Figure 1, but with gains prior to 1 year smoothed so that first point plotted represents average gain from 3 months to 1 year. Gain from 1 to 3 months is off graph (From Tanner, 1962)

By courtesy of Blackwell Scientific Publications

children now in its eighteenth year (*see* Tanner, 1962). A history of the method and a review of previous work has been given by Tanner (1965), together with a full description of the technique and a number of the results described here. More recent results embody the work of many colleagues and will be published in full shortly (Tanner, Goldstein, Hughes, Bell and Whitehouse, in press).

Radiographs of the upper arm, calf and thigh were taken, at an anode-film distance of 2·5 m so as to provide as nearly as possible a parallel beam of x-rays hitting the limb (for illustrations *see* Tanner, 1962). This makes the degree of magnification of the limb outline in the film relatively small, and, more importantly, it minimizes the errors introduced by not posing the limb with the required diameter (calf maximum for example) exactly in a plane parallel to the film.

The left upper arm is positioned laterally so that the two epicondyles appear superimposed on the film, the left calf is positioned with the foot pointing directly forward, weight on both feet evenly, and the thigh radiograph is taken with the subject straddling the casette, which is placed as high into the pubes as possible, with the position lateral so that the two femoral condyles are superimposed. The gonads are protected by the specially designed armadillo lead jock-strap or apron which virtually eliminates all exposure to them (Tanner, Whitehouse and Powell, 1958). The x-rays are coned down exactly to film size. The total skin dose for the three pro-cedures amounts to about 66 mr, which is half the extra annual dose received from natural sources by a child living at 1,000 ft rather than at sea-level; or 1·3 per cent of the dose allowed to a radiographer per year.

Numerous measurements may be made on these radiographs, but as a routine we measure only one diameter on each, using a specially designed needle-point caliper accurate to 0·1 mm over a range of 1,500 mm. In the arm a mark is placed half-way between the tip of the acromion and the head of the radius before the radiograph is taken. A line is drawn on the radiograph passing down the long axis of the arm, that is as nearly parallel to the two skin borders as possible, and a line perpendicular to this is drawn at the marked level. The widths of anterior and posterior subcutaneous tissue and total arm are measured along this line, and width of humerus is measured, cortex and medulla separately, at this level but perpen-dicular to the long axis of the humerus. Muscle width is obtained by subtraction.

The calf is measured at the maximal overall width. We measure total width, lateral and medial fat, tibia and fibula and obtain muscle by subtraction.

The level of the thigh diameter is harder to fix; we have chosen a point one-third of the leg length (taken as stature less sitting height) up from the lower border of the femoral condyles. This coincides approximately with the maximal width of thigh muscle in the majority of people.

The reliability of these measurements is good; during routine and rapid work, with no special effort made, the standard error of measurement was 1–2 per cent of the mean value for muscle and bone in calf and thigh, and 3–4 per cent for fat. In the arm all three components gave figures of 3–4 per cent.

In the great majority of studies only the widths of each tissue have been measured, not the areas. The distinction is important, at

7

least theoretically. Suppose, for example, that the fat width in the calf decreases during a certain period of growth, this does not necessarily imply a decrease in the cross-sectional area of fat at the maximal calf width. The fat is a ring around a muscular-skeletal core which increases continuously during childhood. If the cross-sectional area of fat remained constant over a year during which the muscle and bone core enlarged then the fat ring would inevitably become thinner. It is possible to make an estimate of the cross-sectional areas of bone, muscle and fat on the assumption that the limb is circular and has the bone at its centre, and that the fat rim is equally thick all round. At least in the case of the calf maximal diameter these assumptions are fairly closely met, for the regression of calf circumference on calf width is very nearly π. If b is the radius of the bone, m the width of the muscle ring and f the width of the fat ring then the radius of the limb is $b + m + f$ and the cross-sectional areas are

bone, πb^2,
muscle, $\pi(m^2 + 2bm)$, and
fat, $\pi(f^2 + 2fb + 2fm)$.

By using these formulae it can be seen whether the area of fat is decreasing when the fat diameters are decreasing (and, in fact, we find that during growth they are doing so; the 'ring' effect is not a quantitatively great one).

The volume of tissue in a limb is another matter and can only be approximated with difficulty except in the case of a bone of nearly uniform shape such as the humerus or femur. A decrease in horizontal cross-sectional area of fat does not necessarily mean that the absolute amount of fat in the limb is decreasing (though some of the decrements of cross-section during growth are so great that the volume must clearly be decreasing also).

Results

Figures 3, 4 and *5* show the widths of bone, muscle and fat from age $5\frac{1}{2}$ to 18 years in a sample of the children in the Harpenden Growth Study. By 'total limb bone width' is meant the sum of the widths of humerus, tibia and femur at the levels described above. 'Total limb muscle width' consists of the width of the arm muscle plus the width of calf muscle plus half the width of thigh muscle, the weighting being to reduce the absolute value for the thigh to a figure comparable with arm and calf. The total limb fat width has been

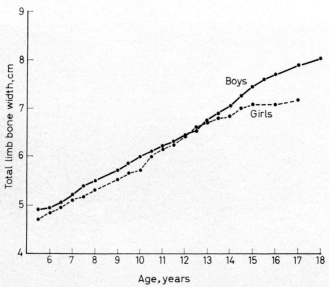

Figure 3. Total limb bone width (humerus plus tibia plus femur) in boys and girls, Harpenden Growth Study. For treatment of data see text

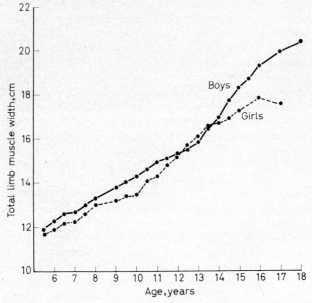

Figure 4. Total limb muscle width (upper arm plus calf plus half thigh) in boys and girls, Harpenden Growth Study. For treatment of data see text

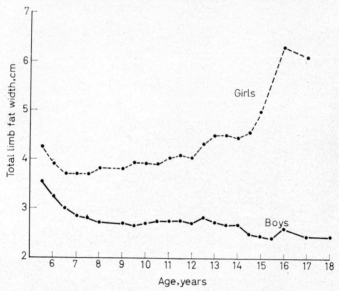

Figure 5. Total limb fat width (upper arm plus calf plus half thigh) in boys and girls, Harpenden Growth Study. For treatment of data see text

calculated similarly to muscle width, using the straight measurement of fat, not a logarithmic one.

The data is mixed longitudinal (*see* Tanner, 1951), that is children entered and left the study at different ages. But the percentage longitudinality from one year to the next is high (about 80) except before age 6 and after 16. The numbers of boys at each age range from 22 to 59, and of girls from 24 to 48. The graphs simply represent the means at each 6 months of age, calculated as though the data were cross-sectional. This is in many circumstances a very inefficient way of estimating the means in a mixed longitudinal series (*see* Tanner, 1951) but under our conditions of high longitudinality it is a close approximation to the best estimates, which we have recently computed and will publish in our full report.

There is a fairly steady rise in limb bone width from 6 years to adolescence, the gain in the combined total being about 2·5 mm/ year. Boys at 6 years have already slightly wider bones than girls, but velocity from 6 years to adolescence differs very little between the sexes. At adolescence both boys and girls have a spurt in bone width. The girls' adolescence being as usual about 2 years earlier than the boys', girls reach briefly a greater mean width than the

10

boys. The boys' spurt then begins and, being larger, carries the boys' mean to well above that of the girls, establishing the adult sex difference. This way of reporting results at adolescence is unsatisfactory, however, in that it does not allow for the large differences between children in the age at which the adolescent spurt occurs. Simply summing each individual's values at each chronological age, ignoring their 'phase differences', results in a mean curve which rises less steeply and for longer than does the curve of the average individual (Tanner, 1962; Tanner, Whitehouse and Takaishi, 1966). The situation at adolescence is better illustrated in the section below.

The mean curve for muscle width (*Figure 4*) resembles quite closely that for bone. The rate of growth is fairly steady from age 6 to adolescence; the boys already have more muscle than the girls at 6 years of age, but both sexes gain at about the same rate until age 10. The girls then have their spurt, which carries them to a higher value than the boys for a year or more. The boys' spurt then occurs and carries them still further beyond the girls than in the case of bone. The details of the adolescent spurt timings and velocities are again shown below.

The fat width curve (*Figure 5*) confirms, in general, the skinfold curve of *Figure 1*. The mean value falls in both boys and girls until about age 8 in the girls and 10 in the boys. It then rises, slowly at first. In the girls the mean value does not actually drop at adolescence (except at one half-yearly interval probably due to sampling errors) but the rise is barely perceptible from $9\frac{1}{2}$ to $12\frac{1}{2}$. After this a large increase takes place, probably magnified by sampling errors and the effects of using a straight scale in a situation where a logarithmic one would really be better (especially at older ages when some girls are getting really fat). The boys' mean curve drops from age $12\frac{1}{2}$ to 15, the drop being flattened and extended by the phase-difference effects; a better view of the adolescence changes appears below.

ADOLESCENCE

The details of tissue changes at adolescence can best be illustrated by curves derived from 28 boys and 21 girls out of the total above. Each individual was measured every 3 months. A smooth curve was drawn through the points representing the height measurement; this was done by eye but using an iterative procedure involving also the height velocity curve, as detailed elsewhere (Tanner, Whitehouse and Takaishi, 1966). From this the point of maximal or peak height velocity (PHV) was determined, and the age at which it occurred. The individuals curves for bone, muscle and fat were then re-aligned according to the point of PHV, after the manner of Boas (1892) and Shuttleworth (1937). Thus, chronological age is replaced by years before or after PHV.

The series covers the period from 3 years before PHV to 2 years after for boys and $2\frac{1}{2}$ years before to 2 years after for girls. The data is pure longitudinal in both sexes from 1 year before to 2 years after PHV. But from 3 years to 1 year before PHV only 18 of the 28 boys were present and from $2\frac{1}{2}$ years to 1 year before PHV only 14 of the 21 girls. The increments of these 18 and 14 have been subtracted from the mean values of the 28 and 21 to arrive at estimated means for the whole group at PHV -3, $-2\frac{1}{2}$, etc. No special selection

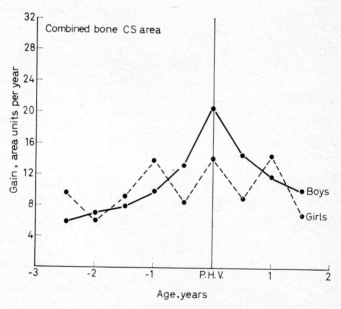

Figure 6. Mean velocity of combined bone cross-sectional area (calf, arm and thigh). Longitudinal data, individual curves aligned on peak height velocity (PHV) (From Tanner, 1965)

By courtesy of Pergamon Press

12

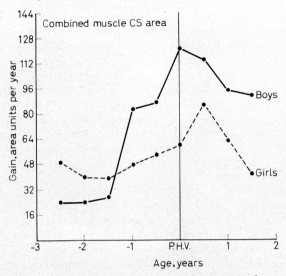

Figure 7. Mean velocity of combined muscle cross-sectional area (calf, arm and thigh). Longitudinal data, individual curves aligned on peak height velocity (PHV) (From Tanner, 1965)

By courtesy of Pergamon Press

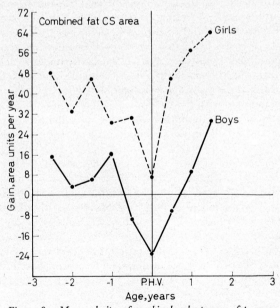

Figure 8. Mean velocity of combined subcutaneous fat cross-sectional area (calf, arm and thigh). Longitudinal data, individual curves aligned on peak height velocity (PHV) (From Tanner, 1965)

By courtesy of Pergamon Press

operated to distinguish the 18 boys who were present throughout the whole 5 years from those present only for 3 years. A number of radiographs were lacking, due to missed attendances or other factors. On these occasions the values were estimated by linear interpolation in the individual's curve.

Figures 6, 7 and *8* illustrate the data in terms of cross-sectional areas rather than the widths so far considered. The areas were calculated on the assumption that at the level measured, bone, muscle and fat constitute concentric circles and using the formulae given above. The areas for each limb were calculated and the three values summed, the values for thigh muscle and fat widths being halved as before.

The figures give the velocity curves rather than the distance, or amount-attained curves. The boys' bone cross-section (*Figure 6*), shows a marked peak coincident as one might expect with peak height velocity. In the same boys, shoulder width (biacromial diameter) and hip width (biiliac diameter) also showed a marked and coincident peak. The girls' bone cross-section on the other hand, shows a flatter less regular curve with less of a spurt. Their biacromial and biiliac diameters also showed the same flat velocity curve in these data, when aligned either on PHV or on menarche. However, the flatness in the limb bone curve is due partly to pooling the three limb areas. A distinct peak can be seen in girls' calf bone, coincident with PHV, and a distinct peak in girls' arm bone, just after PHV. It seems likely that the thigh measurements, always the most difficult, were not always located at exactly the same level and probably introduced considerable measurement error.

Both boys' and girls' muscle cross-sections (*Figure 7*) show a peak, the boys' coincident with PHV (or a trifle after, if smoothed) and the girls' slightly after. The male peak velocity in the arbitrary units used is one and a half times the female peak velocity, as one might expect.

In *Figure 8* the fat cross-sectional areas are shown. There is an absolute loss of fat (negative velocity) in the boys but not in the girls. The individual limb areas all show a loss in the boys; in the girls the upper arm shows a negative velocity coincident with PHV. Calf and thigh in the girls drop to a trough at PHV also, but the trough remains above zero. Both sexes therefore show a considerable deceleration at PHV. The corresponding fat *widths* for girls show very similarly shaped curves, but do reach a mean negative velocity at PHV for calf fat and a zero velocity for the thigh. The fat cross-sectional area mean for calf also shows a negative velocity.

14

CORTEX AND MEDULLARY CAVITY OF HUMERUS

Figures 9 and *10* show that the adolescent spurt in bone width is wholly or mostly attributable to a thickening of the cortical bone, with little if any widening of the medullary cavity, at least in the humerus. Here the alignment is made again on PHV but the curves are of distance not velocity. The data is similar but not identical to that for the whole bone in the section above. The boys' mean curve seems to show a clear-cut spurt in cortex starting a year before PHV and beginning to level off later, about 2 years after PHV. This timing would agree well with the muscle curves and the curves for strength increment, which latter probably have their peak 1 or even 2 years after PHV (*see* Tanner, 1962). The medulla increases very slowly in width and shows no particular change at adolescence.

The girls' curves (*Figure 10*) are harder to interpret. The spurt in cortical thickness is present, with about the same timings. But the medulla shows a curious increase from $2\frac{1}{2}$ years to 1 year before PHV, which has no counterpart in boys. This may be only a

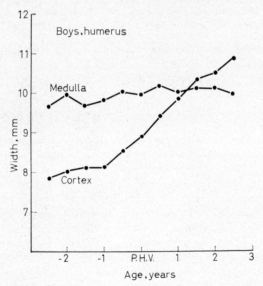

Figure 9. Width of cortex and medullary cavity of humerus, boys, Harpenden Growth Study. Mixed longitudinal data, aligned on peak height velocity

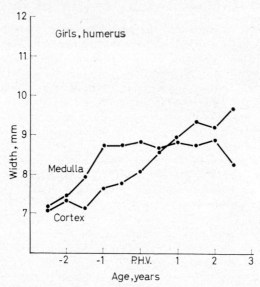

Figure 10. Width of cortex and medullary cavity of humerus, girls, Harpenden Growth Study. Mixed longitudinal data, aligned on peak height velocity

sampling effect; it would seem wise not to consider too seriously whether a repeatable sex difference exists until more extensive data are available.

INTERRELATIONSHIPS OF BONE, MUSCLE AND FAT

It is perhaps surprising that the widths of bone and muscle and fat in the limbs are nearly mutually independent, both in children and adults (Reynolds, 1944; Hewitt, 1958; Tanner, 1965). Fat measurements in arm, calf and thigh correlate with each other at all ages rather highly; in young adults the intercorrelations are usually between 0·7 and 0·8. The three muscle widths have lower but very significant intercorrelations, in adults of the order of 0·45, and the three bone widths intercorrelate to nearly the same extent; the between-tissue coefficients, however, are nearly all below 0·1. At least after puberty, the bone cortical thickness correlates a little more highly with the width of the muscle in the same region, but the association is still not a close one. Widths of bone cortex and medulla are quite independent in adults.

16

SUMMARY

Information on the amount and location of bone, muscle and fat may be obtained by anthropometric, radiographic or various physiological methods. The advantages and disadvantages of each are discussed.

The changes from birth to maturity in subcutaneous fat as measured by skinfold calipers are illustrated; boys have on average an absolute loss of limb fat during adolescence while girls only slow up their fat gain.

The radiographic techniques for measuring bone, muscle and fat widths in the upper arm, calf and thigh are given and the growth of each tissue from age $5\frac{1}{2}$ to 18 years is illustrated in data from the Harpenden Growth Study. At adolescence there is a spurt in growth of muscle and bone widths and this is illustrated by longitudinal data in which individuals' curves are aligned on peak height velocity. The peak of muscle growth velocity is reached in both sexes shortly after the peak height velocity. The decrease in rate of gain of fat widths is also shown, amounting to an absolute loss in boys but not in girls.

Curves for the growth in width of humerus, cortex and medullary cavity are given, with individuals aligned on peak height velocity. It appears that the adolescent spurt in bone breadth is due solely to increase in cortical thickness, the medulla remaining practically unchanged.

ACKNOWLEDGEMENTS

I am much indebted to my colleagues R. H. Whitehouse, P. C. R. Hughes, W. Bell, H. Goldstein and Margaret Manning in collecting and analysing much of the data reported here. I wish also to thank the Medical Research Council and the Nuffield Foundation for grants which have each supported portions of this work.

REFERENCES

Boas, F. (1892). *Science* **19**, 256; **19**, 281; **20**, 351
Brozek, J. (1965). *Biotypologie* **26**, 98
— (1966). *J. Indian anthrop. Soc.* **1**, 27
Edwards, D. A. W., Hammond, W. H., Healy, M. J. R., Tanner, J. M. and Whitehouse, R. H. (1955). *Br. J. Nutr.* **9**, 133
Forbes, G. B. (1964). *J. Pediat.* **64**, 822
Friis-Hansen, B. (1965). In: *Body Composition. Symp. Soc. Study hum. Biol.* **7**, 191

Heald, F. P., Hunt, E. E., Schwartz, R., Cook, C. D., Elliot, O. and Vajdan, B. (1963). *Pediatrics, Springfield* **31**, 226

Hewitt, D. (1958). *Ann. hum. Genet.* **22**, 213

Parizkova, J. (1961). *J. appl. Physiol.* **16**, 173

Reynolds, E. L. (1944). *Child Dev.* **15**, 181

Scott, J. A. (1961). 'Report on heights and weights (and other measurements) of school pupils in the county of London in 1959.' London; County Council

Shuttleworth, F. K. (1937). *Monogr. Soc. Res. Child Development* **2**, No. 5, 253 p.

Tanner, J. M. (1951). *Hum. Biol.* **23**, 93

— (1962). *Growth at Adolescence.* 2nd edn. p. 325. Oxford; Blackwell Sci. Publ.

— (1964). *The Physique of the Olympic Athlete.* London; Allen and Unwin

— (1965). In: *Body Composition, Symp. Soc. Study hum. Biol.* **7**, 211

— Goldstein, H., Hughes, P. C. R., Bell, W. and Whitehouse, R. H. Unpublished

— and Whitehouse, R. H. (1955). *Am. J. phys. Anthrop.* **13**, 743

— — and Powell, J. H. (1958). *Lancet* **2**, 779

— — and Takaishi, M. (1966). *Archs. Dis. Childh.* **41**, 454 and 613

BONE GROWTH AND REMODELLING

H. A. SISSONS

Institute of Orthopaedics, Royal National Orthopaedic Hospital, London

FOR one who is chiefly concerned with disease processes in man, the present symposium is a welcome opportunity to review information on some of the cellular and morphological aspects of bone growth and remodelling. The programme of the present meeting indicates the great extent to which studies by agricultural research workers are contributing information and ideas on basic aspects of growth; it is very desirable to develop and maintain contact between work in the medical and agricultural fields.

TYPES OF BONE GROWTH

The growth processes with which we are concerned in the skeleton are usually considered to be of three types:

(1) *Endochondral ossification*, responsible for the growth of bones in length but also seen in pathological conditions such as fracture repair.

(2) *Membranous ossification*, responsible for periosteal bone growth and for bone growth in other situations where the bone tissue is not preceded by cartilage.

(3) *Bone remodelling*, a process of structural change which involves the adult, as well as the developing, skeleton.

Each of these processes involves the formation of new bone tissue by osteoblasts and this occurs in a basically similar fashion. In endochondral ossification the bone is deposited on a scaffolding of calcified cartilage, while in the other types the new bone is deposited on the surfaces of pre-existing bony structures. The anatomical arrangement of bone tissue, and of its constituent collagen fibres, varies widely in different parts of the skeleton, at different ages, and in different animal species (*see* Enlow, 1966).

In connection with each type of growth process, it is important to realize that calcified bone is a rigid material; it does not show the interstitial expansion which occurs with other types of tissue, and increase in size is brought about only by the addition of new tissue to existing surfaces. In gross terms, this has been known since the eighteenth century, when the early experimental work of Hales, of Duhamel, and of John Hunter (*see* Sissons, 1956), using bone markers, showed that the increase in length of growing bones was due to the addition of new tissue at the ends and not to interstitial growth of the existing shaft. More recently, similar information on the sites of bone growth, and on the rate at which new tissue is formed, has been obtained by radiological measurement following the insertion of metal markers (Haas, 1926; Bisgard and Bisgard, 1935; Sissons, 1953) and by means of a variety of vital bone markers, including alizarin (Brash, 1934), radioactive isotopes (Lacroix, 1960), and tetracyclines (Hulth and Olerud, 1962; Hansson, 1967).

ENDOCHONDRAL OSSIFICATION

The general histological features are well known (*see* Sissons, 1956). The cartilage cells are arranged longitudinally in columns in the growing epiphysial cartilage plate. Cells are continually being added, by mitosis, towards the epiphysial surface of the plate; after growth and maturation of the cells, the 'hypertrophic' cartilage is invaded by blood vessels and connective tissue cells from the metaphysial surface. In contrast to calcified bone, the uncalcified cartilage of the plate grows by interstitial expansion as the cartilage cells multiply and enlarge; vascularization and bony replacement of the cartilage keep pace with its longitudinal growth. In normal endochondral ossification, the 'hypertrophic' cartilage cells do not persist but disintegrate as calcification of the surrounding matrix, and subsequent vascular invasion of the cartilage columns occurs. Studies with ^{3}H thymidine have directly demonstrated the passage through the epiphysial plate of cartilage cells labelled at the time of mitosis (Kember, 1960). Similar investigations with regard to the bone-forming tissue which invades the metaphysial surface of the plate (Young, 1962) have established that the actively proliferating cells are not differentiated osteoblasts, but a morphologically unspecialized 'osteoprogenitor' cell from which the osteoblasts and osteoclasts are both derived.

Under conditions of normal growth, there is close linear relationship between the rate of growth of an epiphysial cartilage plate and its thickness (Fahmy, 1956); this is illustrated in *Figures 1* and *2*. The rate of growth of the plate clearly depends on the rate of formation of cartilage cells in the columns, and on the final size attained

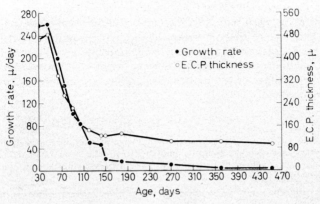

Figure 1. The growth rate and thickness of the upper tibial epiphysial cartilage plate in the optimally-growing male rat. (There is a high degree of correlation between these two variables, significant at the 1:1,000 level.) (From Fahmy, 1956)

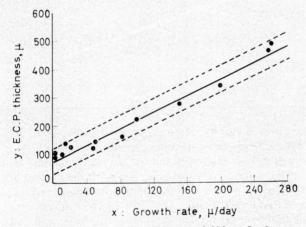

Figure 2. Pairs of values for growth rate and thickness for the upper tibial epiphysial cartilage plate in optimally-growing male rats. The regression line corresponds to the equation:

$$y = 1.415x + 82.9(\pm 23.6)$$

Control limits are drawn at twice the stated standard error, and outline the extent of chance deviation at the 1:20 level (From Fahmy, 1956)

by each of these cells at the metaphysial surface of the plate. The thickness of the plate is also influenced by the time that elapses during the growth and maturation of the cartilage cells, as they

21

progress from one side of the plate to the other. In the young growing rat, information on the number of cells in the columns, and the final size of the cell spaces of the hypertrophic cartilage cells (Table 1), suggests that the linear relationship between the growth rate and thickness of the plate reflects a relatively constant life-span for the cartilage cells, despite considerable variation in their rate of formation and final size.

Table 1

Results for the upper tibial epiphysial cartilage plate in the optimally-growing male rat. For each age group the growth rates are the means of 10 animals and the histological values the means of 10 E.C.P.'s from 5 animals. (In part from Fahmy, 1956)

Age of rat (days)	Growth rate (μ/day)	E.C.P. thickness (μ)	No. of cells in columns	Final cell size (μ)	Rate of cell formation (cells/day)	Life span of cells (days)
30	257	467	36	46	5·6	6·4
45	261	485	33	50	5·2	6·3
60	200	341	27	38	5·3	5·1
90	102	223	20	30	3·4	5·9

The growing epiphysial cartilage plate is affected in a variety of nutritional, hormonal and other circumstances (*see* Sissons, 1956). Its rate of growth is accelerated by increased amounts of pituitary growth hormone (Greenspan, Li, Simpson and Evans, 1949) and virtually ceases in the absence of this hormone (Asling, Walker, Simpson and Evans, 1950). Thyroxin, too, is required for normal endochondral bone growth (Ray, Simpson, Li, Asling and Evans, 1950) and particularly for the closure of epiphyses during skeletal maturation (Ray, Asling, Walker, Simpson, Li and Evans, 1954), while gonadal and adrenal hormones also influence the growth and maturation of the epiphysial plate. Restriction of dietary intake retards endochondral bone growth. In experimental rickets, whether the result of vitamin D deficiency or of abnormal mineral intake, there is failure of calcification of the hypertrophic cartilage of the epiphysial plate, and this is associated with failure of vascularization and bony replacement of the abnormal cartilage. The cartilage plate consequently becomes thickened and irregular; its growth is slowed but its thickness is increased because of an increase in the life-span of the cartilage cells in the absence of normal calcification (Fahmy, 1956; Hjertquist, 1961). Interference with the vascularization of the growing plate can also be produced by surgical incision on its metaphysial surface (Hjertquist and Westerborn, 1962); as in rickets, the hypertrophic cartilage cells continue to accumulate and the thickness of the plate increases. In the condition of lathyrism produced by the experimental administration of certain aminonitriles (Engfeldt, Tegner and Bergquist, 1960) bone growth is affected and the structure of the epiphysial cartilage plate disorganized, apparently as a result of interference with the synthesis of mucopolysaccharide material by the cartilage cells. Administration of the proteolytic enzyme, papain, also results in selective damage to growing cartilage (Westerborn, 1961); the rate of growth of the epiphysial cartilage plate is reduced, there is loss of mucopolysaccharide material from the cartilage, and the plate thickness is greatly reduced.

MEMBRANOUS OSSIFICATION

Here osteoblasts lay down bone on a periosteal, or other, surface without the intervention of a cartilage model. Again, information on membranous bone formation goes back to the time of John Hunter (*see* Sissons, 1956) who used the vital staining produced by madder feeding to demonstrate, in experimental animals, the sites of bone formation and the role of periosteal bone formation in the increase in thickness of the shaft of a long bone during growth. In recent years there has been considerable interest in the cell dynamics of the proliferating osteoblastic tissue as shown by mitotic indicators such as ^{3}H thymidine (Owen, 1963: Owen and MacPherson, 1963).

BONE REMODELLING

During growth, the deposition of new bone tissue is accompanied by resorption of existing bone by means of osteoclasts (*see* Hancox, 1956). It is by the control of these two processes that the size and shape of the developing skeleton is changed. The same processes of bone deposition and bone resorption continue during adult life and are responsible for a continuous remodelling ('turnover') of bone tissue. Normally, the two opposed processes are balanced in any part of the skeleton. As long ago as 1853, Tomes and de Morgan were aware, from histological study of normal bone, of this continuous remodelling of adult bone tissue. They observed resorption cavities and developing Haversian symptoms in cortical bone, and they were the first to realize that the Haversian architecture of this tissue remitted from the deposition of concentric lamellae of bone on the inner walls of resorption cavities.

In recent years, a number of vital markers, including alizarin (the active principle of Hunter's madder), have been used to study the histological localization of the remodelling activity. Of particular interest, in this respect, are the tetracycline antibiotics. When administered to a patient, or an experimental animal, they become attached to microscopic surfaces where bone formation is in progress; they are retained in the tissue when these surfaces become covered by new bone tissue and they can be demonstrated in undecalcified sections by their fluorescence in ultra-violet light. It is possible, by the use of markers of this type, to determine the rate at which bone is deposited on individual surfaces, as well as the rate of turnover for the part of the skeleton concerned (*see* Harris, 1960; Frost, Villanueva and Roth, 1960; Vanderhoeft, Kelly and Peterson, 1962; Harris, Jackson and Jowsey, 1962; Amprino and Marotti, 1964; and Lee, 1964). Such information is of interest in metabolic bone diseases, as well as under normal conditions.

The skeletal turnover of calcium, as opposed to that of actual bone tissue, can be measured by the use of 'bone-seeking' isotopes such as ^{45}Ca and ^{85}Sr (*see* Bauer and Ray, 1958; Heaney and Whedon, 1958); its relationship to bone formation is still under discussion (*see* Lee, Marshall and Sissons, 1965).

REFERENCES

Amprino, R. and Marotti, G. (1964). In: *Bone and Tooth*. Ed. by H. J. J. Blackwood, Proc. First Eur. Symp. Oxford; Pergamon Press

Asling, C. W., Walker, D. G., Simpson, M. E. and Evans, H. M. (1950). *Anat. Rec.* **106**, 555

Bauer, G. C. H. and Ray, R. D. (1958). *J. Bone Jt Surg.* **40A**, 171

Bisgard, J. D. and Bisgard, M. E. (1935). *Arch. Surg.* **31**, 568

Brash, J. C. (1934). *Edinb. med. J.* **41**, 305

Engfeldt, B., Tegner, B. and Bergquist, E. (1960). *Acta path. microbiol. scand.* **49**, 39

Enlow, D. H. (1966). In: *Studies on the Anatomy and Function of Bone and Joints*. Ed. by F. G. Evans. New York; Springer

Fahmy, A. (1956). 'An experimental study of some factors affecting bone growth.' Univ. Lond. *Ph.D. thesis*

Frost, H. M., Villanueva, A. R. and Roth, H. (1960). *Henry Ford Hosp. med. Bull.* **8**, 239

Greenspan, F. S., Li, C. H., Simpson, M. E. and Evans, H. M. (1949). *Endocrinology* **45**, 455

Haas, S. L. (1926). *Arch. Surg.* **12**, 887

Hancox, N. (1956). In: *The Biochemistry and Physiology of Bone.* Ed. by G. H. Bourne. New York; Academic Press

Hansson, L. I. (1967). *Acta orthop. scand.,* Suppl. 101

Harris, W. H. (1960). *Nature, Lond.* **188**, 1038

— Jackson, R. H. and Jowsey, J. (1962). *J. Bone Jt Surg.* **44A**, 1308

Heaney, R. P. and Whedon, G. D. (1958). *J. clin. Endocr. Metab.* **18**, 1246

Hjertquist, S-O (1961). *Acta path. microbiol. scand.* **50**, 163

— and Westerborn, O. (1962). *Acta path. microbiol. scand.* **56**, 266

Hulth, A. and Olerud, S. (1962). *Acta Soc. Med. upsal.* **67**, 219

Kember, N. F. (1960). *J. Bone Jt Surg.* **42B**, 824

Lacroix, P. (1960). *Bone as a Tissue.* Ed. by K. Rodahl, J. T. Nicholson and E. M. Brown. New York; McGraw-Hill

Lee, W. R. (1964). *J. Anat.* **98**, 665

— Marshall, J. H. and Sissons, H. A. (1965). *J. Bone Jt Surg.* **47B**, 157

Owen, M. (1963). *J. cell. Biol.* **19**, 19

— and MacPherson, S. (1963). *J. cell. Biol.* **19**, 33

Ray, R. D., Asling, C. W., Walker, D. G., Simpson, M. E., Li, C. H. and Evans, H. M. (1954). *J. Bone Jt Surg.* **36A**, 94

— Simpson, M. E., Li, C. H., Asling, C. W. and Evans, H. M. (1950). *Am. J. Anat.* **86**, 479

Sissons, H. A. (1953). *J. Anat.* **87**, 228

— (1956). In: *The Biochemistry and Physiology of Bone.* Ed. by G. H. Bourne. New York; Academic Press

Vanderhoeft, P. J., Kelly, P. J. and Peterson, L. F. A. (1962). *Lab. Invest.* **11**, 714

Westerborn, O. (1961). *Acta chir. scand.,* Suppl. 270

Young, R. W. (1962). *J. cell. Biol.* **14**, 357

SOME PROBLEMS ASSOCIATED WITH THE DEVELOPMENT OF THE MAMMALIAN SKELETAL MUSCLE FIBRE

MICHAEL A. MESSAGE

Department of Anatomy, University of Cambridge

IT WAS ONE of the gentlemen named Hippocrates and hailing from the Island of Kos who affirmed that one should never begin a talk with a remark made by another. Nevertheless, the general approach which has been used in planning this paper is summarized most appropriately by an aphorism with which the famous nineteenth century German mathematician Karl Jacobi (1804–1851) is said to have begun his first lecture to freshmen mathematicians:

'Gentlemen, always invert the problem.'

Initially, therefore, and very much in the spirit of Jacobi's advice, I would like to begin by giving a brief outline of what is currently known concerning the structural organization of the established, i.e., adult, mammalian skeletal muscle fibre and then to see to what extent it is possible to answer a few of the many problems which the growth and development of such a fibre and its organization pose. Moreover, it must be emphasized that no attempt will be made to give a detailed, purely descriptive account of the development of the mammalian skeletal muscle fibre. Such a description at the level of the light microscope has been given, for example, by Boyd (1960) and at the level of the electron microscope by Hay (1963) and Allen and Pepe (1965).

STRUCTURAL ORGANIZATION OF ESTABLISHED FIBRES

The established skeletal muscle fibre is, of course, syncytial in nature; it possesses not one but many nuclei, each one lying in an immediately sub-sarcolemmal position.

The sarcolemma, or cell membrane, of the skeletal muscle fibre was first described by Schwann (1839) and Bowman (1840). Both those authorities agreed that it was a smooth and structureless membrane; studies with the electron microscope, however, have now shown that it is structured to the extent that it has the typical trilaminar organization of the so-called 'unit membrane' (Robertson, 1957 and 1966). Thus it has an overall thickness of some 100 Å and is composed of an inner and an outer layer of electron dense material, each layer being some 25 Å in thickness, separated by a space some 40–60 Å in width and which is considerably less electron dense (Robertson, 1957, 1958 and 1966).

That organization exists throughout the sarcolemma but it is now clear that the sarcolemma itself also displays two forms of regional differentiation. Thus in the region of the myoneural junction (m.n.j.) the sarcolemma is invaginated into the underlying junctional sarcoplasm to form a variable number of irregular, in plan view, primary sub-synaptic gutters. These may be seen quite easily with the light microscope (for example, Couteaux, 1955 and 1960) and in each of them runs a terminal arborization of the motor-nerve terminal which supplies the parent muscle fibre. Arising from the sides of each of these primary gutters are further projections into the underlying sarcoplasm. At the level of the light microscope these projections look like solid rods (for example, Couteaux, *op cit.*) but more recent studies with the electron microscope (Palade and Palay, 1954 and Reger, 1957) have shown that in fact they are formed by narrow infoldings of the sarcolemma itself and are now known as 'sub-neural lamellae'. Combination of the data obtained from the two forms of microscopy of the myoneural junction indicates that the sub-neural lamellae are longitudinal outpocketings of the primary sub-synaptic grooves into the underlying junctional sarcoplasm and not tubular recesses; their functional significance is not yet certain.

The second form of regional differentiation which the sarcolemma of the striated skeletal muscle fibre displays occurs at the ends of the fibre. At the level of the light microscope the fibre ends, whether tapering or blunt, appear to be quite smooth. At the level of the electron microscope, however, it may be seen (Couteaux, 1959) that that region actually possesses a very large number of fine sarcolemmal invaginations which penetrate deeply into the underlying sarcoplasm. Into each of those invaginations pass a number of fine collagenous 'micro'-fibrils which eventually insert into the *superficial* aspect of the sarcolemma bounding each invagination. At their

27

other ends the micro-fibrils merge with the tendon of aponeurosis of origin or insertion of the muscle to which the given fibre belongs.

Cytologically, however, the most arresting components of the established skeletal muscle fibre are the myofibrils; the fibre's contractile machinery. They occupy the greater part of a skeletal muscle fibre's volume and are centrally placed within it as they run uninterruptedly, and in parallel with each other, from one end of the fibre to the other. The myofibrils have diameters of between 1 and 2 μ and display a longitudinally repeating pattern with a period of some 2·5 μ. The unit of this pattern is the so-called *sarcomere* which, by definition, is that region of a myofibril lying between successive, so-called Z-lines; the cross-striated appearance of a fibre as a whole derives from the fact that the sarcomeres of adjacent myofibrils are approximately in register with one another.

The detailed organization of the mammalian myofibril is now so well known and documented in the literature as to require, in the present context, no more than the briefest of descriptions.

On the basis of his results from the study of both living and glycerinated skeletal muscle fibres with low angle x-ray diffraction techniques, Huxley (1953a) postulated that each myofibril was itself composed of a double array of filaments and that the characteristic lines and bands of the myofibril resulted from the precise spatial arrangements assumed by these filaments. Subsequent studies with the aid of the electron microscope confirmed those contentions (Huxley, 1953b, 1957).

The individual myofilaments are of two types; the one thick and the other thin. The thick filaments are some 1·5 μ in length, 150–170 Å in diameter and are restricted to the so-called A, or anisotropic, bands of the myofibrils. In transverse section it may be seen that they are arranged in an hexagonal array with a centre to centre spacing of some 450 Å. The thin filaments are about 1·0 μ in length and some 50–60 Å in diameter. These arise from the Z-lines and not only comprise the I, or isotropic, bands but overlap with and interdigitate between the ends of the thick filaments in such a way that in a transverse section through the region of overlap each thin filament is shared, so to speak, by three thick ones and each thick filament is surrounded by six thin ones.

The overall effect of that organization is such that where a longitudinal section passes through the region of overlap between the thick and thin filaments then, depending upon the exact orientation of the plane of section, either one or two thin filaments can appear in the interval between adjacent thick filaments; the interfilament spacings are, of course, different in the two situations.

Since the A-band itself occupies the centre of the sarcomere its constituent filaments are overlapped from both ends by thin filaments. Under normal circumstances, however, these opposing sets of thin filaments do not meet at the centre of the A band; the resulting thin filament free zone is termed the H-zone. At a sarcomere length, i.e., Z to Z distance, of 2·5 μ that zone is approximately 0·5 μ in width. Running across the centre of the H-zone is yet another line; the M-line. This is some 0·06 μ across and, together with a narrow band on either side of it, constitutes the so-called 'pseudo-H-zone' which has an overall width of about 0·1 μ.

In the region of their overlap the thick and thin filaments are linked together by cross-bridges. These 'thick-to-thin' bridges are actually integral parts of the thick filaments and arise spirally along the length of their parent filament; except in the two narrow bands lying immediately adjacent to the M-line. The cross-bridges' spiral of origin has an *axial* length of approximately 400 Å with successive cross-bridges separated by some 70 Å and with, therefore, an angle of approximately 60 degrees relative to the longitudinal axis of the parent filament, between successive cross-bridges.

In the region of the M-line itself are found between three and five sets of cross-bridges which extend from one thick filament to the six nearest such filaments. These 'thick-to-thick' bridges are responsible for most, if not all, of the increased density of the sarcomere in that region.

Finally there is the Z-line itself. From the electron microscopic investigations of Knappeis and Carlsen (1962) and Reedy (1964) it is now quite clear that at the Z-line a thin filament on one side of the line is connected to the four thin filaments which are its nearest neighbours on the other side. The exact way, however, in which that connection is established is a matter of some uncertainty. Thus in a two-dimensional projection of the three-dimensional situation which exists *in vivo*, Knappeis and Carlsen (*op. cit.*) visualize the thin filaments to be lying at the nodes of a square lattice and to be connected to their nearest neighbours by radially orientated cross-links. Reedy (*op. cit.*), however, while agreeing that in such a projection the thin filaments do lie at the nodes of a square lattice, argues that the links to the nearest neighbours leave them tangentially so as to give the appearance of a 'woven', rectangular lattice.

The actual electron micrographs presented in each of those two papers, however, support the appropriate author's contentions. Since, however, Knappeis and Carlsen (1962) worked with frog, and Reedy (1964) with rat and rabbit striated muscle fibres, it may be that their different results and interpretations reflect genuine species differences in the organization of the Z-line, as seen in transverse section, rather than differing interpretations of the same species independent organization.

On the basis of his results with silver impregnation techniques, Veratti (1902) finally resolved a controversy of long standing; he showed quite unequivocally and in material form a wide variety of vertebrates and invertebrates that the skeletal muscle fibre as well as containing myofibrils also possesses a delicate but extensive reticular system within its substance. Moreover, he demonstrated that fundamentally the system was organized into two components; the one running longitudinally and ramifying over the surfaces of the myofibrils and the other running transversely in register with certain of the cross-striations of the myofibrils, which particular cross-striations depending upon the species, and linking the components of the longitudinal system.

Despite its clarity and definitive character, however, that paper fell into obscurity and it was not until some 50 years later that interest in reticular systems within skeletal muscle fibres was revived when Bennett and Porter (1953) described a 'sarcoplasmic reticulum' in electron micrographs of chicken breast muscle. In a subsequent investigation (Porter and Palade, 1957) the common identity of the 'sarcoplasmic reticulum' and Veratti's reticulum was firmly established.

The results from Porter and Palade's (1957) study suggested that, in fact, the sarcoplasmic reticulum consisted of a longitudinally running system comprising dilated cisterns and tubules which ramified over the surfaces of the myofibrils and which were interrupted at regular intervals to allow the interpolation of transversely orientated strings of 'vesicles'. Immediately adjacent to these vesicles the longitudinal cisterns were characteristically dilated; each vesicle and its two flanking cisternal dilatations forming what subsequently came to be known as a 'sarco-tubular triad'. In the case of mammalian skeletal muscle fibres (*m. sartorius*; *op. cit.*) those triads were found to be opposite the A/I junctions while in the case of the amphibian *Amblystoma* (myotomal muscle; *op. cit.*) they occurred opposite the Z-lines.

Since the publication of Porter and Palade's (1957) paper there has been an ever increasing number of studies with the electron microscope (for a recent review *vide* Smith, 1966) directed towards the further elucidation of the detailed organization of the sarcoplasmic reticulum. Thus Andersson-Cedergren (1959), on the basis of serial section techniques, established that the central component of the sarco-tubular triad was derived, in fact, from a highly convoluted tubule which ran radially within the parent muscle fibre; the subsystem formed by these tubules became known as the transverse, or '*T*-system'. From her studies, however, Andersson-Cedergren (*op. cit.*) could find no evidence of continuity between the *T*-system and either the sarcolemma or the cisternal system.

Subsequently, however, Franzini-Armstrong and Porter (1964), in a purely morphological study, concluded that the *T*-system did become continuous with the sarcolemma but not with the cisternal system. In the same year Huxley (1964a) and Page (1964a) showed experimentally that that must be the case. Thus, these latter workers soaked frog skeletal muscle fibres in solutions containing ferritin, a material which has a characteristic appearance in the electron microscope, and showed that it could be found only in the *T*-system of a fibre; whether or not the continuity of that system with the sarcolemma is permanent or phasic is as yet undecided.

Clearly the deep penetration of the *T*-system into muscle fibres, its association there with the myofibrils and its peripheral continuity with the sarcolemma render it an ideal pathway for the inward spread of excitation. Experimental corroboration for this view is provided by the findings of Huxley and Taylor (1958). With the aid of electrode micropipettes these workers applied localized stimuli to the surfaces of isolated skeletal muscle fibres (crab, lizard and frog) and observed the fibres under the polarizing microscope. They found that as the micro-electrodes were moved over the surface of a fibre they evoked localized contractions of the underlying myofibrils at discrete loci only (the strengths of the applied stimuli were too small to evoke a propagated response). Moreover, in the case of the frog those foci were invariably opposite the *Z*-line and for the lizard invariably opposite the *A/I* junctions; exactly the levels at which the *T*-systems are known to lie in those species.

SOME ASPECTS OF THE DEVELOPMENT OF THE MUSCLE FIBRE

So much then for the basic structural organization of the mature mammalian skeletal muscle fibre; what of the developmental processes involved in its establishment? Clearly, within the present limitations of space, there are many more problems than can be considered, so attention will be restricted to but a few of the relevant questions and problems with the aim of considering them in detail, both methodologically and in terms of the available experimental evidence, and of deciding to what extent they have been resolved and, in cases of incomplete resolution, to speculate a little.

Origin of the multinucleate state

To date the problem which has provoked, perhaps, the most investigation and speculation, but which now may be considered to have been resolved, is that of the mode of origin of the established skeletal muscle fibre's multinucleate state. *A priori* there are four mechanisms which could, in principle at least, account for that condition:

(1) The nucleus of a single myoblast undergoes mitotic division but the cell as a whole fails to divide; this process of karyokinesis without cytokinesis is then repeated several times for each daughter nucleus.

(2) The nucleus of a single myoblast undergoes amitotic division but the cell as a whole fails to divide; then as in (1).

(3) The fusion of a large number of mononucleate myoblasts.

(4) Some prescribed amalgamation of possibilities (1) to (4).

On the grounds, however, that even after the administration of colchicine one rarely sees any evidence of mitotic activity within a *developing* skeletal muscle fibre once it has attained the multinucleate state, the first possibility (1) given above will be disregarded and attention given primarily to possibilities (2) and (3) which, for convenience, will be termed henceforth the '*amitotic*' and the '*fusion*' hypotheses respectively.

In its essentials the case of the amitotic school of thought is that whereas there is no evidence of mitotic activity in developing skeletal muscle fibres, not only is there considerable variation in nuclear shape but also it is possible, on the basis of their shapes, to arrange the nuclei of such fibres into a continuous series ranging from well separated, smooth ellipsoidal nuclei at one end of the series through nuclei with just the suspicion of a waist and then through clearly dumbbell shaped nuclei to, finally, smooth ellipsoidal nuclei whose adjacent poles are very close together. The existence of such a clearly defined series of nuclear shapes establishes, for some, that the nuclei of a developing skeletal muscle fibre undergo amitotic division. Moreover, this school claims, that careful examination of sectioned material fails to reveal any evidence of fusion either between two or more mononucleate myoblasts or between a myoblast and an already multinucleate skeletal muscle fibre.

The basic position of the protagonists of the fusion hypothesis is that the pleiomorphism, as seen in sectioned material, displayed by the shapes of the nuclei of developing skeletal muscle fibre reflects either differences in the functional activity and maturity of the nuclei and/or fixation artefacts and does not show that during development such nuclei undergo amitotic division. Moreover, in a more positive sense, this school asserts that a careful examination of sections of developing skeletal muscle shows unequivocal evidence of the two types of cellular fusion (*vide supra*) which, *ex hypothesi*, must exist.

In attempting to assess the relative merits of these rival hypotheses it must be kept in mind that historically the controversy arose directly from attempts to delineate the fine details of a developmental process, which is necessarily dynamic, by the unsatisfactory method of the interpretation of the necessarily static appearances presented by fixed, sectioned and stained material. At best such an approach can provide only a circumstantial interpretation, irrespective of whether one is using light or electron microscopy. Clearly, therefore, to achieve a definite decision recourse must be made to techniques which allow one to follow the process of myogenesis, either directly or indirectly over a reasonable period of time.

The simplest method of achieving that aim directly would be to study the behaviour of developing skeletal muscle in tissue culture by either direct microscopic observation or, preferably, time-lapse cinematography. In fact the latter technique has been used recently by a number of investigators in attempts to resolve the conflict between the fusion and amitotic schools of thought for the establishment of the multinucleate state of the skeletal muscle fibre; the results obtained have been, effectively, identical.

Thus Capers (1960), Cooper and Konigsberg (1961) and Bassleer (1962) cultivated chick myoblasts *in vitro* and were able to show unequivocally that a mononuclear myoblast could fuse either with another mononuclear myoblast or with an already formed, multinucleate myostrap (or ribbon) and that myostraps could themselves fuse with one another. Moreover, there was a complete absence of mitotic activity in established myoblasts and myostraps.

In addition, the results from all three investigations showed that within the multinucleated myostraps the nuclei were in constant motion, both rotatory and translatory, and that associated with these movements the nuclei themselves frequently underwent extreme alterations in shape with time, often approaching dumbbell configurations. In no case, however, was a nucleus found to become divided into two although, as a result of their movements up and down a myostrap, two nuclei could become arbitrarily close, or indeed collide, when if one was ignorant of their past histories one might consider that they were in fact the daughter nuclei resulting from an amitotic division of another nucleus.

In view of the fact that *in vitro* the shapes of individual nuclei in a myostrap are so variable with time, it is clear that if such variation also occurs *in vivo* then the examination of fixed and stained material would undoubtedly reveal the existence of nuclear pleiomorphism; the significance of that pleiomorphism, however, would be entirely wrongly assessed. It would be erroneously attributed, certainly by the upholders of the amitotic school of thought, to the occurrence of amitosis.

Although purely observational in character the preceding *in vitro* observations cast considerable doubt upon the validity of the amitotic hypothesis. The evidence to be recounted in the immediately succeeding paragraphs is held to resolve the situation unequivocally.

Its nuclei mean clearly DNA, an alternative approach to the problem of the mode of origin of the multinucleate state of the established skeletal muscle fibre, would be to study the DNA content of individual fibre nuclei during myogenesis. The premises underlying such studies and the results to be expected on the basis of each of the possible explanations which might be advanced to account for that state are clear cut and easily summarized.

Thus if the multinucleate state derives solely from the fusion of mononucleate myoblasts then each of the fibre nuclei would contain the diploid $(2n)$ amount of DNA; the fibre as a whole would then be K-ploid where K is the total number of nuclei in the fibre. If, however, that state results from *amitotic* nuclear division *preceded* by the replication of the nuclear DNA then during the phase of nuclear proliferation any individual fibre nucleus could have a DNA content anywhere within the range of values corresponding to the diploid $(2n)$ and the tetraploid $(4n)$ states. Thus during the proliferative phase the mean nuclear DNA level would be between those for the (2) and $(4n)$ states while the final outcome would again be a K-ploid fibre with each nucleus possessing the $(2n)$ level of DNA.

Alternatively it might be argued that the replication of nuclear

DNA occurs in the daughter nuclei immediately after the process of amitosis. In this situation, therefore, during the phase of nuclear proliferation any given fibre nucleus would have a DNA content lying between the haploid (n) and diploid $(2n)$ levels so that during this phase the mean nuclear DNA would necessarily lie between these limits. Nevertheless, the final outcome would again be a K-ploid cell with each nucleus possessing the $2n$ value of DNA.

But if amitotic division were to be neither preceded nor succeeded by the replication of nuclear DNA then the situation becomes entirely different. If, at amitosis, there was an equipartition of DNA between the daughter nuclei and there were J generations of nuclei, giving a total of $2^J = K$ nuclei, then at the end of the phase of nuclear proliferation each nucleus would have a DNA content equal to the $(2n/K)$ state. During that phase, therefore, any given nucleus would have a DNA content between those equivalent to the haploid (n) and the $(2n/2^J-)$ levels, where J is the generation of the nucleus; the mean level would necessarily lie within those limits. Should, however, the partition of DNA between the daughter nuclei be random, then during the phase of nuclear proliferation any individual nucleus could have a DNA content within the range of values equivalent to the, effectively, zeroploid $(0n)$ and the diploid states. At all times, therefore, the mean nuclear DNA content would be less than that equivalent to the diploid $(2n)$ level and once nuclear division had ceased would be equal to the $(2n/K)$ level. However, for both of these last two cases the total fibre DNA would still be equal to the diploid $(2n)$ value and so, presumptively, the fibre would still possess a copy of its owner's complete genotype.

Experimentally two complementary cytological techniques have been used in attempts to decide between those various possibilities for the establishment of the skeletal muscle fibre's multinucleate state; cytophotometry of Feulgen stained nuclei and autoradiography with tritiated thymidine. Of these techniques the former gives static information, in that it provides an accurate measure of the DNA content of a nucleus at the time of its fixation, while the latter gives more dynamic information. Thus that technique not only defines those nuclei which contain DNA synthesized between the administration of the isotopically labelled compound and fixation of the parent tissue but also permits the extent of that synthesis to be quantitated.

Feulgen cytophotometry has been used to measure the DNA content of nuclei in regenerating mouse muscle (Lash, Holtzer and

Swift, 1957) and in chick striated muscle growing in tissue culture (Firket, 1958; Bassleer, 1962 and Strehler, Konigsberg and Kelley, 1963). The results from all these studies were identical. Thus the nuclei of the mononucleate cells surrounding the multinucleate ones showed a bimodal distribution of DNA content with the peaks at values equivalent to the diploid $(2n)$ and tetraploid $(4n)$ states. In complete contrast to this the nuclei of the multinucleate cells showed a unimodal distribution of DNA content with the peak at the value equivalent to the $(2n)$ state. Furthermore, the spread about this peak was extremely small with no nucleus even approaching the haploid (n) or the $\frac{3}{4}$-ploid levels. These results suggest very strongly indeed that the fusion hypothesis is correct.

Further support for this view derives from the results obtained with tritium labelled thymidine; thymidine itself, of course, is a specific precursor of DNA. Thus (Bassleer, 1962), if a 'pulse' dose of tritiated thymidine is given to a culture of developing muscle and the culture sacrificed shortly after, it is found that whereas the culture contains a very large number of mononucleate myoblasts with labelled nuclei any multinucleate myostraps present in the culture contain only unlabelled nuclei. The latter finding would not have been expected if those nuclei had been synthesizing DNA. If, however, a culture is allowed to continue for some appreciable time after the administration of the labelled thymidine, then it is found that the multinucleate myostraps in the culture contain both labelled and unlabelled nuclei.

Taken together, therefore, the results obtained from the two types of labelling experiment (Bassleer, *op. cit.*) demonstrate quite unequivocally that the multinucleate state of the striated muscle fibre derives from the *fusion* of mononucleate myoblast with mononucleate myoblast and/or multinucleate myostrap. This, of course, has the necessary corollary that, in the development of the striated muscle fibre, synthesis of nuclear DNA and myoblastic fusion are independent processes.

Finally, if indeed nuclear replication is not a necessary concomitant of the establishment of the multinucleate state then it should be possible to dissociate the one from the other. That prediction is upheld by the results obtained by Konigsberg, McElwain, Tootle and Herrmann (1960) and Bassleer, Collignon and Matagne-Dhoosche (1963). Both groups of workers used radio-mimetic drugs to inhibit DNA synthesis; the former group in the case of muscle developing in culture and the latter group in the cases of the chick embryo *in ovo* and of *Amblystoma* during regeneration of their tails after operative interference. In all three series of experiments, notwithstanding the fact that DNA synthesis was inhibited, multinucleate skeletal muscle fibres were formed.

Taken together, therefore, all the preceding data indicate that the fusion hypothesis is the correct interpretation for the mode of origin of the multinuclearity of the established skeletal muscle fibre.

At the present time, and in complete contrast to the situation concerning the origin of the multinucleate state of the established mammalian skeletal muscle fibre, our knowledge of the way in which the complex structural organization of the myofibril is established is rudimentary, despite not only the many studies of myogenesis at the level of the light microscope but also the few so far undertaken at the level of the electron microscope.

It will be remembered that during the description of myofibrillar organization given at the outset of the present paper the primary structural elements of a myofibril were referred to simply as thick and thin filaments; no attempt was made to define their precise natures. Now, however, it is well established (*vide* the review by Huxley and Hanson, 1960) that the thick filaments are composed of the protein myosin while the thin ones consist of the protein actin, probably, with some 10 per cent of tropomyosin. Moreover, it is clear that each filament is not composed of a single molecule of the relevant protein but in fact is a polymer of that protein. Thus, from a consideration of the total amount of myosin per gramme of muscle, the molecular weight of myosin and the dimensions of individual myosin filaments, it may be calculated that each such filament is composed of some 350 myosin molecules. In this context it should be noted that brief tryptic digestion of myosin causes its dissociation into two sub-units named light and heavy meromyosin; the significance of this phenomenon will be considered later. Similar computations for the thin filaments indicate that they contain approximately 350 to 400 actin monomers.

Establishment of the internal organization of the myofibril

In the light of what has been said already it would seem that the problems associated with the establishment of the internal organization of the individual myofibril may be grouped conveniently under four, not necessarily independent headings; namely, the nature of the mechanisms concerned with:

(1) the production of the primary building blocks, i.e., of both the meromyosins, actin and tropomyosin,

(2) the control of the polymerization of those building blocks into thick and thin filaments of the correct lengths,

(3) the arrangement of the individual filaments into the correct configuration as seen in transverse sections of mature myofibrils, and

(4) the establishment of the sarcomeric segmentation of the myofibril.

The first of these problems is clearly that of the definition of the mechanisms underlying the synthesis of specific proteins; the present view of that process may be formulated briefly as follows. According to contemporary molecular biological thought, the structure of any given protein, as reflected in the amino acid sequence(s) of its constituent polypeptide chain(s), is defined by a specific, so-called structural gene composed of DNA and situated in the parent cell's nucleus. The primary product of the structural gene is an RNA copy of it which is known as the specific messenger RNA (mRNA) for the given polypeptide chain. This mRNA copy is released into the cytoplasm where it associates with pre-existing particles, also containing RNA, known as ribosomes. It is on the surfaces of these ribosomes that the polypeptide chains are actually synthesized, amino acid by amino acid, under mRNA control.

More recently, however, it has been established that the primary protein synthetic unit is not a single ribosome but rather a group of them; the so-called 'polysome' in which a number of ribosomes are bound together by the mRNA molecule upon which they are operating (Warner, Rich and Hall, 1962; Rich, 1963). Recently such polysomes have been observed both in electron microscopic sections (for example Allen and Pepe, 1965; Heuson-Stiennon, 1964; Message, 1967a and Waddington and Perry, 1963) and homogenates (Allen and Pepe, *op. cit.*) of developing skeletal muscle fibres. Since it may be taken as established that *all* proteins are synthesized in the manner outlined above, the first problem posed above (1) in the context of myofibrillogenesis may be taken as having been answered in terms of biological principal but not, as yet, in terms of biochemical detail.

When, however, one considers problems (2) to (4) inclusive the extent of our knowledge is considerably less. Nevertheless, the view that *in vivo* the construction of the primary myofilaments, both thick and thin, could be accomplished by a process of *controlled* polymerization of simple, primary building blocks would seem to receive strong circumstantial support from the results of certain recent *in vitro* studies of the properties of highly purified preparations of myosin and actin (for example Huxley, 1963; Hanson, 1967).

Although both heavy and light meromyosins are readily soluble in solutions of high ionic strength, if the ionic strength of the solution is reduced then the individual molecules form aggregates. If such aggregates are negatively stained and examined with the electron microscope, then it is found that the appearances presented by aggregates which have resulted from lowering the ionic strength of a solution containing *both* heavy and light meromyosins are strikingly similar to those of thick myofilaments. Thus, such synthetic aggregates are some 0·3–2·0 μ or more in length, 150–200 Å in diameter, have tapering ends and have distinct projections over most of their surfaces which are very reminiscent of the thick-to-thin cross-bridges found *in vivo*. Moreover, and even more significantly, the central 0·1–0·2 μ of such synthetic filaments are frequently devoid of any such projections; cf., the appearances of thick filaments *in vivo*. If, however, the aggregates have come from a solution which contained only light meromyosin, then although the aggregates are filamentous, have similar dimensions and taper towards their ends, their surfaces are entirely smooth.

The current view upon the form of the myosin molecule is that its overall shape is very like that of a child's lollipop, composed of a 'stem' of light meromyosin some 1300 Å in length and 20 Å in diameter to one end of which is attached a 'head' which is globular in character with a length of approximately 250 Å and a diameter of some 40 Å (Huxley, 1963).

Thus to construct a thick filament, either *in vivo* or *in vitro*, apparently all that need be done is to join the stems of two myosin molecules in anti-parallel, i.e., so that the heads of these two molecules point in opposite directions, and then to add further myosins in parallel with each of the two starter molecules but with succeeding molecules displaced some 70 Å along the growing filament and with their sites of attachment arranged spirally around the filament's long axis, with six new molecules added per turn of the spiral. This process would produce thick filaments with only their central regions devoid of projections, with tapering ends and with each end of a filament capable of forming cross-bridges with up to six thin filaments.

Under suitable conditions it is also possible to bring actin out of solution; again with the formation of aggregates. When these aggregates are negatively stained and examined with the electron microscope they are found to have indefinite lengths, sometimes tens of microns, diameters of 60–70 Å and a beaded appearance. There is now general agreement that in fact these synthetic actin filaments have the form of two chains, each composed of a string of rather globular sub-units some 55 Å in diameter, twisted around each other. The pitch followed by each individual chain has a period of some 700 Å but since in a filament the two chains are out of step with each other by half the diameter of one of the sub-units, the double helical pattern repeats approximately every 350 Å (Huxley, 1963; Hanson, 1967).

These synthetic actin filaments also possess the property of binding heavy meromyosin (Huxley, 1963); if then such filaments are exposed to a solution containing heavy meromyosin and subsequently negatively stained a striking electron microscopic picture emerges. The diameters of the stained filaments are some 200–300 Å and the axial repeat of 350 Å of the actin filament itself is now very pronounced. Moreover, and much more significantly, it is clear that each synthetic actin filament possesses a distinct polarity throughout its length; a polarity which is best described in terms of the filament's internal organization giving rise to an appearance of a series of 'arrowheads' all pointing in the same direction. The electron microscopic appearances of thin filaments isolated from established myofibrils and then negatively stained are identical with those of the synthetic actin filaments in terms of diameter, internal organization and polarity but not in terms of length; the significance of that difference will be considered later.

It would seem, therefore, that all that need be done to construct a thin filament, either *in vivo* or *in vitro*, is for individual actin monomers to join with each other to form chains and when two such chains come into contact for them to wind around each other to form a double helix with the two chains half the width of one monomer out of step with each other and giving, thereby, a thin filament with an internally polarized structure.

From the preceding paragraphs it might seem that *in vivo* the operation of only the simplest of processes, that of the polymerization of simple and pre-existing building blocks, is necessary in order to produce both the thick and the thin myofilaments ready for incorporation into a developing myofibril. In fact, however, such a process can be no more than a part, albeit an important part, of the whole mechanism.

Thus it is quite clear that *in vivo* a process of simple polymerization would be quite capable of producing myofilaments with organizations which appear to be very similar, if not identical, to those of the thick and thin filaments composing real myofibrils. Nevertheless, *per se* as a mechanism for the *in vivo* production of such filaments, it suffers from a crucial disability in that it would exercise no control over the lengths of the filaments which it produced whereas *in vivo* their lengths are precisely controlled. Thus Page (1964b), in an electron microscopic study of mature chicken and frog skeletal muscle fibres, estimated the lengths of the thick filaments to be 1·6 μ with very little variation about the mean and of two thin filaments and the interposed Z-line to be 2·06 μ, again with very little variation about the mean, so that a single thin filament must have a length of very nearly 1·0 μ.

The results obtained by Allen and Pepe (1965) from an examination of myofilaments isolated from developing skeletal muscle fibres (chick myotome) are therefore of considerable interest. Those workers found that the thin filaments had a maximum length of 1·0 μ although, due either to breakage during preparation and/or incomplete development, their individual lengths could be considerably less than that. The thick filaments, however, from the stage at which they could first be separated intact, invariably had a length of 1·5 μ and diameters of 105–120 Å (cf., the dimensions of the thick filaments in established myofibrils). It was concluded from this finding that the assembly of the thick filaments is accomplished very rapidly.

From a study of routine electron microscopic preparations of their material, Allen and Pepe (1965) concluded that the thin filaments appear at sites scattered (? randomly) throughout the sarcoplasm and in advance of the first appearance of thick filaments. The latter view has been challenged by Dessouky and Hibbs (1965) on the basis of their electron microscopic study of the myogenesis of chick myotomal muscle fibres but corroborated upon immunological grounds by Ogawa (1962). Both of Allen and Pepe's (1965) interpretations have received corroboration from a preliminary study of human foetal muscle (Message, 1967a).

If, of course, production of one type of myofilament were to be started throughout the sarcoplasm of a developing skeletal muscle fibre in advance of the commencement of production of the other, then as soon as the second type of primary myofilament began to appear there would be a rapid formation of thick/thin complexes at numerous sites scattered, effectively, at random throughout the sarcoplasm of the parent muscle fibre. Such a situation would necessarily result in similar complexes appearing over an appreciable period of time, for them to undergo further development in parallel and, in the early stages at least, to be orientated at random relative to the long axis of their parent fibre. As the complexes increased in length and in girth by the incorporation of further thick and thin filaments and/or by the coalescence of pre-existing thick/thin complexes, the number of completely new complexes appearing would tend with time to decrease while the complexes themselves would come to lay preferentially in parallel with the long axis of the fibre.

Moreover, since each end of an individual thick filament is so constructed that it can form cross-bridges with, maximally, six thin filaments and, by inference from the situation in the established myofibril, the structure of each thin filament is that it can come to be shared by up to three thick filaments, then in the first instance it might seem that a simple process of co-polymerization between the two filament types is all that is required in order to produce myofibrils.

Certainly such a process would yield structures in which the appearances of transverse sections taken through regions of thick-to-thin cross-bridging would be identical with those of sections taken through the region of overlap between the actin and myosin filaments of undoubted myofibrils. Furthermore, if initially the longitudinal extent of the overlap between thick and thin filaments is random then in longitudinal section the thick/thin complexes would, certainly initially display no evidence of cross-striation.

Now both the light and electron microscopic evidence (for example Hay, 1963; Allen and Pepe, 1965; Dessouky and Hibbs, 1965 and Message, 1967a) suggests that *in vivo* the first sign of myofibrillogenesis proper is the appearance, at sites which are seemingly scattered at random throughout the sarcoplasm of the parent fibre of apparently randomly orientated complexes between thick and thin myofilaments which are not only unstriated in the longitudinal sense but also conform in the transverse sense to the Huxley (*vide supra*) pattern. At first sight, therefore, it might seem that the initial stages of myofibrillogenesis proper are indeed accomplished by a process of *simple* co-polymerization between the thick and thin primary myofilaments. Again, however, despite its *a priori* appeal that view is certainly an oversimplified one.

Thus, as it stands, it would permit a thin filament to become attached to the *same* longitudinal row of projections on both sides of the central bare area of a thick filament; a situation which certainly does not exist in mature myofibrils and which, on the basis of a preliminary study of the development of human skeletal muscle fibres (Message, 1967a), would seem not to occur in developing

42

myofibrils. *A priori*, of course, the most probable reason for the non-occurrence *in vivo* of the phenomenon just predicted, on the basis of a process of simple co-polymerization, would seem to be that in order that a thin filament may become cross-linked to one end only of a thick filament the internal polarity of the thin filament must be appropriately orientated relative to the long axis of the thick filament. There is evidence from the structure of the established myofibril that, in fact, that condition does obtain *in vivo*.

In the established myofibril the thin filaments attached on one side of a Z-line have a polarity which is opposite to that of those attached on the other. Thus under appropriate conditions it is possible to isolate from such myofibrils units which are termed *I*-segments (*vide supra*) which consist of a Z-line and the two sets of thin filaments attached to it. Examination of negatively stained *I*-segments with the electron microscope (Huxley, 1963) shows that the arrowheads of all the thin filaments on one side of a Z-line invariably point away from that line; how does this selectivity of orientation of the thin filaments arise?

An individual thin filament undoubtedly possesses its characteristic internal, arrowhead polarity *before* it is incorporated into a developing myofibril. Furthermore, the internal organization of each thick filament is essentially bipolar in that its constituent myosin molecules are subdivided into two sub-populations, each one lying in *anti*-parallel to the other. In view of this organization it would seem very probable that the local chemical topography of the tips of the projections found towards each end of a thick filament, and which form the actual cross-bridges with the thin filaments, is such that for a thin filament to become attached to a given longitudinal row of such projections it must be orientated in a given direction. This view receives support from the facts that, on the one hand, in the established myofibril the thick to thin cross-bridges do not stick out from the thick filaments at right angles but are actually directed towards the nearer Z-line, while on the other, the thin filaments themselves are internally polarized. Taken together those conditions would seem to be both necessary and sufficient to ensure that the thin filaments attaching at the same end of a given set of thick filaments all point in the same direction and in the opposite direction to those attaching at the other.

Such a situation has, of course, two critical corollaries. The first is that adjacent sets of thin filaments, i.e., those which will come to lay on opposite sides of the same Z-line, must point in opposite directions (cf., *in vivo*). The second is that during the formation of thick/thin complexes no one thin filament may become attached to a given longitudinal row of the projections found at the two ends of a given thick filament *although* the centrally directed end of a thin filament might overlap the entirety of the thick filament's central, bare area.

In summary, therefore, it would seem that provided the process operates according to the topographical restrictions defined in the immediately preceding discussion, *selective* co-polymerization between thick and thin myofilaments could well underlay at least the initial phases of myofibrillogenesis.

However, in its modified form and notwithstanding the fact that it can produce a thick/thin complex with an appropriate structure in the transverse sense a process of selective co-polymerization of thick and thin fibres, even combined with a longitudinal shuffling of the elements of such a complex, is insufficient to produce a *stable* form of the characteristic cross-striations which are possessed by the mature myofibril. How then does an initially unstriated myofibril acquire its characteristic cross-striations?

Developmentally, the first of the striations to appear is the Z-line (Allen and Pepe, 1965; Message, 1967a) and a consideration of the organization of the established Z-line suggests how this might be achieved in a simple fashion. It will be remembered (*vide supra*) that a thin filament on one side of a Z-line is connected to its four nearest neighbours on the other. But this poses the question of how a double helix, the configuration of an actin filament, forms four cross-links and Knappeis and Carlsen (1962) showed that the juxta Z-line end of each thin filament is provided with four tail filaments and suggested that they might consist of tropomyosin. That suggestion receives support from the facts that, on the one hand, the electron micrographic appearances of tropomyosin (Huxley, 1963) are very similar to those of transverse sections of Z-lines and, on the other, that actin preparations always seem to have a rather constant proportion of tropomyosin associated with them before they are extensively purified (Perry and Corsi, 1958; Corsi and Perry, 1958).

However, irrespective of their exact chemical nature, given that the tails of the thin myofilaments do underlay the primary structure of the mature Z-line then the next question is whether the thin filaments acquire their tails before or after they become incorporated into a developing myofibril. The only published photographs of thin filaments isolated from developing skeletal muscle fibres (chick myotomal muscle; Allen and Pepe, 1965) do not permit of an answer to that question. Allen and Pepe (*op. cit.*) did make the intriguing

44

observation, however, that the first sign of a longitudinal segmentation of a developing myofibril was the appearance at intervals of approximately 1·5 μ, along an otherwise unstriated myofibril, of fine tubules which ran perpendicularly across the developing myofibril. In some cases these tubules appeared to be replaced by a band of dense material reminiscent of Z-line material. A preliminary study of electron micrographs of human foetal muscle suggests that similar phenomena occur (Message, 1967a). Whether, of course, there is any causal relationship between such tubules and the establishment of myofibrillar cross-striations is, in the present state of knowledge, clearly problematical. However, *if in practice* those tubules are associated with the formation of Z-lines then since the surfaces of the tubules themselves are smooth and protein synthesizing membrane systems are usually rough it would seem to be more plausible to assert that in fact the thin filaments acquire their tails before they are incorporated into a developing myofibril and that the tubules merely form a surface upon which those tails may come into contact with each other and establish a Z-line.

By reason of the restrictions placed upon the way in which thick and filaments may become cross-linked, it is clear that the opposed (or laterally adjacent) ends of the thin filaments attached to the adjacent ends of any two successive thick filaments along (or across) a developing myofibril will be the tailed ends of the thin filaments. By a process of longitudinal shuffling of thick and thin filaments the opposed (laterally adjacent) tailed ends of the thin filaments could come into contact with each other when, *ex hypothesi*, they would become cross-linked with each other and, by inference from the properties of those cross-links in the mature myofibril, these cross-linkages would be most unlikely to be broken subsequently. The proposed process of shuffling could easily be accomplished by the (? random) operation of the mechanisms which are involved in the making and breaking of thick-to-thin cross-bridges during the contractile activity of established myofibrils. The operation of such processes throughout a developing myofibril would rapidly and easily establish the essential basis of the longitudinal, sarcomeric segmentation displayed by established myofibrils, i.e., the Z-lines.

Of course the processes considered so far would also suffice to bring the thin filaments into their correct relationship to the Z-lines and into transverse register with each other. *Per se*, however, those processes would be insufficient to subdivide the region between successive Z-lines into the so-called '*I*' and '*A*' regions which are also characteristic of the established sarcomere. Nevertheless, that end could be gained if those processes were to be augmented by one further simple process which indeed must operate *in vivo*.

The additional mechanism turns upon the fact that the synthetic myosin filaments produced *in vitro* (Huxley, 1963) and separated from developing chick muscle fibres (Allen and Pepe, 1965) have a bare central region whereas *in situ* in an established myofibril that region of the thick filament possesses projections which form very stable thick-to-thick cross-bridges and form, thereby, the structural basis of the *M*-line.

Thus if after, more probably than before, the establishment of any two successive *Z*-lines the thick filaments lying between them not only continue to be shuffled backwards and forwards but also acquire such centrally placed projections, then when adjacent thick filaments came into an appropriate register they would form stable cross-bridges with each other. As this process spreads across the width of a sarcomere the sarcomere would necessarily become subdivided into its other characteristic regions; *H*-zone, *A*-band and *I*-band etc. Repetition of these processes along the length of the developing myofibril would cause it to acquire its characteristic longitudinal organization.

Briefly to summarize and synthesize the preceding discussion of myofibrillogenesis, it would seem to be highly probable that the detailed organization of a myofibril results from the operation of a so-called 'self-organizing process' which uses essentially very simple building blocks, the primary thick and thin myofilaments, and works according to a small number of elementary rules. Similar systems have been proposed increasingly in explanation of the behaviour of biological systems (for example Penrose, 1958 and 1960).

Necessarily, in the light of the gaps in contemporary knowledge, assumptions have had to be introduced in the development of such an interpretation of the process of myofibrillogenesis but those assumptions seem currently not unreasonable. It must be emphasized, however, that the thesis developed here is both preliminary and speculative in nature; it is currently undergoing further detailed evaluation but obviously only time and further *in vivo* study will decide whether or not its development and presentation here have been worth while. It is hoped that the answer will be in the affirmative and it is perhaps worth noting in passing that the model appears to have considerable potential in the field of the processes involved in the further growth, both in length and diameter, of an established myofibril (Message, 1967b).

46

Development of the sarcoplasmic reticulum

Whereas it would seem that we can define the mechanisms responsible for the establishment of the multinucleate state of the skeletal muscle fibre with almost complete certainty and at least make an informed guess as to the mechanisms involved in the construction of myofibrils, our knowledge of the processes concerned in forming and correctly positioning not only the cisternal but also the T-system components of the sarcoplasmic reticulum is rudimentary, indeed almost non-existent.

It is quite clear (Allen and Pepe, 1965; Dessouky and Hibbs, 1965 and Message, 1967a) that the sarcoplasm of developing skeletal muscle fibres contains large numbers of irregularly shaped vesicles of varying sizes and one could easily envisage that the cisternal component of the sarcoplasmic reticulum is established by those vesicles undergoing anastomatic fusion. The resulting complex(es) would become progressively restricted to the inter-myofibrillar spaces and, thereby, come to occupy their characteristic position of ramifying between and over the surfaces of the myofibrils. Again, however, this is an interpretation of over simplicity; it completely ignores the fact that *per se* such a process would necessarily produce a cisternal system which would run *uninterruptedly* throughout the sarcoplasm, whereas *in vivo* the cisternal system is interrupted (*vide supra*) at the levels at which the T-system traverses the parent fibre. How this interruption is accomplished is not known but it may be that the T-system itself imposes it upon an initially continuous cisternal system.

Turning now to the T-system component of the sarcoplasmic reticulum, the fundamental question is how it comes to lie at the correct level of each sarcomere; the A/I junction in the mammalian skeletal muscle fibre (for other species *vide* the review of Smith, 1966). Apart from the fact that it does not seem to appear until myofibrillogenesis is extremely well advanced (Message, 1967a), there is at present no evidence on this question at all.

Essentially there would seem to be two mechanisms which could, in principle at least, account for the topographic specificity of the positioning of the T-system relative to a sarcomere; both mechanisms, however, depend upon assuming that at the appropriate level of the sarcomere the local properties of the sarcomere itself are highly chemo-differentiated. It must be emphasized from the outset, however, that *the author regards both possibilities as being at the moment no more than the purest speculation.*

47

According to the first, the local chemical properties of each sarcomere would cause the T-system for that sarcomere to differentiate *in situ*; interrupting the cisternal system if necessary. Then, because the distances between adjacent myofibrils are small and the sarcomeres in adjacent myofibrils are extremely well in register, a progressive linkage of the T-system across the parent fibre could be easily and quickly accomplished. The final step would be for it to fuse with the sarcolemma.

The second possibility derives from the fact that in the early phases of myogenesis established myofibrils come to lie immediately subjacent to the sarcolemma while the nuclei are lined up along the fibre axis; the so-called 'myotube' phase of muscle fibre development. Subsequently, the relative positions of myofibrils and nuclei are reversed. Now it will be remembered that the tubules of the T-system appear to be, in effect, ingrowths of the sarcolemma (cf., Franzini-Armstrong and Porter, 1964). Thus it might be argued that during their sub-sarcolemmal sojourn the early myofibrils, again by virtue of some local chemo-differentiation within each of their sarcomeres, induce the formation of sarcolemmal ingrowths; the tubules of the T-system. When the nuclei and myofibrils exchange position then, *ex hypothesi*, those ingrowths would be carried into the depths of the parent fibre. Such a process raises major problems, however, *viz à viz* the addition of further myofibrils during the remainder of the developmental period and, possibly, also during post-natal growth.

Those problems, however, are not necessarily insuperable. Thus initially each newly added myofibril will be slender and it could 'thread its way' between the T-tubules as they run to the surface. Subsequently, as the myofibril increases in girth the tubules would be gently displaced sideways about the developing myofibril; a process which certainly must happen during fibre growth anyway. Since, moreover, adjacent myofibrils are very much in register with one another, the correct positioning of the T-tubules relative to the sarcomeres of the new myofibrils would be almost automatic. In passing, however, it is worth emphasizing that lurking under the commonplace observation that the sarcomeres of adjacent myofibrils are in register there is yet another set of, as yet, undefined development processes and problems.

In summary, therefore, it is quite clear from the preceding, albeit brief, consideration of mammalian myogenesis that notwithstanding the extensive data currently available concerning the structure of the established skeletal muscle fibre and of some of the aspects of the development of that organization, we are still profoundly ignorant of the precise nature of most of the mechanisms involved and how the mechanisms themselves are controlled.

Necessarily, therefore, much of this paper has been deliberately speculative in an attempt to underline the shortcomings of our knowledge of myogenesis and to suggest some new ways of approaching that phenomenon. This may be justified by the old Spanish proverb:

'If a man has never been known to be wrong it is probably because he has never said anything worth while.'

REFERENCES

Andersson-Cedergren, E. (1959). *J. Ultrastruct. Res.* **1**, Suppl. 1
Allen, E. W. and Pepe, F. A. (1965). *Am. J. Anat.* **116**, 115
Bassleer, R. (1962). *Z. Anat. EntwGesch* **123**, 184
— Collignon, P. and Matagne-Dhoosche, Fr. (1963). *Archs. Biol., Paris* **74**, 79
Bennett, H. S. and Porter, K. R. (1953). *Am. J. Anat.* **93**, 61
Bowman, W. (1840). *Phil. Trans. R. Soc.* **130**, 457
Boyd, J. D. (1960). In: *The Structure and Function of Muscle*, Vol. 1, p. 63. (Ed. by G. H. Bourne.) London; Academic Press
Capers, C. R. (1960). *J. biophys. biochem. Cytol.* **7**, 559
Cooper, W. G. and Konigsberg, I. R. (1961). *Anat. Rec.* **140**, 195
Corsi, A. and Perry, S. V. (1958). *Biochem J.* **68**, 12
Couteaux, R. (1955). *Int. Rev. Cytol.* **4**, 335
— (1959). *C.r. hebd. Séanc. Acad. Sci., Paris* **249**, 307
— (1960). In: *The Structure and Function of Muscle*, Vol. 1, p. 337. (Ed. by G. H. Bourne.) London; Academic Press
Dessouky, D. A. and Hibbs, R. G. (1965). *Am. J. Anat.* **116**, 523
Firket, H. (1958). *Archs. Biol., Paris* **69**, 1
Franzini-Armstrong, C. and Porter, K. R. (1964). *J. Cell Biol.* **22**, 675
Hanson, J. (1967). *Nature, Lond.*
Hay, E. D. (1963). *Z. Zellforsch. mikrosk. Anat.* **59**, 6
Heuson-Stiennon, J. A. (1964). *J. Microsc.* **3**, 229
Huxley, A. F. and Taylor, R. E. (1958). *J. Physiol., Lond.* **144**, 426
— (1953a). *Proc. R. Soc.* **141B**, 59
— (1953b). *Biochim. biophys. Acta* **12**, 387
— (1957). *J. biophys. biochem. Cytol.* **3**, 631
— (1963). *J. molec. Biol.* **7**, 281
— (1964a). *Nature, Lond.* **202**, 1067
— (1964b). *Proc. R. Soc.* **160B**, 442
— and Hanson, J. (1960). In: *The Structure and Function of Muscle*, Vol. 1, p. 183. (Ed. by G. H. Bourne.) London ; Academic Press

Knappeis, G. G. and Carlsen, F. (1962). *J. Cell Biol.* **13**, 323

Konigsberg, I. R., McElwain, N., Tootle, M. and Herrmann, H. (1960). *J. biophys. biochem. Cytol.* **8**, 333

Lash, J. W., Holtzer, H. and Swift, H. (1957). *Anat. Rec.* **128**, 679

Message, M. A. (1967a). In preparation

— (1967b). In preparation

Ogawa, Y. (1962). *Expl Cell Res.* **26**, 269

Page, S. (1964a). *J. Physiol., Lond.* **175**, 10P

— (1964b). *Proc. R. Soc.* **160B**, 460

Palade, G. E. and Palay, S. L. (1954). *Anat. Rec.* **118**, 335

Penrose, L. S. (1958). *Ann. hum. Genet.* **23**, 59

— (1960). *New Biol.* **31**, 57

Perry, S. V. and Corsi, A. (1958). *Biochem. J.* **68**, 5

Porter, K. R. and Palade, G. E. (1957). *J. biophys. biochem. Cytol.* **3**, 269

Reedy, M. K. (1964). *Proc. R. Soc.* **160B**, 458

Reger, J. F. (1957). *Expl Cell Res.* **12**, 661

Robertson, J. D. (1957). *J. Physiol., Lond.* **140**, 58

— (1958). *Anat. Rec.* **130**, 440

— (1966). In: *Principles of Biomolecular Organisation.* p. 357 (Ed. by G. E. W. Wolstenholme and M. O'Connor.) London; Churchill

Rich, A. (1963). *Scient. Am.* **209**, 44

Schwann, T. (1839). *Mikroskopische Untersuchungen über die Uebereinstimmung in der Struktur und dem Wachstum der Thiere und Pflanzen.* Berlin; Reimer

Smith, D. S. (1966). *Prog. Biophys. biophys. Chem.* **16**, 107

Strehler, B. L., Konigsberg, I. R. and Kelley, J. E. T. (1963). *Expl Cell Res.* **32**, 232

Waddington, C. H. and Perry, S. V. (1963). *Expl Cell Res.* **30**, 599

Warner, J. R., Rich, A. and Hall, C. E. (1962). *Science, N.Y.* **138**, 1399

Veratti, E. (1902). *Memorie Ist lomb. Sci. Lett.* **19**, 87. (Translation: *J. biophys. biochem. Cytol.* **10**/Suppl.)

ASPECTS OF FAT DEPOSITION AND MOBILIZATION IN ADIPOSE TISSUE

J. H. ADLER and H. E. WERTHEIMER

Hebrew University Hadassah Medical School, Jerusalem

IN THE RECENT PAST, adipose tissue was not regarded as a dynamic organ, and the accumulation of triglycerides in the fat cells was considered to be a purely passive process. Indeed, microscopic examination of the tissue reveals a mass of cells incorporating 95 per cent triglycerides, approximately 3 per cent connective tissue and 2 per cent protoplasm. The fat mass was thought of, primarily, as an energy reservoir characterized by a very low metabolic rate, served by a meagre vascular system and having a negligible degree of innervation. The passive protection against the dissipation of heat and, in some instances, mechanical tasks were recognized as additional functions of fat.

Within the last two decades a series of publications (Schoenheimer, 1942; Wertheimer and Shapiro, 1948; Dole, 1956; Gordon and Cherkes, 1956; Wertheimer and Shafrir, 1960; Kinsell, 1962; Jeanrenaud, 1963 and Randle, Garland, Hales, Newsholme, Denton and Pogson, 1966) have radically revised our concepts of the subject. New methods have made possible the demonstration of rich vascularization (Gersh and Still, 1945) and developed innervation in adipose tissue (Wirsen, 1965). Today, adipose tissue is included among the metabolically active organs, together with muscle and liver. The tissue is regarded as a dynamic centre of energy regulation, one which is integrated with physiological, neural and humoral mechanisms. These mechanisms synchronize the balance between fat deposition and mobilization with the changing requirements of the organism.

Research on the physiological activity of adipose tissue has been carried out both *in vivo* and *in vitro*. The *in vitro* methods include perfusion of fat pads (Robert and Scow, 1963), the use of adipose tissue slices, isolated fat cells and fat cell membranes (Rodbell, 1964). These investigations revealed various effects of hormones and metabolites on the preparations studied, on the organism, the tissue and the cell. The logical consequence of these studies was the investigation of cell-free systems, the enzymes and their kinetics; these subjects are reviewed in *The Handbook of Physiology*, Section V (1965), American Physiological Society.

FAT DEPOSITION IN ADIPOSE TISSUE

Prior to 1948, the conversion of carbohydrates to fat was thought to be carried out only in the liver. Shapiro and Wertheimer (1948) demonstrated the transformation of glucose and acetate to free fatty acids in adipose tissue with the aid of deuterium. Favarger and Gerlach (1954) showed that only a minor fraction of the total lipogenesis occurred in the liver of mice, while most of the triglycerides were found to be synthesized in adipose tissue.

Renold and Cahill (1965) showed that insulin was the major hormone promoting lipogenesis and may be regarded, therefore, as the anabolic hormone of adipose tissue. In addition to promoting lipogenesis, insulin also increased adipose tissue glycogen, protein and RNA synthesis. These effects have been summarized by Randle *et al.* (1966). The effect of physiological doses of insulin on the epidydimal fat pad is quantitatively graded and serves as one of the most useful bio-assays for small amounts of insulin in serum. In diabetes, adipose tissue anabolism is decreased and is associated with lipolysis and ketogenesis. These trends can be reversed by the administration of insulin. The relationship of insulin to fat anabolism and lipolysis can be demonstrated *in vivo* and *in vitro*. Many investigators today believe that, quantitatively, the most important function of insulin is its anabolic effect on adipose tissue.

The manner in which insulin links carbohydrate and fat metabolism and participates in the regulation of energy metabolism can be readily appreciated, since lipogenesis in adipose tissue is a function of the concentration of circulating insulin, and the concentration of circulating insulin depends on the blood glucose level. When glucose is available, insulin secretion is promoted, leading to glycogen and triglyceride storage. When glucose is deficient, adipose tissue anabolism is inhibited. Hyperglycaemia, in the absence of insulin, does not promote fat anabolism, as can be seen in pancreatic diabetes.

FAT MOBILIZATION FROM ADIPOSE TISSUE

The active mobilization of depot fat is a function of the sympathetic division of the autonomic nervous system; this has been demonstrated both *in vivo* and *in vitro*. Adipose tissue is known to store norepinephrine, which is the transmitter substance of the autonomic nervous system. Paoletti, Smith, Maickel and Brodie (1961) showed that these stores were depleted following transection of the autonomic nerve supply to adipose tissue. Direct stimulation of the autonomic nerve supply to the incubated epidydimal fat pad of rats *in vitro* has been shown by Correll (1963) to result in lipolysis and the release of fatty acids. The addition of anti-adrenergic drugs caused partial or complete inhibition of the response. Rossel (1966) demonstrated the release of free fatty acids and glycerol from canine subcutaneous adipose tissue, by stimulation of sympathetic nerves supplying the region. The release of endogenous adrenalin and noradrenalin from the adrenal medulla also causes depot fat mobilization and is characterized by an increase in circulating free fatty acids and glycerol. In normal circumstances, tonic activity of the sympathetic nerve supply to adipose tissue is, apparently, the major adrenergic factor contributing to the balance between lipolysis and lipogenesis. Plasma levels of free fatty acids are not affected in adrenalectomized dogs maintained on cortisone, but when sympathetic activity is inhibited by ganglion-blocking agents, plasma free fatty acid levels drop (Havel and Goldfein, 1959).

Exogenous noradrenalin, both *in vivo* and *in vitro*, causes lipolysis in adipose tissue leading to the release of free fatty acids and glycerol. The catecholamines act on a special catecholamine-sensitive lipase; insulin counteracts this activity by specifically decreasing lipolysis.

Thanks to the basic contribution of Dole, at the Rockefeller Institute, and Gordon and Cherkes, at the N.I.H., we now know that the transport of energy from adipose tissue occurs in the form of free fatty acids which are circulated in the blood in an albumin-bound form. These free fatty acids (FFA) were referred to as NEFA (non-esterified fatty acids) by Dole and as UFA (un-esterified fatty acids) by Gordon and Cherkes. The free fatty acids provide a major source of energy to the tissues of the organism, particularly skeletal muscle, cardiac muscle and kidney. The mobilization of free fatty acids is activated by means of the sympathetic nervous system; either by local release of noradrenalin in the nerve endings serving the adipose tissue or by the systemic release of adrenalin and noradrenalin from the adrenal medulla in the case of emergency. When this occurs, increased catecholamines can be demonstrated in the serum and urine together with elevated levels of circulating free fatty acids.

SPECIES VARIATIONS

The physiological role of glucose, insulin and catecholamines relative to adipose tissue regulation, is well-established. In a number of mammalian species, lipolysis may be enhanced by the absence of insulin and fat deposition can be decreased by adrenergic activity; in both cases circulating free fatty acids will be elevated.

The physiological role of other lipolytic hormones such as ACTH, growth hormone and glucagon is not clear; their effect varies in different species. Shafrir and Wertheimer (1965) have pointed out that the physiological activity of adipose tissue in different species may vary, not only quantitatively, but also qualitatively.

Carlson, Liljedahl, Verdy and Wirsen (1964), found that administration of norepinephrine to the domestic fowl did not bring about a significant elevation in the level of plasma free fatty acids. He measured lipolysis in fowl and rat adipose tissue *in vitro* and found that poultry adipose tissue was not responsive to norepinephrine or porcine ACTH, while rat tissue was. Glucagon stimulated glycerol release from both rat and poultry adipose tissue. Avian adipose tissue differs, apparently, from that of mammals in a number of respects and migratory birds have special photoperiodic dependent patterns of fat deposition.

In addition to species differences, strain differences also exist. Larson (1966) found that five different strains of mice varied in carcass composition and fat content, when fed the same diet. After a 24 h fast, all strains decreased in weight, but not equally. The amount of fat lost was a strain characteristic; some strains have a higher fat-mobilizing capacity than others and lose more fat. Plasma UFA and glucose levels also differed in different strains.

Adipose tissues taken from different parts of the same animal may reveal significant differences in their metabolic activity. If, for example, the activity of brown fat is compared with that of the occular fat body, it can be seen that brown fat may have one of the highest metabolic rates of all tissues, relative to its dry weight (Joel, Treble and Ball, 1964) and it can serve as an hermogenic centre, whereas the function of occular fat is primarily mechanical and its metabolic rate is low. Subcutaneous and intraperitoneal fat have an intermediate position, relative to their rate of metabolism.

SOME ASPECTS OF FAT DEPOSITION
AND MOBILIZATION IN RUMINANTS

The mechanisms of adipose tissue regulation, discussed above, are based on studies in non-ruminant mammals, primarily the rat, dog and man. In the case of ruminants, somewhat different, or additional, regulatory mechanisms and metabolic pathways are to be expected, although some of these differences may be quantitative rather than qualitative. As mentioned previously, the major quantitative role of insulin in non-ruminant mammals is the conversion of glucose to triglycerides in adipose tissue. This does not appear to be the case in ruminants because of a relative glucose deficiency on one hand and the presence of ample sources of metabolites, such as acetic acid for the production of long chain fatty acids, on the other. Armstrong (1965) showed that if sufficient glucose is administered intravenously to sheep, its conversion efficiency for lipogenesis is equal to that found in non-ruminants, but he considered this unlikely to occur in ruminants in normal circumstances.

When considering energy metabolism in ruminants, their physiological and metabolic peculiarities must be borne in mind. Annison and White (1961) and Bergman (1963) have demonstrated that sheep oxidize less glucose than non-ruminant mammals. Ruminants showed decreased glucose tolerance, as demonstrated by McCandless and Dye (1950) in sheep. Reid (1950) and Jasper (1953) demonstrated a diminished rate of fall of blood glucose in response to insulin injection in sheep, as compared with non-ruminant animals. Jasper (1953) showed that a similar phenomenon occurred in the cow and that the cow had a marked resistance to insulin-induced hypoglycaemia.

Ruminants, however, do not lack insulin; insulin of ruminant origin serves as a major pharmaceutical source of the hormone. McCandless, Woodward and Dye (1948), Jarret (1946) and Lindsay (1961) demonstrated alloxan diabetes in sheep. Cunningham (1962a and b) demonstrated the presence of circulating insulin in ovine and bovine plasma. Cunningham et al. (1963) induced hyperglycaemia in both cows and sheep by the injection of guinea-pig insulin antiserum. They reported that the rate of increase of glucose, following the injection, was slower than that observed in non-ruminant animals under similar experimental conditions. Boda (1964), employing an immunological assay for serum insulin, showed that both starved and fed sheep responded with a tenfold increase in insulin to exogenous glucose and that there was a positive relationship between insulin secretion and the utilization rate of injected glucose. He concluded that 'certain characteristics of sheep, such as reduced glucose tolerance and low glucose oxidation, are not due to a lack of circulating insulin or to an inability of hyperglycaemia to mobilize insulin stores'. The glycaemic levels induced by Boda in his experiments exceeded 200 mg per cent. These values are far higher than those which normally occur in sheep.

Glucose catabolism in ruminants begins in the digestive tract, by virtue of the action of rumen micro-organisms, and the amount of glucose absorbed is negligible. The portal circulation of the ruminant host is presented with acetic, butyric and propionic acids, or their derivatives, in lieu of glucose. Of these, only proprionic acid can serve as a source of glycerol and it comprises 20–30 per cent of the total VFA in the rumen; acetic and butyric acids account for approximately 60 per cent and 20 per cent respectively. The mass of raw material for the synthesis of triglycerides in non-ruminant animals includes significant quantities of glucose. In ruminants, the materials available for depot fat synthesis do not include absorbed glucose but consist largely of acetic, propionic and butyric acids, while ample sources exist for the production of the fatty acid moiety of the adipose tissue triglyceride, including acetic and butyric acids.

Comparison of the triglyceride precursors available to ruminants with those available to non-ruminant mammals, reveals the relative paucity of exogenous ruminant sources of glycerol. Quantitatively, propionate appears to be the logical source of glycerol. The known metabolic pathways suggest that it can be converted to glycerol in an insulin independent manner. When considering gluconeogenic pathways from protein as sources of glycerol production, insulin dependence is not conditional.

The utilization of acetic acid, for the purpose of adipose tissue triglyceride synthesis, depends, apparently, on the presence of propionate or glucose. Armstrong and Blaxter (1957a and b) and Armstrong, Blaxter, Graham and Wainman (1958) studied different mixtures of VFA in proportions which are likely to occur in the rumen, even when extreme diets are given. They found that the efficiency of utilization of the different mixtures for maintenance of sheep was between 80 and 85 per cent, even when acetic acid comprised 90 per cent of the total. In studying the utilization of different VFA mixtures for fat production, the least efficient was acetic acid given alone. In these experiments, it was shown that as the level of acetic acid increased, the efficiency of utilization for fat production decreased; the reverse was true for propionic acid or glucose. Efficiency of utilization of acetic acid was shown to be similar to that of propionic acid for milk production in goats. It is probable that the inefficiency of acetic acid as a source of fat is due to the lack of glycerol precursors.

Some factors which affect circulating UFA levels in ruminants have been established. Increased plasma UFA levels have been demonstrated in starved sheep by Annison (1960) and Lindsay (1961). Reid and Hinks (1962) found that the plasma UFA level was a most sensitive indication of the degree of undernutrition in pregnant ewes. Glucose infusion reduced the plasma UFA level of starved sheep (Annison, 1960). This effect of glucose was shown to be quantitative, depending on the amount of glucose injected (Patterson, 1964). It has been shown also that the increased plasma UFA resulting from starvation is associated with hypoglycaemia and hyperketonaemia (Annison, 1960; Lindsay, 1961). Alloxan diabetes in sheep was associated with elevated plasma UFA levels (Lindsay, 1961). Cunningham *et al.* (1963) showed that the inactivation of circulating insulin by the injection of guinea-pig insulin antiserum, led to elevated UFA levels in sheep and cows.

Adler and Wertheimer (1962) studied 35 normal lactating cows and found that there was a reciprocal relationship between blood glucose and plasma UFA; the latter increased as blood glucose diminished. Kronfeld (1965) showed that plasma UFA diminished when insulin or glucose was injected intravenously and that plasma UFA increased with starvation and adrenalin injection in cows.

If some of the factors which affect the level of circulatory UFA in ruminants are summarized, it appears that their responses to hypoglycaemia, hyperglycaemia, starvation, adrenalin, insulin or its deficiency, are similar, at least qualitatively, to those of non-ruminant mammals. Plasma UFA levels of sheep were also shown to be affected by psychological factors (Reid and Mills, 1962; Patterson, 1964).

The quantitative aspects of these relationships are concerned with the control of energy metabolism. The concept of caloric homeostasis presents a useful approach for the evaluation of this problem. Fredrickson and Gordon (1958) suggested that the interrelationship between circulating UFA and carbohydrates and the mechanisms that regulate them could be explained, at least tentatively, in terms of a concept of caloric homeostasis. The assumption behind this concept is that physiological regulatory mechanisms organize the supply of tissue energy requirements; for example, hypoglycaemia is compensated for by an increase in plasma UFA. The level of circulating energy available to the tissues cannot be calculated simply by multiplying the concentration of circulating metabolites by their caloric value and rate of utilization, because additional variables are involved. Tissues differ in their ability to utilize certain metabolites. Changes in the levels of circulating metabolites may affect each other's rates of utilization. The endocrine dependence of different tissues regarding metabolic activity varies and interaction between hormones may modify their metabolic effect. Mayfield, Bensadoun and Johnson (1966) studied the incorporation of acetate into homogenates of different tissues of sheep. Under their experimental conditions, acetate oxidation and incorporation into long chain fatty acids and neutral lipids occurred at a greater rate in adipose tissue homogenates than in those of kidney, muscle, heart, lung, liver and brain. Weber, Convery, Lea and Stamm (1966) have shown that increasing the level of octanoate *in vitro* inhibited key glycolytic enzymes of the liver. They interpreted this effect in terms of a negative feed-back mechanism which could switch on gluconeogenesis. Positive feed-back mechanisms of this nature also occur, for example, the utilization of acetate by mammary gland slices of non-ruminants is enhanced by the presence of glucose (Weber *et al.*, 1966).

Comparison of ruminant with non-ruminant mammalian mammary gland slices *in vitro* provides a good example of species variation and tissue specificity relative to metabolite utilization and hormone dependence. Foley (1956) summarized a series of publications on metabolism in the mammary gland; these showed that the ruminant mammary gland utilized acetate in preference to glucose for the purpose of fat production, while the opposite is true for the mammary gland slices of non-ruminants. Insulin was shown to increase fat production and glucose utilization in non-ruminant mammary tissue slices while similar preparations, of ruminant origin, were indifferent to the presence of insulin. Insulin independence of the ruminant mammary gland may account for the fact, previously mentioned, that increased dietary acetate is associated with high milk fat levels and that poor depot fat production, in ruminants, is insulin dependent. Since propionate is glucogenic, its presence may be related to increased insulin secretion. To our knowledge, no one has demonstrated the degree of insulin dependence of depot fat in ruminants.

The central role of insulin in the control of caloric homeostasis in non-ruminant mammals is apparent in view of the fact that exogenous glucose is a major source of energy. As 80–90 per cent of ruminant energy requirements are derived from VFA, any concept of caloric homeostasis in these species has to account for the integration of these acids within the framework of a physiological control mechanism.

Bergman, Roe and Kon (1966) found that 50 per cent of the absorbed propionate in sheep is converted to glucose and that it supplies 20–40 per cent of total glucose of the adult ruminant.

Lindsay (1961) discussed the question of endocrine control of energy metabolism in ruminants and suggested the probable significance of adipose tissue in relation to acetate metabolism, since acetate contributes as much as 50 per cent of the ruminants' energy supply. The data he presented showed that the intravenous injection of acetate in intact sheep leads to the reduction of blood glucose and plasma UFA. We confirmed this observation by perfusion of acetate through a duodenal fistula. Additional evidence for the existence of a reciprocal relationship between absorbed ketogenic fatty acids and circulating glucose and UFA is available. Reid and Hinks (1962) showed that plasma UFA levels of sheep declined after feeding, while blood ketone levels increased. Radloff, Schultz and Hoekstra (1966) found that the plasma UFA and blood glucose of the cow declined and blood ketones increased after feeding. While it appears that reciprocal relationships exist between absorbed acetic and butyric acids or their products, on the one hand, and circulating glucose and UFA, on the other, we do not yet understand the regulatory mechanisms that control them. Horino, Machlin, Hertelendy and Kipnis (1966) showed an increase in circulating insulin following the administration of butyric or propionic acid, but acetate infusion did not stimulate insulin secretion.

PRACTICAL APPLICATIONS TO RUMINANT PROBLEMS

The significance of ketosis in ruminants has presented a theoretical and clinical problem ever since Jöhnk (1911, 1912) found ketonuria in bovine clinical ketosis. The occurrence of ketonuria was recognized also in normal cows and its clinical significance is equivocal. Jöhnk considered that the steady secretion of ketone bodies occurred normally in cows. Udall (1954) stated that the appearance of ketone bodies in the urine of cows was not necessarily associated with clinical ketosis.

The views of Soskin and Levine (1952) that 'the fundamental disturbance underlying all ketosis is a relative or absolute lack of carbohydrates in the liver leading to an excess breakdown of fat' is not necessarily correct in the case of ruminants, since they absorb considerable quantities of ketogenic VFA. Pennington (1952) found that rumen epithelium *in vitro* converts butyric acid to β-hydoxybutyric acid and Annison, Hill and Lewis (1957) showed that the concentration of ketones in portal blood exceeds that of carotid blood in sheep. Furthermore, it must be borne in mind that absorbed acetic and butyric acids may be converted into ketones in the liver. Dole (1958) showed that the increased plasma UFA in diabetes was correlated with ketonaemia and considered that the ketone bodies originated from the plasma UFA.

These considerations led us to believe that the study of circulating free fatty acids in normal cows and in bovine clinical ketosis would illuminate the question of the dual origin of ketones in ruminants (Adler, Wertheimer, Bartana and Flesh, 1963). We found that in typical cases of clinical ketosis, plasma UFA levels were elevated and exceeded 0·6 m-equiv./l plasma.

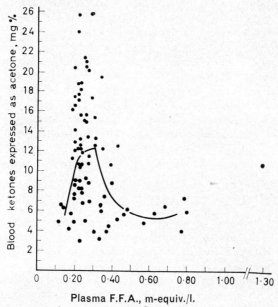

Figure 1. Blood ketones expressed as milligrammes per cent acetone plotted against FFA m-equiv./l plasma. Data from 77 ewes

The UFA levels of normal cows were lower and, in addition, were not correlated with blood ketone levels. The finding of elevated UFA in bovine clinical ketosis was confirmed by Kronfeld (1965). We concluded, therefore, that ketonaemia of normal cows originated primarily from the rumen, while in cases of clinical ketosis the ketones may have a dual origin, or, if the cow is not eating, keto-naemia will be due entirely to free fatty acid mobilization from adipose tissue.

The second project of this nature we undertook involved Awassi sheep (Adler and Lotan, 1967). Awassi sheep, in Israel, have been selected for high milk production. Unselected native sheep yield 80–100 kg of milk annually, while selected herds with an average

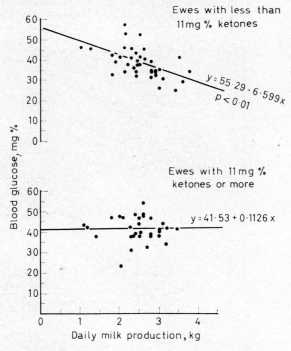

Figure 2. Blood glucose in milligrammes per cent plotted against daily milk production in kilogrammes for two groups of ewes. One regression line was calculated from data on 38 ewes with blood ketone levels of less than 11 mg per cent (expressed as acetone); the other regression line was calculated from data on 31 ewes with blood ketones of 11 mg per cent or more (expressed as acetone)

annual milk production of 250 kg or more, are common. It was reasoned that these ewes were candidates for lactation ketosis, in analogy to high-producing cows. The study undertaken involved 84 lactating ewes; blood glucose and plasma UFA concentrations were plotted against milk production. A significant decrease in blood glucose and increase in plasma UFA levels was found to be associated with increased milk production. When total blood ketones, expressed as milligrammes per cent acetone, were plotted against plasma UFA, the highest ketone values coincided with relatively low UFA concentrations of 0·2 and 0·4 m-equiv./l (*Figure 1*). It was considered improbable, therefore, that these ketones resulted from depot fat mobilization. These findings suggested that the observed ketones, or their precursors, were suppressing plasma UFA. In view of this interpretation, it was decided to analyse separately the data on ewes with high blood ketone levels and low blood ketone

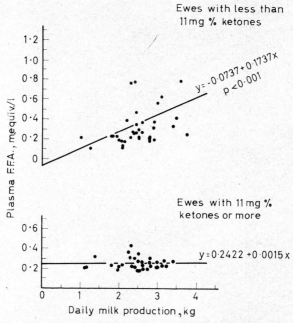

Figure 3. FFA in m-equiv./l plasma plotted against daily milk production in kilogrammes for two groups of ewes. One regression line was calculated from data obtained from 38 ewes whose blood ketone levels were below 11 mg per cent; the other regression line was calculated from data on 31 ewes whose blood ketone levels were 11 mg per cent or more

levels. Ewes whose blood ketone level exceeded 11 mg were included in the high ketone group, the remainder were regarded as the low ketone group; 11 mg was selected as the dividing value as it was the mean blood acetone value. *Figure 2* shows that the blood glucose of the low ketone groups of ewes diminished as milk production increased, while the blood glucose level of the high ketone group of ewes was independent of milk production. When plasma ketones were plotted against milk production (*Figure 3*), it could be seen that UFA levels increased with milk production in the low ketone group of ewes while the UFA level of the high ketone group was independent of milk production. These results suggest that the blood ketones, or their precursors, are glucose sparing and that they participate in caloric homeostasis. By correlating blood ketone levels with plasma UFA in ruminants, it is possible to glean some information as to the origin of circulating ketones; those that are associated with high UFA levels probably derive from adipose tissue. In recognition of the importance of the dual origin of ketones in ruminants, it is suggested that rumen derived ketones be designated RDK and adipose derived ketones as ADK. Some of the relationships discussed here are summarized in *Figure 4.*

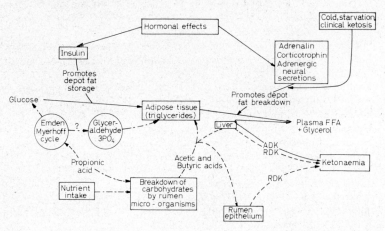

Figure 4.   Schematic outline of suggested pathways associated with ADK (Adipose derived ketones) and RDK (Rumen derived ketones) (based on Adler and Wertheimer, 1962)

Continuous lines ——— represent known hormonal control mechanisms of adipose tissue deposition and breakdown in nonruminant mammals.   Those mechanisms are probably common to ruminants with reservations related to ruminant insulin resistance and special digestive processes

Dots and dashes −·−·−·− show special ruminant digestive mechanisms, high VFA absorption available for formation of long chain fatty acids in adipose tissue.   The suggestion that propionate may be converted to glycerol without glucose degradation is incorporated

Dashes − − − − indicate route of RDK formation in rumen epithelium and liver

REFERENCES

Adler, J. H., Wertheimer, E. H., Bartana, U. and Flesh, J. (1963). *Vet. Rec.* **75**, 304

— and Lotan, E. (1967). *J. agric. Sci., Camb.* **69**, 349

— and Wertheimer, E. H. (1962). *Bull. Res. Coun. Israel*, E. **10**, 97

Annison, E. F., Hill, K. J. and Lewis, D. (1957). *Biochem. J.* **66**, 592

— and White, R. R. (1961). *Biochem. J.* **80**, 162

— (1960). *Aust. J. agric. Res.* **11**, 58

Armstrong, D. G (1965). In: *Physiology of Digestion in the Ruminant*, p. 272 (Ed. by R. W. Dougherty). London; Butterworths

— and Blaxter, K. L. (1957a). *Br. J. Nutr.* **11**, 247

— — (1957b). *Br. J. Nutr.* **11**, 413

— — Graham, N. McC. and Wainman, F. W. (1958). *Br. J. Nutr.* **12**, 177

Bergman, E. N. (1963). *Am. J. Physiol.* **204**, 147

— Roe, W. E. and Kon, K. (1966). *Am. J. Physiol.* **211**, 793

Boda, J. M. (1964). *Am. J. Physiol.* **206**, 419

Carlson, L. A., Liljedahl, S. O., Verdy, M. and Wirsen, C. (1964). *Metabolism* **13**, 227

Correll, J. W. (1963). *Science, N.Y.* **140**, 387

Cunningham, N. F. (1962a). *J. Endocr.* **25**, 35

— (1962b). *J. Endocr.* **25**, 43

— Patterson, D. S. P. and Wright, P. H. (1963). *J. Physiol.* **169**, 137

Dole, V. P. (1956). *J. clin. Invest.* **35**, 150

— (1958). *Bull. N.Y. Acad. Med.* **34**, 21

Favarger, P. and Gerlach, J. (1954). *Helv. physiol. pharmac. Acta* **12**, C15

Foley, S. J. (1956). *The Physiology of Lactation.* Springfield, Ill.; Charles Thomas

Fredrickson, D. S. and Gordon, R. S. (1958). *Physiol. Rev.* **38**, 585

Gersh, I. and Still, M. A. (1945). *J. exp. Med.* **81**, 219

Gordon, R. S. Jr. and Cherkes, A. (1956). *J. clin. Invest.* **35**, 206

Havel, R. J. and Goldfein, A. (1959). *J. Lipid Res.* **1**, 102

Horino, M., Machlin, L. J., Hertelendy, F. A. and Kipnis, D. M. (1966). *Proc. Endocr. Soc. 48th Meeting.* p. 32, Abstr. 11

Jarret, I. G. (1946). *Aust. J. exp., Biol. med. Sci.* **24**, 95

Jasper, D. E. (1953). *Am. J. vet. Res.* **14**, 184

Jeanrenaud, B. (1963). *Helv. med. Acta* **30**, 1

Joel, C. D., Treble, D. H. and Ball, E. G. (1964). *Fedn Proc. Fedn Am. Socs. exp. Biol.* **23**, Abstr. 48th Meeting. p. 271

Jöhnk, M. (1911). *Wschr. Tierheilk. Viehz.* **55**, 301

— (1911). *Wschr. Tierheilk. Viehz.* **55**, 383

— (1912). *Wschr. Tierheilk. Viehz.* **56**, 257

Kinsell, L. W. (Editor) (1962). *Adipose Tissue an Organ.* Springfield, Ill.; Charles Thomas

Kronfeld, D. (1965). *Vet. Rec.* **77**, 30

Larson, S. (1966). *Acta physiol. scand.* **68**, 43

Lindsay, D. B. (1961). *Digestive Physiology and Nutrition of the Ruminant.* p. 235. (Ed. by D. Lewis.) London; Butterworths

Mayfield, E. D., Bensadoun, A. and Johnson, B. C. (1966). *J. Nutr.* **89**, 189

McCandless, E. L. and Dye, J. A. (1950). *Am. J. Physiol.* **162**, 434

— Woodward, B. A. and Dye, J. A. (1948). *Am. J. Physiol.* **154**, 94

Paoletti, R., Smith, R. L., Maickel, R. P. and Brodie, B. B. (1961). *Biochem. biophys. Res. Commun.* **5**, 424

Patterson, D. S. P. (1964). *Res. vet. Sci.* **5**, 286

Pennington, R. S. (1952). *Biochem. J.* **51**, 251

Radloff, H. D., Schultz, L. H. and Hoekstra, W. G. (1966). *J. Dairy. Sci.* **49**, 179

Randle, P. J., Garland, P. B., Hales, C. N., Newsholme, F. A., Denton, R. M. and Pogson, C. I. (1966). *Recent Prog. Horm. Res.* **22**, 1

Reid, R. L. and Mills, S. C. (1962). *Aust. J. agric. Res.* **13**, 282

— (1950). *Aust. J. Res.* **1**, 182

— and Hinks, N. T. (1962). *Aust. J. agric. Res.* **13**, 1092

Renold, A. G. and Cahill, Jr. J. F. (1965). *Handbook of Physiology. Section 5. Adipose Tissue.* p. 824. Washington D.C.; Am. Physiol. Soc.

— Crofford, O. B., Stanffacher, W. and Jeanrenaud, B. (1965). *Diabetologia* **1**, 4

Robert, A. and Scow, R. O. (1963). *Am. J. Physiol.* **205**, 405

Rodbell, M. (1964). *J. biol. Chem.* **239**, 375

Rossel, S. (1966). *Acta physiol. scand.,* **67**, 343

Schoenheimer, R. (1942). *The Dynamic State of Body Constituents.* New York; Hafner Pub. Co. Inc.

Shafrir, E. and Wertheimer, E. H. (1965). *Handbook of Physiology. Section 5. Adipose Tissue.* p. 417. Washington, D.C.; Am. Physiol. Soc.

Shapiro, B. and Wertheimer, E. H. (1948). *J. biol. Chem.* **173**, 725

Soskin, S. and Levine, R. (1952). *Carbohydrate Metabolism.* 2nd Edn, Univ. Chicago Press

Udall, D. H. (1954). *The Practice of Veterinary Medicine.* 6th edn. Ithaca, N.Y.; Baillière, Tindall and Cox, Ltd

Weber, G., Convery, H. J. H., Lea, M. A. and Stamm, M. B. (1966). *Science, N.Y.* **154**, 1357

Wertheimer, E. H. and Shapiro, B. (1948). *Physiol. Rev.,* **28**, 451

— and Shafrir, E. (1960). *Recent Prog. Horm. Res.* **16**, 447

Wirsen, C. (1965). *Handbook of Physiology. Section 5. Adipose Tissue.* p. 97. Washington D.C.; Am. Physiol. Soc.

DISCUSSIONS

PROFESSOR LAMMING (Nottingham) observed that Professor Tanner's data showed clearly that the peaks in development of the tissues bone, muscle and fat were less distinct than indicated by earlier data and enquired whether PROFESSOR TANNER could offer any explanation for the mechanism of this spacing of development peaks. PROFESSOR TANNER replied that the growth peaks were linked to increased output of anabolic hormones from the adrenal cortex and the increase in growth of bone and muscle was due, presumably, to increased output of androgens. It was not at all clear, however, why peaks should be at slightly different times for different tissues but even within tissues there were different peaks for different parts of the body; for example, peak bone growth in the legs occurred 3–6 months before it did in the trunk. If there were these gradients within tissues it was not surprising that they occurred also between tissues.

DR. TIMON (Dublin) enquired whether the growth of fat tissue occurred independently of bone and muscle growth, to which PROFESSOR TANNER replied that there appeared to be a negative correlation between them.

DR. GREENWOOD (London) enquired about the result of force feeding during the period of the 'growth spurt'. PROFESSOR TANNER observed that children sometimes force-fed themselves and in obese children fat may not be lost during the adolescence period. To a query from DR. HUNTER (Edinburgh) on the effect of force feeding on muscle growth, PROFESSOR TANNER expressed uncertainty but confirmed that there was an adolescent spurt in bone growth. DR. MESSAGE (Cambridge) enquired what were the mechanisms of the rapid increase in growth rate after adolescence and could it be related to the fact that the new-born baby was subjected to gravitational force for the first time? He mentioned that mice which were centrifuged had lower growth rates than those which were not. PROFESSOR TANNER replied that the mechanisms of these changes were not known, nor were those of the fat waves which occurred up to 9 months of age and then declined. He did not consider that subjection to gravitational force was of significance, as birth appeared to exert little change on the pattern of the growth curves.

DR. GREENWOOD agreed with this point and expressed the view that nutrition was the major factor involved; the unborn child received a maximum food supply and hormonal levels would have little effect on prenatal growth.

To a comment by Dr. Rerat (Jouy-en-Josas) on the use of Ca45 in studies on bone growth and remodelling, DR. SISSONS agreed that radioisotopes were valuable for this purpose and revealed that Ca was taken up partly in the formation of new tissue and partly in exchange processes in tissue already formed.

The discussion then centred on the accumulation of tetracycline in bone tissue. DR. SISSONS elaborated on this and stated that uptake was initially on the bone surface but remained only where bone tissue was being formed; it thus acted as an indicator of areas of active bone growth. The amounts retained were not great and would have neither a beneficial nor adverse effect on bone tissue.

PROFESSOR TANNER enquired about the mechanism of increase in muscle width as, for example, during adolescence in the human. Was this an increase in the number or in the breadth of myofibrils? DR. MESSAGE then elaborated on what was known about postnatal increase in muscle fibre number and cast doubts on the usual interpretation of some of these data. Thus, it was generally presumed that fibre number increased to some extent following birth and then stabilized but there was increasing evidence from studies on mice and cattle that the true situation was a postnatal increase followed by a *decrease*. Thus, in British Friesian cattle there was no change in muscle fibre number up to 12 months of age but between 12 and 24 months there was a dramatic fall of 40 per cent in m. *longissimus dorsi* and of 50 per cent in m. *semitendinosus*. In this respect there were probably both species and muscle differences. Muscle development depended both on nutrient level and exercise; initially there was an increase in fibre number but subsequent development depended on these two factors.

MR. HOUSTON (Leeds) queried the validity of muscle fibre counts if the fibres were formed interstitially and did not extend from origin to insertion. DR. MESSAGE agreed that this was valid comment on the technique, because in some muscles of some species the fibrils extended the full length of the muscle, but others they did not. It was necessary to look for a more reliable parameter; for example, the total cholinesterase activity in the muscle divided by mean activity for individual myoneural junctions. Breed differences also influenced experimental results of fibre counts; between breeds of rat fibre counts for the same muscle were 2,500 as against 6,000.

To a query by MR. THURLEY (Weybridge) on how extra myofibrils were laid down with increase in muscle size, DR. MESSAGE replied that they were synthesized at random by a collision process in which the combination of one actin and one myocin provided the basis of a new myofibril, and that once the process had started it must continue to produce a complete myofibril.

DR. ARMSTRONG (Newcastle upon Tyne) commented that the ruminant animal was evolved to handle cellulose with the necessary enzymes for this and the usual diet provided little in the way of glucose or glucose precursors. The metabolism of glucose in the adipose tissue of the ruminant animal might be very different from that in the non-ruminant. It had been shown that in the mammary gland of the ruminant there were very low concentrations of citrate cleavage enzyme and therefore protection against the carbon of glucose being used to provide fatty acids for lactation. These same enzymes may be lacking in adipose tissue, in which case the role of glucose in live weight gain of the non-lactating ruminant was speculative.

DR. ADLER remarked that Blaxter and Armstrong had found acetate to be less efficiently utilized for fat production than mixtures of acetate with either propionate or glucose. It was not known how acetic acid was metabolized at the endocrine level, where the glycerol came from or what happened to acetic acid if it wasn't utilized.

DR. ARMSTRONG replied that glucose was a major supplier of the carbons of glycerol in the mammary gland and, therefore, perhaps also in the non-lactating animal. Furthermore, glucose supplied the necessary reduced coenzymes for fat synthesis even from acetate.

DR. GREENWOOD asked whether plasma insulin had been measured in, for example, the goat, to which DR. ADLER replied that it had been measured in the sheep and the cow by both serological and bioassay methods; he went on to observe that butyric or acetic acids may interfere with glucose utilization in the presence of insulin. Recent work on starved and non-starved sheep showed that insulin was released in response to glucose and to butyrate injection. DR. GREENWOOD stated that there was strong evidence that GH had a lipid mobilizing activity under a variety of stimuli, such as stress, exercise or fasting. Insulin had the action suggested by DR. ADLER but there were a multiplicity of lipid mobilizing factors the interplay of which, at least in the human, was not clear.

DR. FOWLER (Aberdeen) asked Professor Tanner whether he thought the need to use a log transformation in his data was because the linear measurement was really a cube root measurement of weight and the factor that was varying normally was weight of tissue, so that what was being measured was the cube root of weight. PROFESSOR TANNER replied that he thought it wrong to regard weight as more important than other measures, it was *least* important and was biologically very unspecific.

67

DR. FOWLER also asked Professor Tanner for his view on determinants in the growth of fat tissue. Thus, while some fat tissue could be regarded as essential to the fat-free body mass, it would seem that most of the fat tissue was simply the result of energy intake in excess of actual requirement. PROFESSOR TANNER agreed that excess energy *was* stored as fat but that, at least in the human, there were appetite regulating mechanisms which could prevent excess fat deposition in 'genetically thin' subjects. DR. ADLER commented that caloric homeostasis was a very complex situation and there were insufficient data at the present time to allow simple mathematical calculation of fat deposition from energy turnover.

DR. ARMSTRONG referred to DR. SISSONS' comments on the presence of tetra-cyclines in bone tissue and asked whether these could have any long term effect on function or strength. DR. SISSONS thought that the concentrations were too low to have any effect, deleterious or beneficial. Tooth discoloration was the only adverse effect of which he was aware; this was seen in children receiving large doses of tetracycline for long periods.

DR. BRAUDE (Reading) asked whether the isotopic methods for measuring bone growth mentioned by Dr. Sissons, could be applied to the measurement of muscle growth, to which DR. MESSAGE replied that no-one appeared to have attempted this with isotopically labelled amino acids; indirectly, tritiated thymidine had been used but the results of this, showing the presence of labelled nuclei, were very difficult to interpret because of the fact that nuclei moved along muscle fibres. The presence of labelled nuclei also raised the interesting question of whether there was a true turnover of DNA or whether interstitial nuclei were being incorporated into existing fibres.

II. HORMONAL INFLUENCES

A DIMINISHED ROLE FOR GROWTH HORMONE IN THE REGULATION OF GROWTH

W. M. HUNTER

Medical Research Council Clinical Endocrinology Research Unit, Edinburgh

INTRODUCTION

THE LAST 4 years have seen great strides in our understanding of the physiology of growth hormone in man. This advance has come about largely through the application of a new immunological assay procedure (Hunter and Greenwood, 1962, 1964; Glick, Roth, Yalow and Berson, 1963) which permits the precise estimation of the hormone in the small concentrations present in the bloodstream. I shall pursue the question of what part growth hormone plays in the regulation of growth largely by the use of our own data, partly because it is more familiar to me but also because, in some cases, our experiments were designed to yield evidence on this particular subject of growth regulation.

It should be understood that the physiology to be described applies to man and probably to the other primates since, chemically, primate growth hormones are very similar to each other. Because growth hormone from lower mammals are chemically and immunologically different, the physiology may not be the same and, further, the assay as used for measuring GH in man cannot be directly applied for measurement in other species.

TIME SCALE OF SECRETION AND BIOLOGICAL ACTIONS

Most people when confronted with the problem of hormonal regulation of growth would think in terms of some sort of direct relationship which could be expressed in the form:

$$\text{growth velocity} \propto \text{rate of growth hormone secretion.}$$

However, growth hormone is still produced when growth ceases, indeed extraction procedures find as much in the pituitary glands of adults as in those of children. This, of course, presented a major point of interest for the investigation of GH physiology.

The first surprise resulting from the application of the radio-immunoassay to human growth hormone (HGH) was the remarkably quick disappearance of the hormone from the plasma following its I.V. injection; a half-life of 20–30 min has been obtained in three laboratories (Parker, Utiger and Daughaday, 1962; Hunter and Greenwood, 1964; Glick, Roth and Lonergan, 1964). This was surprising, of course, because as growth is a slow process it might be thought not to require a hormone capable of showing rapidly fluctuating plasma levels. It might have been expected that plasma levels would remain fairly constant and one's first thought about the short half-life was that the body was simply rather wasteful in this instance. It soon became apparent, however, that plasma GH levels *do* modulate very markedly on an hour-to-hour time-scale. Once this was seen, interest quickly turned to an examination of the acute effects known to follow the experimental administration of the hormone. These fall into two groups, those affecting nitrogen metabolism and those affecting energy-providing mechanisms, i.e. fat and carbohydrate metabolism.

Table 1 shows the principal immediate effects of GH on N metabolism.

Table 1

Immediate effects of growth hormone on nitrogen metabolism

Fall in plasma amino acids	Russell, 1955
Increased tissue uptake of amino acids *in vivo*	Riggs and Walker, 1960
Increased transport of amino acids into rat diaphragm *in vitro*	Noall *et al.*, 1957
Increased incorporation of amino acids into protein of rat diaphragm *in vitro*	Kostyo, 1964

There is a fall in plasma amino acids in a nephrectomized animal whose plasma amino acid level has been raised by prior administration of a protein hydrolysate (Russell, 1955). There is at the same time an increased tissue uptake of amino acids (Riggs and Walker, 1960). Increased transport of amino acids into rat diaphragm preparations *in vitro* have been demonstrated (Noall, Riggs, Walker and Christensen, 1957). There is a single instance (Kostyo, 1964), as yet unconfirmed, of GH added to the medium causing an increase in the incorporation of amino acids into protein by the rat diaphragm *in vitro* even when Na^+ was replaced by choline to stop the sodium pump upon which active transport of amino acids depends.

Table 2 shows the immediate effects upon fat and carbohydrate metabolism known to follow administration of GH. The principal effect here is undoubtedly the marked rise in plasma non-esterified fatty acids (NEFA) which reaches a maximum at 4–6 h (Raben and

Table 2

Immediate effects of growth hormone on fat and carbohydrate metabolism

Rise in plasma non-esterified fatty acids	Raben and Hollenberg, 1959
Stimulation of lipolysis in isolated fat cells	Fain, Kovacev and Scow, 1965
Increased uptake of NEFA by muscle followed by increased NEFA release by adipose tissue	Rabinowitz, Klassen and Zierler, 1965

Hollenburg, 1959). This effect was readily demonstrated and it merited particular attention since it required only very small doses of hormone. More recently Fain, Kovacev and Scow (1965) have been able to demonstrate an action of growth hormone in stimulating lipolysis in isolated fat cells. Concentrations of 10–100 μmg/ml GH were effective in this respect and this is precisely the physiological range of the hormone's plasma concentration. Forearm perfusion studies by Rabinowitz, Klassen and Zierler (1965) have shown that the first action of growth hormone is an increased uptake of NEFA by muscle and that this is followed by increased NEFA released from adipose tissues. It is perhaps worth emphasizing the enormous and rather recently established importance of NEFA in energy provision. That fat provides vastly more stored fuel than does carbohydrate has long been recognized. What is relatively new is the knowledge that NEFA is used directly, in many instances in preference to glucose, and that it is quantitatively much more important than glucose. Thus one of the most notable findings from the forearm perfusion experiments mentioned above was that the RQ of resting muscle was around 0·7 indicating an almost total reliance on fat. Further, much of the mobilization and combustion occurred quite locally without use of the blood compartment. The intermingling of fat in muscle tissue makes this easily possible and the mechanism further emphasizes how hormonal regulation of lipolysis can play a major role in energy regulation. The longer term effects of GH in decreasing body fat and causing insulin resistance, ketosis and transient or permanent diabetes can now be seen as the consequences of these primary actions.

PATTERN OF CHANGE IN PLASMA HGH LEVELS

Having now reviewed the direct actions of growth hormone, we can examine its plasma concentrations by means of the radio-immunoassay and later draw the two groups of data together.

The starting point came in 1963 when Roth, Glick, Yalow and Berson showed that insulin-induced hypoglycaemia or fasting, each stimulated the secretion of human growth hormone and that the plasma levels fell promptly following the administration of glucose. These responses were not given in stalk-sectioned subjects so the feed-back which abolishes HGH secretion probably has its receptors in the hypothalamus. There is considerable evidence for a GH or somatotrophin releasing-factor (SRF) which passes to the pituitary by way of the portal circulation. If we look at the effect of glucose more closely we see that the rate of fall in plasma HGH which follows intravenously administered glucose is exactly the same (20–30 min) as that which follows the intravenous injection of the hormone itself. This, of course, indicates that glucose abolishes HGH secretion. No substance present in the blood stream other than glucose has been shown to reduce HGH secretion.

To turn to stimulation of HGH secretion, we find that growth hormone rises in fasting. *Figure 1* shows HGH values for three healthy adult males on complete rest. Intermittent bursts of secretion occur from about noon onwards after an overnight fast (Hunter, Willoughby and Strong, 1968). Exercise in the fasting state is associated with more marked increases in plasma HGH levels. *Figure 2* shows data from experiments during which 6 adults walking at 4 mile/h for 2 h showed rises of 20- to 50-fold. These were clearly associated with a marked elevation of plasma NEFA and fall in RQ. In the next experiment (*Figure 3*), four walks by the same subject (a 29 year-old man) are represented. On two occasions he walked in the fasting state, the data are represented as means by the full lines, and the pattern is similar to that shown in *Figure 2*. The subject walked twice more, this time taking 25 g glucose before and at 30 min intervals while walking. The broken lines show the mean values for these experiments and it is clear that glucose abolishes alike the fall in RQ, the rise in NEFA and the rise

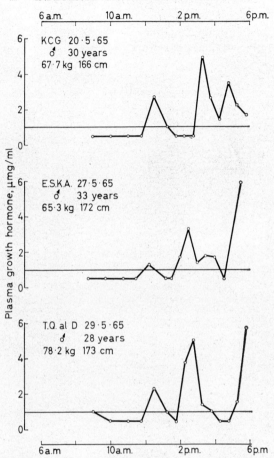

Figure 1. HGH during 24 h fasts. Plasma HGH levels in three healthy men during the latter part of the 24 h spent fasting in bed. The solid base-line represents the threshold of sensitivity of the HGH assay (Hunter, Willoughby and Strong, 1968)

in growth hormone. This provides good evidence that it is the need for mobilization of stored fuel which calls for the increased growth hormone secretion during exercise in the fasting state. *Figure 4* shows data from a much longer walk, this time 7 h at 4 mile/h without food, when the fall in RQ and the rise in NEFA became very marked and the growth hormone pattern in response

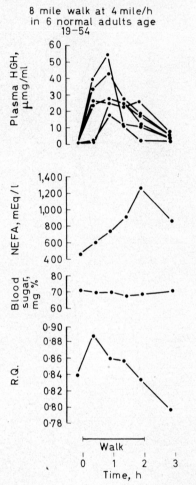

8 mile walk at 4 mile/h in 6 normal adults age 19–54

Figure 2. Plasma HGH and NEFA, blood glucose and RQ during 8 mile walks at 4 mile/h in six healthy adults aged 19–37. Figures for NEFA glucose and RQ are means+S.E.M. (From Hunter, Fonseka and Passmore, 1965)

By courtesy of the American Society for the Advancement of Science

Figure 3. Measurements during 8 mile walks at 4 mile/h by a 29 year-old healthy male in fasting state —·—·— *and while ingesting glucose (25 g immediately before and half-hourly during the walk)* – – –○– – –, *or protein (casein) in similar amounts* ·····△····· (*From Hunter, Fonseka and Passmore, 1965*)

By courtesy of The Editor, *The Quarterly Journal of Experimental Physiology*

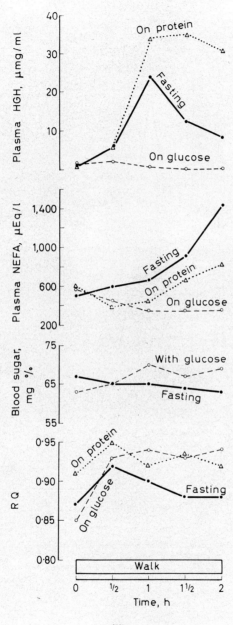

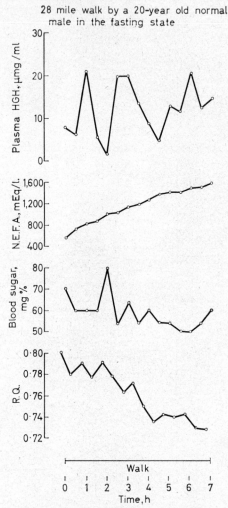

Figure 4. Measurements during a 28 mile walk at 4 mile/h by a 20 year-old male, in the fasting state (From Hunter, Fonseka and Passmore, 1965)

By courtesy of the American Society for the Advancement of Science

78

to the prolonged need for fat mobilization consists of a series of bursts of secretion. Finally, a third situation, in which there is an increase in plasma NEFA. *Figure 5* gives blood glucose and plasma NEFA, insulin and growth hormone levels during the 6 h following an oral glucose load. Here, growth hormone rises with the secondary rise in NEFA which is associated with the 'switch over' in fuel supply from the exogenous carbohydrate to endogenous fat.

These then represent the main lines of growth hormone physiology in the adult with no secretion occurring during the absorptive phase of carbohydrate-containing meals, the levels rising immediately

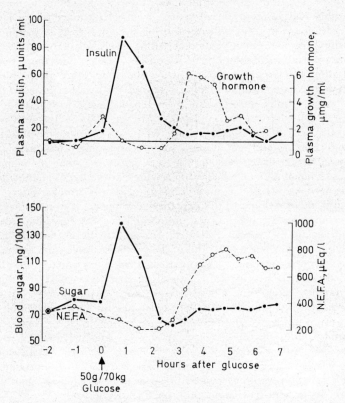

Figure 5. Plasma HGH insulin and NEFA and blood glucose for 2 h before and 7 h following an oral glucose load of 50 g/70 kg. Plasma insulin was determined by the method of Yalow and Berson (1960). The solid base-line in the upper figure represents the threshold of sensitivity of the two radio-immunoassays (From Hunter, Willoughby and Strong, 1968)

thereafter to help in the initiation of fat mobilization and showing further bursts of secretion later to maintain this if fuel demands are continued or increased. To this may now be added some more recent indications that, in some situations, amino acids or protein may stimulate HGH secretion. Infusion of rather large aphysiological doses of amino acids cause a rise in plasma HGH (Merimee, Lillicrap and Rabinowitz, 1965 and Knopf, Conn, Fajans, Floyd, Guntsche and Rull, 1965). We also have noted (*Figure 3*) that if protein (casein) is eaten in 25 g amounts before and at 30 min intervals during exercise, an enhanced secretion of HGH occurs above that found in the same subject when walking in the fasting state. I do not believe that these situations involving protein or amino acids necessarily represent a new mechanism calling for HGH secretion, as it is not difficult to bring the findings back within our original system. We know that NEFA levels fall slightly following administration of amino acids (Gordon, 1957). The more recent finding that insulin is secreted when amino acids are given (Floyd, Fajans, Knopf, Rull and Conn, 1965) offers an explanation for this. Doses of insulin too small to lower blood glucose will nevertheless cause well marked falls in NEFA (Mirsky, 1963). It seems most likely that in the situations referred to above ingestion of amino acids (or eating protein in some subjects and some situations) causes an increased secretion of insulin, this causes the small or incipient fall in blood glucose which is represented at the hypothalamic level as carbohydrate deficiency, and this in turn is the stimulus for growth hormone secretion. If this explanation is correct it would of course be quite good economics for the organism. Insulin is capable of enhancing the active transport of amino acids across cell walls (Kipnis and Noall, 1958), and in the post-absorptive state the intracellular concentration of amino acids is some ten times higher than their plasma concentration (van Slyke and Meyer, 1912). When protein is eaten, insulin is secreted to assist in this process. If unresisted this would seriously deplete supplies of both fuels; of fat by its powerful blocking of lipolysis and of carbohydrate by driving the equilibrium towards glycogen synthesis. This, of course, would have serious consequences for the newly ingested amino acids which would be in immediate demand as mere fuel. Growth hormone in this situation would then still appear as a fat mobilizing agent and here its protein-sparing action is particularly direct.

The foregoing account of that portion of growth hormone physiology which is quite clearly not concerned with the regulation of growth, adults having in Professor Tanner's terminology a growth velocity of zero, must be excused as a necessary prelude to an understanding of the function of HGH in children.

PLASMA HGH LEVELS IN CHILDREN

HGH levels in cord blood are always high and the hormone derives from the foetal pituitary (Greenwood, Hunter and Klopper, 1964). In neonates also HGH is continuously high, the physiological modulation which has been described above for adults is developed during the first few days of extra-uterine life (Cornblath, Parker, Reisner, Forbes and Daughaday, 1965). The important point here is that glucose abolishes HGH secretion in children above the age of a few days just as it does in adults. The pattern again is one of no HGH during the period when exogenous carbohydrate is being used followed by a marked rise assisting in the conversion to fat utilization after about 3 h. *Figures 6, 7* and *8* show the sort of diurnal pattern found in children. It may be noted that HGH was always low after meals, as was found in similar studies on 9 children. They emphasize clearly that if we have to assess the relative importance of (*1*) the simple feed-back based upon blood glucose (or something closely dependent upon this) and (*2*) the rise in HGH secretion which may follow ingestion of protein, then clearly the former is generally overriding. Thus, the late morning, late afternoon, and the evening peaks of HGH appear some 3–4 h after meals

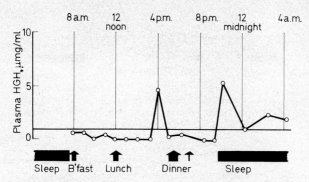

Figure 6. *Plasma HGH levels in an* 11 *year* 7 *month-old boy without endocrinopathy. Bone age* 11 *years 6 months. Sexual maturation score* (*SMS*) 1 (*Tanner, 1962*)
The times of meals are shown by the arrows (From Hunter and Rigal, 1966)

By courtesy of Cambridge University Press

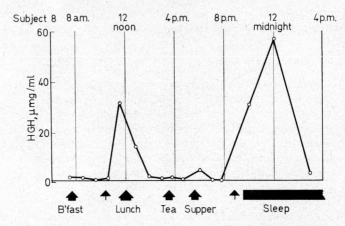

Figure 7.   Plasma HGH in a 14 year 8 month-old boy without endocrinopathy. Bone age 15 years, SMS 1 (From Hunter and Rigal, 1966)

By courtesy of Cambridge University Press

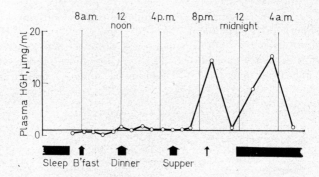

Figure 8.   Plasma HGH from a 12 year 8 month-old boy without endocrinopathy.   Bone age 11½ years, SMS 1 (From Hunter and Rigal, 1966)

By courtesy of Cambridge University Press

and, as such, are analogous to the secondary rises associated with fat mobilization which occur after the absorption of a glucose load.   If stimulation by protein were more important, then lunch (which contained, for example roast beef, *Figure 6*) and supper (which included fish and milk, *Figure 6*) would have been followed by higher and not lower plasma HGH levels.   Although the sampling is not frequent enough to show the full pattern there is evidence of raised HGH levels during the night. *Figure 8* shows that if meals are frequent enough and large enough there may be no secretion of HGH during the day-time. Its first appearance here is in the form of the secondary rise after supper. Again levels are high at night.

This night-time activity prompted more detailed sampling and *Figure 9* illustrates the pattern which is reminiscent of that seen in, adults during prolonged exercise.

Very recently, Quabbe, Schilling and Helge (1966) have published the results of HGH estimations on similar night studies in 6 adults. The pattern in their subjects is similar to this but the levels are much lower. How should we interpret this night-time secretion of HGH in children? It seems, quite simply, that it represents a response to fasting which is both earlier and bigger than that seen in adults. If

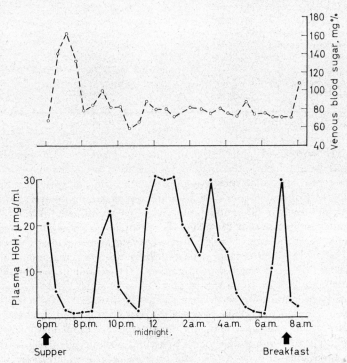

Figure 9. Plasma HGH and blood glucose during the night in a 13 year 6 month-old boy without endocrinopathy. Bone age 12½ years, SMS 2. Arrows indicate times of meals (From Hunter and Rigal, 1965)

By courtesy of The Editor, *Acta Endocrinology, Copenhagen*

this is the case then the easiest quantitative comparison between the two groups would be made by examining the HGH levels in children and adults who merely omit breakfast and remain in bed for sampling during the morning. Table 3 shows such a comparison and it will be seen that the children show clearly higher levels. Almost all of the measured HGH in the adult group in this series was due in fact to one high-responding subject.

Table 3

Growth hormone in adults and children at rest after an overnight fast

Age	Adults 25–35	Children 8–15
Number of subjects	16	28
Number of samples	79	78
Number of estimations < 2 μmg/ml	53	7
Mean $\pm$ S.D. $\pm$ SEM	$4.04 \pm 8.37 \pm 0.90$	$10.04 \pm 9.57 \pm 1.08$
Significance	$P < 0.001$	

The pattern of HGH levels in children (*Figures 6–9*) is precisely what would be wanted if lipid mobilization were the hormone's main purpose, since increases occur at the time of switching fuel supplies from exogenous carbohydrate to endogenous fat and during the night when fat mobilization must be continued. What use would an increase in tissue uptake of amino acids serve at these times? At the time of the secondary rises the amino acids are nearly all located inside the tissue cells and the quantities remaining in the plasma are very small by comparison. It is difficult to visualize much advantage from this action. The evidence then is quite clearly in favour of growth hormone acting primarily (and possibly solely) as an agent which increases the mobilization and combustion of fat. The assurance of a continuous supply of fat as fuel, permits the body to regulate the use of incoming protein according to the requirements of growth. In the absence of growth hormone this economy is lost and proteins are largely wasted in use as fuel. It seems quite possible that this action could account for the whole of the growth supporting action of this hormone.

RELATIONSHIP BETWEEN GROWTH VELOCITY
AND HGH SECRETION

That growth hormone is necessary for the support of growth is not in question.   The recent highly successful use of administered HGH to restore normal height in hypopituitary children (Raben, 1962; Medical Research Council, 1959; Henneman, Forbes, Moldawer, Dempsey and Carroll, 1960), provides excellent evidence that this is true in man and the foregoing is a possible mechanism which could account for this.   We may now examine the second question of how (or indeed whether) growth hormone has a role in regulating growth velocity.   It seems to be quite clear that we must abandon the sort of simple concept, quoted at the outset, which implies:

$$\text{growth velocity} \propto \text{HGH secreted}$$

since clearly the right-hand side can be varied at will by altering feeding habits without affecting the left-hand side.   If we try to retain this equation, it becomes evident in the light of the above findings that food, by decreasing the amount of HGH secreted, actually would reduce the growth rate!

The question now arises 'Is growth hormone secretion increased during growth?'.   As we have already seen that in one situation, the 12 h fast which night represents, there is greater GH activity in children than in adults, GH secretion *is* adjusted in order to support growth.   The next question then becomes: 'Is GH secretion increased first with increased growth following, or alternatively, does an increase in growth rate call at the periphery for greater energy supplies which are merely added to other energy requirements and so represented (by means of the blood sugar feed-back) at the hypothalamus, which responds with enhanced HGH secretion?'. Certainly this question cannot be answered on the basis of currently available data.   However, it is worth examining the special demands on a lipid mobilizing hormone which may be expected during growth.   If we compare the situation of an average (70 kg) man and a 13-year old boy who would weigh on average about 40 kg, the tables of calorie requirements (Manual of Nutrition, 1961) are 2,500 for the man and 3,150 for the boy.   Correcting for body weight this becomes 36 cal/kg for the man and 79 cal/kg or about twice as much energy per unit weight for the boy.   We now examine the timing of food intake and find that since the boy probably sleeps for more hours than the man he will most likely pack his calorie intake into a shorter period of time, and will probably have three meals a day, the same as the man. On average, then, these meals will be twice as big per unit of body weight in the boy. Of course, these expansive habits which are characteristic of adolescence are well known and not all of this voracious eating is attributable to greater exercise. The records we kept of the food intake of our orthopaedic cases, who were on complete bed-rest during the diurnal studies (*Figures 6–9*), make horrifying reading and as *Figure 10* indicates there does indeed appear to be a higher basal metabolic rate in children than in adults. The BMR is, of course, very difficult to determine in children and I know of no attempt at correlating BMR with growth velocity. During adolescence then the demands put upon growth hormone secretion are considerably more than those in adult life. It is possible that all of the differences between children and adults so far described can be related solely to these increased demands; that is, we do not necessarily have to postulate a 'resetting' of the hypothalamus pituitary level.

However, growth hormone may play a part in regulating growth velocity, as opposed merely to responding itself to changes in growth rate but evidence on this question will be very difficult to obtain. We need first a standardized test for HGH secretion which must be based upon some well defined and standardized stimulus. There are three possibilities:

(*1*) A standardized period of exercise. This has the merits of being a primary effect and of being physiological. It is a possible

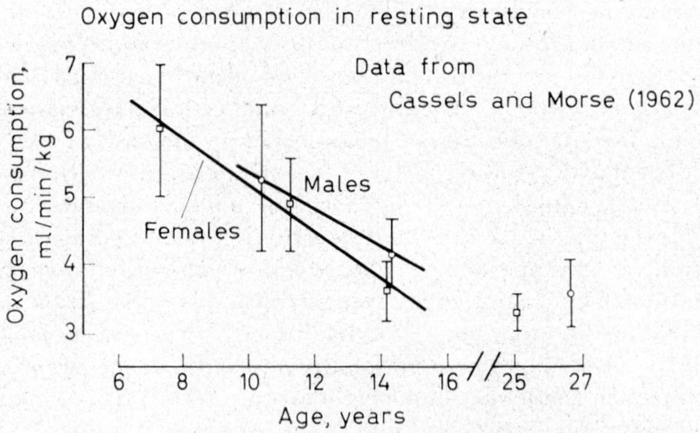

Figure 10. Regression of BMR on age. Oxygen consumption in children and adults. Each point represents mean $\pm$ S.D. and there were 11–23 subjects in each group (Data from Cassels and Morse, 1962)

test in adults because most adult subjects have immeasurable HGH levels during the morning if they rest and fast.   There is an approximately tenfold scatter between adult subjects in this test (Hunter, Fonseka and Passmore, 1965).   However, as we have seen (Table 3), children show increased levels which undulate during the night (*Figure 9*).   It is very difficult to show a well defined increment due to exercise.

(*2*) An extended glucose tolerance test in which the secondary rise in HGH secretion is measured, either as peak height or as the area under the curve, following a glucose load which is adjusted to body weight.   This test has the merits of being very gentle and of representing one of the main physiological functions of the hormone. However, its disadvantage is that the growth hormone secretion

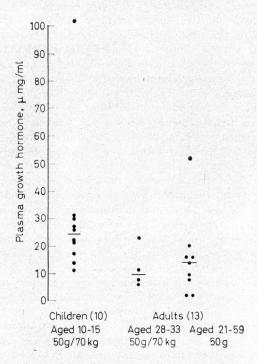

Figure 11.   Comparison in adults and children of magnitude of secondary rise in HGH which follows 3–4 h after a glucose load.   The points represent peak HGH levels from half-hourly plasma sampling.   The bars indicate median values

occurs after and is dependent upon a whole series of earlier events; absorption of glucose by the alimentary tract, and its handling by the many energy consuming and storing tissues. These processes are themselves profoundly dependent upon the secretion of insulin. *Figure 11* gives peak heights for the secondary rise in plasma HGH levels in a group of endocrinologically normal children in comparison with those in healthy adults. It will be immediately apparent that there is a very wide scatter within each group. Though the median value for the children is higher than for adults there is much over-lapping and clearly correlation with growth velocity (again adults = zero) is quite impossible.

(*3*) A standardized insulin sensitivity test. This procedure is not physiological and though not hazardous in practiced hands it is too unpleasant to be used widely for normal volunteers. However, it is the only available direct test for HGH secretion. Greenwood (1967) has found no clear difference between adults and children when tested by this means. Further, these workers (Greenwood, Landon and Stamp, 1966) have found that HGH values from duplicate tests on the same individual subjects do not agree well. More recently, Stimmler and Brown (1967) have found a marked correlation between growth velocity and growth hormone response to insulin in a cross-sectional study of some 30–40 children.

It is evident then that no single test is wholly satisfactory. Either (*2*) or (*3*) might be used, however, in longitudinal studies through the adolescent growth spurt during which very marked growth rate changes occur. Such studies would at least avoid the problem noted above of wide between-subject fluctuation in response. However, the responses themselves may very well be dependent upon BMR and would only be of value in showing a primary function for growth hormone if their variations exceeded those due to changes in BMR. This would be extremely difficult to demonstrate. We must therefore conclude that the experimental procedure for answering this question is not at hand.

Figure 12 shows the possible mechanisms involved in the regulation of growth by growth hormone. The pathways whereby growth hormone could regulate growth velocity are (*1*) in increasing the sensitivity of the receptors to stimuli in the hypothalamus, and (*2*) increasing the response in the hypothalamus or in the pituitary to incoming stimuli. Alternatively, the increase in growth hormone secretion which occurs during growth might result purely from an increased demand at the periphery which finds expression at the hypothalamic level solely in terms of the blood glucose feed-back.

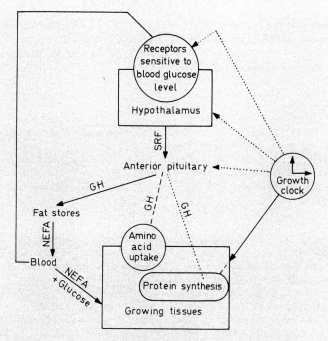

Figure 12. Role of growth hormone in growth regulation. Established mechanisms are shown in full line, probable ones dashed and possible ones are shown dotted

SUMMARY

In summary, the only well-established function for growth hormone is in the maintenance of fuel supplies for the tissues by means of increasing lipolysis. The signal for this necessity is, or is closely related to, the blood glucose level. The secretion of HGH is enhanced in children but the pattern of secretion is not different from that in adults. It is not yet possible to say whether during growth the pituitary is reset in such a way as to help in the dictation of growth velocity. Alternatively, changes in HGH secretion might represent the dispensations of a pituitary which is receiving in the adult, the ordinary demands of energy expenditure, and in the child, these with the addition of those necessary for growth.

ACKNOWLEDGEMENTS

I am grateful to my colleagues; to Dr. F. C. Greenwood for well enjoyed collaboration earlier, and in Edinburgh especially to Professor J. A. Strong and Drs. M. Willoughby and W. M. Rigal for permission to use unpublished material.

REFERENCES

Cassels, D. E. and Morse, M. (1962). *Cardiopulmonary Data for Children and Young Adults.* Springfield; Thomas

Cornblath, M., Parker, M. L., Reisner, S. M., Forbes, A. E. and Daughaday, W. H. (1965). *J. clin. Endocr. Metab.* **25**, 209

Fain, J. N., Kovacev, V. P. and Scow, R. O. (1965). *J. biol. Chem.* **240**, 3522

Floyd, J. C., Fajans, S. S., Knopf, R. F., Rull, J. and Conn, J. W. (1965). *Clin. Res.* **13**, 323

Glick, S. M., Roth, J. and Lonergan, E. T. (1964). *J. clin. Endocr. Metab.* **24**, 501

— — Yalow, R. S. and Berson, S. A. (1963). *Nature, Lond.* **199**, 784

Gordon, R. S. (1957). *J. clin. Invest.* **36**, 810

Greenwood, F. C. (1967). Personal communication

— Hunter, W. M. and Klopper, A. (1964). *Br. med. J.*, **i**, 22

— Landon, J. and Stamp, J. C. B. (1966). *J. clin. Invest.* **45**, 429

Henneman, R. M., Forbes, A. P., Moldawer, M., Dempsay, E. F. and Carroll, E. L. (1960). *J. clin. Invest.* **39**, 1223

Hunter, W. M., Fonseka, C. C. and Passmore, R. (1965). *Science, N.Y.* **150**, 1051

— — — (1965a). *Q. Jl exp. Physiol.* **50**, 406

— and Greenwood, F. C. (1962). *Biochem. J.* **85**, 39P

— — (1964). *Biochem. J.* **91**, 43

— — (1964a). *Br. med. J.* **i**, 804

— and Rigal, W. M. (1965). *Acta endocr., Copenh.* Suppl. 100, **50**, 121

— — (1966). *J. Endocr.* **34**, 147

— Willoughby, M. and Strong, J. A. (1968). *J. Endocr.* (In press)

Kipnis, D. M. and Noall, M. W. (1958). *Biochim. biophys. Acta* **28**, 226

Knopf, R. F., Conn, J. W., Fajans, S. S., Floyd, J. C., Guntsche, E. M. and Rull, J. A. (1965). *J. clin. Endocr.* **25**, 1140

Kostyo, J. L. (1964). *Endocrinology* **75**, 113

Manual of Nutrition (1961). London; H.M.S.O.

Medical Research Council (1959). *Lancet* **i**, 7

Merimee, T. J., Lillicrap, D. A. and Rabinowitz, D. (1965). *Lancet* **ii**, 668

Mirsky, I. A. (1963). *Endocrinology* **73**, 613

Noall, M. W., Riggs, T. R., Walker, L. M. and Christensen, H. N. (1957). *Science, N.Y.* **126**, 1002

Parker, M. L., Utiger, R. D. and Daughaday, W. H. (1962). *J. clin. Invest.* **41**, 262

Quabbe, H-J., Schilling, E. and Helge, H. (1966). *J. clin. Endocr. Metab.* **26**, 1173

Raben, M. S. (1962). *New Engl. J. Med.* **266**, 82

— and Hollenberg, C. H. (1959). *J. clin. Invest.* **38**, 484

Rabinowitz, D., Klassen, G. A. and Zierler, K. L. (1965). *J. clin. Invest.* **44**, 51

Riggs, T. R. and Walker, L. M. (1960). *J. biol. Chem.* **235**, 3603

Roth, J., Glick, S. M., Yalow, R. S. and Berson, S. A. (1963). *Metabolism* **12**, 577

Russell, J. A. (1955). 'Methods of detection and assay of growth hormone.' *Int. Symp. on Hypophyseal Growth Hormone*, p. 17. (Ed. by R. W. Smith, O. H. Gaebler and C. N. H. Long.) New York; McGraw-Hill

van Slyke, D. D. and Meyer, G. M. (1912). *J. biol. Chem.* **12**, 399

Stimmler, L. and Brown, G. (1967). *Archs. Dis. Childh.* (In the press)

Tanner, J. M. (1962). In: *Growth and Adolescence*, Ch. 2. 2nd Edition. Oxford; Blackwell

Yalow, R. S. and Berson, S. A. (1960). *J. clin. Invest.* **39**, 1157

SEX DIFFERENCES IN THE GROWTH
OF SHEEP

P. G. E. BRADFIELD

Grassland Research Institute, Hurley, Berks.

INTRODUCTION

IN ORDER to increase the output of animal protein it is necessary to define the rate of reproduction and growth of our domestic animals. Considerable research has been carried out already on various factors which influence an animal's development, such as genetic potential, nutritional status and environmental influences. The physiological changes which occur within the body during the growing period are still not well understood, however, and our ability to control the physiological development of our domesticated animals by controlling these processes is only possible within the bounds of current knowledge. Very little work has been directed towards relating growth rate and sex differences in development to the endocrine status of the animal.

Undoubtedly, sex and sex hormones affect the growth of species used to produce animal protein and this knowledge has been used to try and produce more animal protein for the feed utilized. Research on the effects of sex and sex hormones has resulted in modifications of feeding systems to accommodate the metabolic changes known to result from the use of exogenous sex hormones. Segregation of males and females to facilitate an appropriate adjustment of their nutrient intake is becoming more common in intensive enterprises. With cattle and sheep there is a trend towards the use of the intact male because it is known that, given a suitable nutrient allowance, it can grow faster and more efficiently and so use food nitrogen to better advantage than castrated males or females.

The necessity of castration to make animals docile is now less important as they are usually slaughtered before the age where testicular activity promotes aggressiveness. So, apart from the theoretical aspects of how sex and sex hormones affect growth, there is an increasing realization that these effects could be used to produce meat more efficiently.

The pituitary gland, the gonads, the thyroids and the adrenals are the endocrine glands which predominantly influence the growth of the animal. The impetus for growth has been assumed to be influenced by the endocrine system as a whole, with the influence of growth hormone being greatest in the young animal and with the sex steroids of minor importance at this stage.

Lamond (1963), while discussing the role of growth hormone, cited the work of Baird, Nalbandov and Norton (1952) in which they determined the relationship between the amount of growth hormone in the anterior pituitary gland and body weight. In their experiments with pigs the amount of growth hormone per unit of pituitary tissue remained essentially the same as the animal matured while at the same time the anterior pituitary became larger with advancing age. The growth of the body tissues, however, became proportionately greater in relation to the weight of the pituitary, and to the amount of growth hormone. Therefore, the growth hormone available per unit of body tissue decreased as the animal approached maturity. The work is based on the supposition that secretion of growth hormone is directly proportional to concentration measured in the gland.

Work by Armstrong and Hansel (1956) with cattle, and Simpson, Asling and Evans (1950) with rats produced very similar results on the levels of growth hormone. Simpson *et al.* (1950) administered growth hormone at 0·4 mg/rat per day after growth stasis had occurred and induced further weight gain. Once the level of growth hormone had become ineffective, the dose was further increased to 0·6 mg/day and weight gain was once more induced. This suggests that the growth hormone levels in the body become ineffective in inducing body weight increase once the level per unit of body weight falls below a certain point. This forms the basis of the hypothesis proposed by Nalbandov (1963) that the availability of growth hormone per unit of tissue is the basic determinant of growth.

Sex has long been recognized as a major source of variation in the growth of domesticated meat-producing animals and as more attention is now being paid to the use of different 'sexes' in meat production, an understanding of the somatic and endocrine effects produced by endogenous and exogenous sex hormones has become necessary. In the various meat-producing animals the male is generally born heavier than the female (Starke, Smith and Joubert,

1958; Donald, 1962). Joubert (1956) showed that in sheep the differences in mature size between males and females can be attributed partly to muscle cell number. This number is determined by about the 100th day of pregnancy and is greater in males than females. Whether an animal then reaches this ultimate size depends upon factors such as environment, nutrition and endocrine effects.

Assuming that eventually it will be possible to control the genetic potential of an animal so that a desirable predetermined mature weight can be established, with the animal following a pre-set growth curve, interest then focuses on the physiological mechanisms which control the rate at which the animal reaches this weight. The relationship between the developmental hormones of the pituitary gland and the sex steroids, and particularly the influence that the latter have on the pituitary gland, are therefore of considerable importance. It has been shown, for instance, that the administration of pituitary growth hormone increased growth and efficiency more in females than males (Nelson, Palmer and Kennedy, 1935), which suggests that the lower growth rate found in females may be due to lower inherent levels of growth hormone in the female. There is considerable evidence to show that a direct action of sex hormone on tissues generally exists (Villee, 1961) but the work of Shroder and Hansard (1958) and Struempler and Burroughs (1959) on the effect of stilboestrol on the size of the anterior pituitary gland in sheep and cattle indicates indirect action on tissues via the pituitary. They demonstrated that the pituitary gland was larger and the total growth hormone content greater in treated animals than in controls. This is supported by the report of Martin and Lamming (1958) who showed that in normal stilboestrol-treated hoggets there was a significant relationship between growth rate and the ribonucleic acid/deoxyribonucleic acid ratio in the pituitary gland. The level of RNA and the ratio between RNA/DNA have been used in other experimental fields as an indication of the rate of protein synthesis in different tissues. This ratio was adopted therefore by Martin and Lamming to provide an estimate of pituitary activity.

THE EFFECTS OF SEX HORMONES ON GROWTH

Androgens stimulate the growth of body tissues as a whole, as well as organs specific to the male animal (Brody, 1945; Gaunt, 1954). Under normal conditions the fusion of the epiphyses in the long bones is brought about by the increase in androgen secretion at the time of puberty. Castration should lead therefore to a disproportionate increase in the length of these bones because of the delayed closure of the epiphyses, but although this has been shown to be true, the overall reduction in growth rate means that these animals do not attain the mature skeletal size of their entire counterparts.

Androgens influence the tissues directly while oestrogens affect the growth of tissues via stimulation of the developmental hormones of the pituitary gland. More specifically, androgens are known to stimulate certain muscle groups and these are mainly associated with the neck and head regions. This explains the characteristic growth of the entire male with the greater development of the forequarters. At the same time, treatment with androgens or a comparison between entire and castrate males shows that the hormones promote the development of the early maturing tissues as well as the early maturing regions. This means that in an animal under the influence of androgen there is a greater proportion of bone and muscle and less fat than in the corresponding castrate.

Thus, as would be expected, it has been shown recently that androgens have a profound effect on protein anabolism. Administration of testosterone proprionate in small doses to rats stimulates RNA polymerase activity and results in an increased synthesis of RNA (Wicks and Kenney, 1964; Widnell and Tata, 1966). Hamilton (1948) also reported that there was a decrease in muscle cell size and protein content of the sarcoplasm following castration. These findings would explain, in part, the effect of castration where there is a reduced growth of bone and muscle with a concurrent increase in fat deposition.

The oestrogens, apart from stimulation of the organs specific to the female, inhibit the linear growth of long and flat bones and tend to accelerate skeletal maturation by bringing about epiphyseal closure in the growing animal. As far as protein metabolism is concerned there are specific differences which have not been clarified. Synthetic oestrogens have been widely used in animal production to promote increased live-weight gain. Many reports have shown that there is an increase in the early developing tissues in ruminants and pigs (Wilkinson, O'Mary, Wilson, Bray, Pope and Casida, 1955; Wilkinson, Carter and Copenhaver, 1955; Preston and Burroughs, 1958). This supports the contention that these steroids may be anabolic agents to a limited degree (Kochakian, 1946). It has not been conclusively explained, however, whether this is a direct influence or whether it is manifested through increased growth hormone secretion. This is probably because little direct comparison has been made between intact and gonadectomized females. Hammond (1932) originally suggested that, in sheep, inhibition of growth by the ovary is greater than the stimulation by the testes. Everitt and Jury (1966a, b) in a detailed experiment on the effect of gonadectomy on growth and development showed, however, that ovariectomy reduced growth rate, although they only took their animals through to puberty and so any possible retarding influence of the oestrous cycle was not revealed. There is limited evidence in other farm species that ovariectomy does reduce weight gain but, in contrast to this, the growth rate of rats is increased by the removal of the ovaries because of the removal of the inhibiting influence of oestrogens. Ovariectomy leads to an early maturation of the carcass with a greater percentage of fat and also an increased area of eye muscle when compared with an intact animal of the same weight (Everitt and Jury, 1966a, b).

Recent work has shown that oestrogens may have a similar effect on protein metabolism to that of androgens and other developmental hormones. Oestradiol treatment produces a rapid acceleration of synthetic reactions in the rat uterus which leads to an accumulation of phospholipids, ribonucleic acid and protein (Aizawa and Mueller, 1961; Noteboom and Gorski, 1963; Talwar and Segal, 1963; Ui and Mueller, 1963). The latter suggested that oestrogen stimulates an initial step in the synthesis of RNA with the subsequent stimulation of overall protein metabolism. Whether this reaction of the uterus to oestrogen is unique or whether it holds true for the tissues of the body as a whole has yet to be answered.

SEX DIFFERENCES IN SHEEP

In order to provide more exact information on the influence of sex and sex hormones on the growth of young animals, experiments were carried out on the influence of both orchidectomy and ovariectomy on the rate of growth and characteristics of live weight increment in the growing sheep. These experiments indicated that the point of inflection in the growth curve is reached first by ovariectomized females, then intact females, castrated males and finally intact males in that order and that the characteristics of live weight increment made at any particular time coincides with the position on the growth curve.

Attempts to determine the influence of sex hormones on the growth hormone content of the anterior pituitary gland have been seriously hampered by the lack of suitable assays for growth hormone. The original work carried out by Martin and Lamming (1958) showed that there might be a correlation between pituitary activity and growth rate. Consequently it was decided to examine further the influence of sex, gonadectomy and sex hormone treatment on the RNA/DNA ratio of the anterior pituitary gland and on the growth rate and body conformation of growing sheep.

The first experiment was designed to provide information on the relationship between growth rate and pituitary activity in the different 'sexes' and how this may change with time in lambs slaughtered serially before and after puberty. Entire and gonadectomized lambs were used so there were, therefore, four groups or 'sexes'; entire males, castrate males, entire females and ovariectomized females.

The four slaughter weights used were 60, 80, 100 and 120 lb live weight. Lambs were assigned to their respective groups early in each experiment. A proportion of the male lambs were castrated within 3 days of birth and the appropriate females were ovariectomized at 40 lb live weight. Although Everitt and Jury (1966a, b) have since operated on lambs at a much earlier age, it was felt that by leaving animals until this weight it would minimize the postoperative set-back they might possibly suffer. The lambs ran at pasture with the ewes until weaned at 14 weeks of age. They continued grazing until the available grass was considered insufficient to give satisfactory weight gain even when supplemented with a concentrate ration. They were then housed during the finishing period and fed *ad lib.* on a pelleted ration based on ground straw and barley.

Weekly weight records were taken throughout the experiment and the overall growth rates were calculated from birth to time of slaughter in kg/week (Table 1). Because of the uneven numbers in this first experiment the results were analysed to take into account any possible interactions between sex and weight. As these were not apparent, weighted marginal means have been quoted in the respective tables for growth rate and nucleic acid ratios for each sex and each slaughter weight and the statistical analyses refer to the differences between these marginal means. Analysis of variance showed highly significant differences in the growth rate between the 'sexes' and also at the different slaughter weights. Ovariectomy caused a significant reduction in growth rate which was evident even at the lowest slaughter weight. The growth rates of the entire and

Table 1 (Experiment I)

Mean growth rates (kg/week from birth to slaughter)

	Slaughter weight (lb live weight)				
	60	80	100	120	Weighted mean
Entire males	1·79	1·74	1·56	1·45	1·640
Castrate males	1·78	1·66	1·23	1·23	1·479
Entire females	1·68	1·56	1·03	1·14	1·350
Ovariectomized females	1·48	1·40	1·00	1·07	1·242
					Overall mean
Weighted mean	1·679	1·591	1·197	1·222	1·425
					S.E. ± 0·0384

The average standard error of the difference between the marginal means was 0·109. The standard errors of each difference have not been quoted because of the very small variation found.
At $P = 0·05$ the least difference required for significance between the marginal means is 0·218.

castrate males and the entire females were similar until the lambs reached 80 lb live weight but thereafter the entire males maintained a relatively faster growth rate while there were marked reductions in the growth rates of both the castrated males and the entire females.

Figure 1 shows the same data in graph form and reveals clearly the characteristic growth pattern of the four 'sexes'.

At slaughter, the lambs were dressed as quickly as possible and after the skull had been cleaved behind the horn buds the pituitary gland was dissected out. It was cleaned and homogenized in ice-cold 10 per cent trichloracetic acid. Extraction of the RNA and DNA was performed by a modification of the Schmidt Thannhauser method (1945) which is based on the fact that RNA is susceptible to alkaline hydrolysis while DNA is stable under these conditions. The nucleic acid content of the two separate fractions was estimated by ultra-violet absorption. The intense and characteristic absorption of ultra-violet light by the purine and pyrimidine bases of the nucleic acids in the region of 260 μ and 268 μ provides a very good method for the determination of RNA and DNA respectively. It is complicated to some extent by the fact that proteins and peptides absorb ultra-violet light at similar wave-lengths and to correct this the two wave-length method reported by Tsanev and Markov (1960) was adopted.

Table 2 shows the pituitary nucleic acid ratio of each group together with the sex and weight means and their respective standard

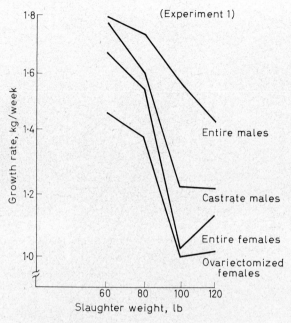

Figure 1. Growth rate of lambs in kg/week *(Experiment I)*

Table 2 (Experiment I)
Pituitary nucleic acid ratios RNA/DNA

	Slaughter weight (lb live weight)				
	60	80	100	120	*Weighted mean*
Entire males	1·17	1·10	1·04	0·97	1·073
Castrate males	1·08	1·15	0·92	0·92	1·015
Entire females	1·12	1·08	0·92	0·84	0·991
Ovariectomized females	1·11	0·98	0·81	0·73	0·912
					Overall mean
Weighted mean	1·118	1·077	0·920	0·864	0·997
					S.E. ± 0·010

The average standard error of the difference between the marginal means was 0·0282.
At $P = 0·05$ the least difference required for significance between the marginal means is 0·0564.

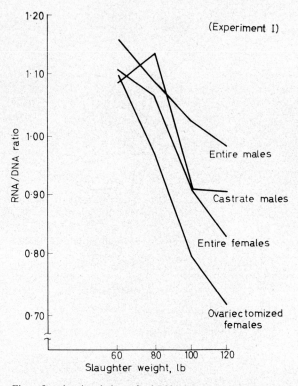

Figure 2. Anterior pituitary gland RNA/DNA ratios (Experiment I)

errors. In *Figure 2* the same data is presented in graph form and the close similarity between the growth rate patterns (*Figure 1*) and the nucleic acid ratios is evident. The entire males showed a significantly better growth rate at the higher weight and this was reflected in a higher nucleic acid ratio at these weights. The intact females and the castrate males showed little difference in growth rate at the lower weights when compared with the intact males but subsequently both sexes took a much longer period to reach the higher weights and as their overall growth rate fell so did their nucleic acid ratios. The ovariectomized lambs grew significantly more slowly than the other three sexes, while also showing a significantly lower nucleic acid ratio.

Regression analysis showed a highly significant correlation between growth rate and pituitary activity in all four 'sexes'. The

100

regression lines are shown in *Figure 3* together with the equations for the four lines and the determination coefficients. These express the proportion of the variation in growth rate which can be attributed to the nucleic acid levels. In this experiment they varied from between 43·9 to 60·5 per cent in the four 'sexes'. The correlation coefficients between growth rate and pituitary nucleic acid ratio were 66·2, 77·2, 77·9 and 72·1 respectively for the four 'sexes'.

Having outlined the basic differences in pattern of growth between the four 'sexes' and shown a positive correlation between growth rate and pituitary activity, a second experiment was conducted to examine the influence of exogenous oestrogens on the growth rate, development and pituitary activity of the four 'sexes'. The slaughter groups were confined to 80 and 100 lb live weight as the value of the response to oestrogens is greatest between these weights. Endogenous sex steroids, or more probably the ratio between growth hormone and the sex steroids, contribute to the

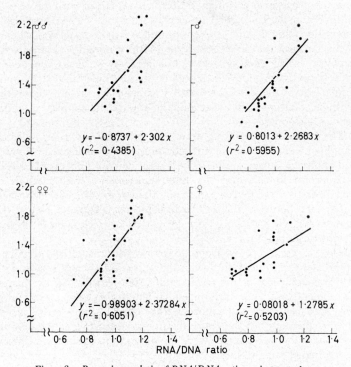

Figure 3. Regression analysis of RNA/DNA ratio against growth rate

101

reduction in the growth rate of the animal after puberty and therefore influence mature size by bringing about epiphyseal closure. Exogenous oestrogens in the form of stilboestrol and hexoestrol have been used widely to promote growth, particularly in the castrate male. The effects produced by the use of these hormones in ruminants are well known; there is an increase in growth rate and feed conversion efficiency while at the same time there is an increase in the nitrogen and water content of the carcass with a concurrent decrease in the fat content. The overall effect is to confer a more juvenile appearance on the carcass. The response in the castrate male is not typical, therefore, of the developmental influence of endogenous oestrogen.

Several workers have shown that this effect is mediated via the pituitary gland which shows an increased activity after oestrogen treatment. The present experiment was designed to study further the physiological effects of these hormones. The lambs were treated as in the previous experiment in that they remained with the ewes at pasture and were weighed at weekly intervals. The appropriate proportion of male lambs were castrated at 3 days of age and a group of ewe lambs was ovariectomized at 40 lb live weight. Lambs were implanted with 12 mg of oestradiol benzoate at 60 lb live weight while a control group of untreated lambs continued under the same conditions. Oestradiol benzoate was used in preference to the synthetic oestrogens in an attempt to simulate the effects of endogenous hormones.

The animals were slaughtered as they reached their selected slaughter weights and the anterior pituitary glands were removed and analysed as previously. There were equal numbers of lambs in each group in this and the final experiment. The analysis of variance was simplified therefore and the standard errors refer to the differences between the individual groups. Table 3 shows the growth rate data of the sixteen groups in kg/week from 60 lb live weight to slaughter.

The four 'sexes' showed similar growth patterns to those obtained in the first experiment, in that in the untreated animals best growth rate was achieved by the entire males, followed by castrate males, entire females and ovariectomized females in that order. Secondly, oestrogen treatment had a more marked effect at the lower slaughter weight where all but the entire males responded to treatment while the ovariectomized females were the only group to show a response at the higher weight.

The growth rates of the untreated entire males were higher at both

Table 3 (Experiment II)

Mean growth rates (kg/week from 60 lb live weight to slaughter)

Slaughter weight	80 lb		100 lb	
	Untreated	*Implanted with oestradiol*	*Untreated*	*Implanted with oestradiol*
Entire males	1·11	1·17	1·22	0·89
Castrate males	1·01	1·19	1·14	1·08
Entire females	0·98.	1·48	1·00	0·77
Ovariectomized females	0·75	1·11	0·79	0·91

Standard error of difference between any two means $= \pm 0·181$
At $P = 0·05$ the least difference required for significance $= 0·361$.

weights when compared with the other 'sexes' and although oestrogen treatment had no effect on the growth rate at 80 lb live weight it depressed it significantly at the highest slaughter weight. On the other hand, oestrogen treatment had a beneficial effect on the growth rate of the castrate male at the lower slaughter weight. This was not apparent at 100 lb live weight and a possible explanation is that the potency of the hormone may have been reduced during the relatively long period between implantation and slaughter.

The entire female responded to oestrogen treatment at the lower weight and the growth rates of the animals in this group were significantly higher than any other. There was, however, a significant reduction in growth rate achieved by the lambs slaughtered at 100 lb. This may have resulted indirectly from a hastening of sexual maturity and a natural slowing in growth as the animals passed puberty. The untreated ovariectomized females again grew significantly more slowly than the other 'sexes'. Treatment with oestrogen increased the growth rate at both slaughter weights although the difference was only significant at the lower weight.

Table 4 shows the nucleic acid ratios found in the anterior pituitary gland and again the ratios follow a very similar pattern to the growth rate data. In the entire and castrate male lambs which were not implanted there were no significant differences in the growth rates or nucleic acid ratios at either slaughter weight, although in each group the results were higher for the entire animals for both growth rate and nucleic acid ratio. The nucleic acid ratios of the untreated ovariectomized lambs at the higher slaughter weight were significantly lower than all groups, as was their growth rate, while at the lower weight the nucleic acid ratios were lower but not significantly so.

Table 4 (Experiment II)

Pituitary nucleic acid ratios (RNA/DNA)

Slaughter weight	80 lb		100 lb	
	Untreated	Implanted with oestradiol	Untreated	Implanted with oestradiol
Entire males	1·08	1·05	1·09	1·05
Castrate males	1·04	1·12	1·05	1·01
Entire females	1·02	1·15	1·04	0·96
Ovariectomized females	0·98	0·99	0·89	0·96

Standard error of difference between any two means = $\pm$ 0·043
At P = 0·05 the least difference required for significance = 0·086

Oestrogen treatment had a beneficial effect on the growth rate of the castrate males at the lower weight and an increase in the nucleic acid ratio was also evident. The growth rate of the group slaughtered at 100 lb live weight was lower however and this was reflected in a decrease in the nucleic acid ratio. Although there were differences in growth rates between treated and untreated intact males there was no significant difference in the nucleic acid ratios at either weight. Implantation of the ovariectomized lambs significantly increased the growth rate at the lower slaughter weight although this was not maintained to 100 lb live weight. There was, however, no significant effect on the nucleic acid ratios at either weight. In the entire females the highly significant increase in the growth rate in the lower slaughter group was reflected in a significant increase in the nucleic acid ratio while at the higher weight there was a corresponding reduction in both the growth rate and nucleic acid ratio.

The final experiment was designed to study the influence of exogenous male sex steroids in the form of testosterone proprionate on the pituitary activity of intact and castrate males. No females were included in this trial because little commercial use of testosterone has been shown in the literature. Two slaughter groups were used as in the previous experiment and the appropriate lambs were

Table 5 (Experiment III)

Mean growth rates (kg/week from 60 lb live weight to slaughter)

Slaughter weight	80 lb		100 lb	
	Untreated	Implanted with testosterone	Untreated	Implanted with testosterone
Entire males	1·10	1·09	1·58	1·32
Castrate males	1·23	1·09	1·15	1·12

Standard error of difference between any two means = ± 0·141
At P = 0·05 the least difference required for significance = 0·284

castrated at 3 days of age. The lambs were implanted with 23·5 mg of testosterone proprionate at 60 lb live weight.

Table 5 shows the growth rate of the four groups from 60 lb live weight to the time of slaughter. Similar growth rates and differences between the 'sexes' were achieved in this experiment when compared with the two previous experiments. At the lower weight, the growth rate of the two 'sexes' was not significantly different but at 100 lb the entire males again showed a prolonged active growth phase while the growth rate of the castrated lambs had decreased. Testosterone implantation caused some reduction in the growth rate of all groups when compared with their untreated counterparts, although significant differences were only apparent in the heavier entire males and the castrated lambs slaughtered at 80 lb live weight.

Table 6 shows the nucleic acid levels of the pituitary glands. In

Table 6 (Experiment III)

Pituitary nucleic acid ratios (RNA/DNA)

Slaughter weight	80 lb		100 lb	
	Untreated	Implanted with testosterone	Untreated	Implanted with testosterone
Entire males	1·05	1·02	1·15	1·05
Castrate males	1·05	1·00	1·07	1·02

Standard error of difference between any two means = ± 0·044
At P = 0·05 the least difference required for significance = 0·088

4*　　　105

the entire males, testosterone treatment produced a reduction in growth rate at the higher weight and this was reflected in a significant reduction in the nucleic acid ratios. The differences shown in the nucleic acid levels in both the entire and castrated males which were treated were not as great as those produced in the previous experiments, which confirms the opinion that testosterone has little effect on the hormonal activity of the pituitary. Any changes which testosterone implantation may have therefore on carcass composition are probably produced through a direct effect on tissue metabolism and are not manifested through the pituitary.

CONCLUSIONS

This series of experiments has demonstrated the very different growth patterns of the four 'sexes' with the increments of gain in each sex being characteristic of the point reached on their particular growth curve. Although not discussed in this paper, very obvious carcass differences were shown between the 'sexes' and these were closely related to the stage of maturity reached by each sex at the different slaughter weights. A high correlation between pituitary nucleic acid ratio and growth rate has been established over a considerable range of slaughter weights in both untreated and treated lambs. The correlation coefficients in these three trials varied from 59·58 to 84·09 per cent.

The effect that can be expected from the use of exogenous sex hormones would seem to depend on the basic metabolic activity of the animal, the hormone used and the age and sex of the animal concerned. The influence of exogenous hormones in entire animals is identical to that of endogenous hormones, in that oestrogens hasten maturity and sexual activity in the female. There is evidence that testosterone has a direct effect on tissue metabolism and little influence on the pituitary, as there were only small changes in the pituitary nucleic acid ratios in testosterone treated animals while there were significant changes in carcass characteristics and growth rate. Prescott (1963) showed a sex and dietary protein relationship in the pig and there may well be a similar effect in the sheep. One trial, which has not been reported here, was conducted using only intensively reared animals and the growth rates in the ram lambs were then much greater than those obtained in the field while the differences in the growth of the other 'sexes' was much less marked. If the available dietary protein was limiting to the ram lambs this may explain why the growth rate of entire and castrate males and entire females was very similar until the higher weights were reached, by which time the lambs were receiving supplementary feed.

Oestrogens have been used in recent years to promote growth in meat-producing enterprises. The physiological changes are not fully understood but these experiments support the contention that there is a direct stimulation of pituitary activity with no sex reaction being manifested and that the changes in carcass production are the result of increased hormone secretion from the anterior pituitary.

REFERENCES

Aizawa, Y. and Mueller, G. C. (1961). *J. biol. Chem.* **236**, 381

Armstrong, D. T. and Hansel, W. (1956). *J. Anim. Sci.* **15**, 640

Baird, D. M., Nalbandov, A. V. and Norton, H. W. (1952). *J. Anim. Sci.* **11**, 292

Brody, S. (1945). *Bioenergetics and Growth.* 2nd edn. 1964. New York; Reinhold

Donald, H. P. (1962). *Anim. Prod.* **4**, 369

Everitt, G. C. and Jury, K. E. (1966a). *J. agric. Sci., Camb.* **66**, 1

—— (1966b). *J. agric. Sci., Camb.* **66**, 15

Gaunt, R. (1954). In: *Dynamics of Growth Processes.* p. 183. (Ed. E. J. Boell.) Princeton Univ. Press

Hamilton, J. B. (1948). *Recent Prog. Horm. Res.* **3**, 257

Hammond, J. (1932). *Growth and Development of Mutton Qualities in the Sheep.* Edinburgh; Oliver and Boyd

Joubert, D. M. (1956). *J. agric. Sci., Camb.* **47**, 382

Kochakian, C. D. (1946). *Vitams. Horm.* **4**, 255

Lamond, D. R. (1963). *Symp. Carcase Composition and Appraisal of Meat Animals.* Paper 8-1. (Ed. by D. E. Tribe.) Melbourne; C.S.I.R.O.

Martin, E. M. and Lamming, G. E. (1958). *Proc. Nutr. Soc.* **17**, XLVIII

Nalbandov, A. V. (1963). *J. Anim. Sci.* **22**, 558

Nelson, H. W., Palmer, L. S. and Kennedy, C. (1935). *Am. J. Physiol.* **III**, 341

Noteboom, W. D. and Gorski, J. (1963). *Proc. natn. Acad. Sci. U.S.A.* **50**, 250

Prescott, J. H. D. (1963). 'The effects of sex and of certain steroids on the growth of the pig.' *Ph.D. Thesis*, Univ. Nottingham

Preston, R. L. and Burroughs, W. (1958). *J. Anim. Sci.* **17**, 140

Schmidt, G. and Thannhauser, S. J. (1945). *J. biol. Chem.* **161**, 83

Shroder, J. D. and Hansard, S. L. (1958). *J. Anim. Sci.* **17**, 569

Simpson, M. E., Asling, C. W. and Evans, W. M. (1950). *Yale J. Biol. Med.* **23**, 1

GROWTH AND DEVELOPMENT OF MAMMALS

Starke, J. S., Smith, J. B. and Joubert, D. M. (1958). *Sci. Bull. Dep. Agric. S. Afr.* No. 382, 28 pp.

Struempler, A. W. and Burroughs, W. (1959). *J. Anim. Sci.* **18**, 427

Talwar, G. P. and Segal, S. J. (1963). *Proc. natn. Acad. Sci.* **50**, 226

Tsanev, R. and Markov, G. G. (1960). *Biochim. biophys. Acta* **42**, 442

Ui, H. and Mueller, G. C. (1963). *Proc. natn. Acad. Sci.* **50**, 256

Villee, C. A. (1961). In: *Sex and Internal Secretions.* Vol. I. (Ed. by W. C. Young.) London; Baillière, Tindall & Cox

Wicks, W. D. and Kenney, F. T. (1964). *Science, N.Y.* **144**, 1346

Widnell, C. C. and Tata, J. R. (1966). *Biochem. J.* **98**, 621

Wilkinson, W. S., Carter, R. C. and Copenhaver, J. S. (1955). *J. Anim. Sci.* **14**, 1260

— O'Mary, C. C., Wilson, G. D., Bray, R. W., Pope, A. L. and Casida, L. E. (1955). *J. Anim. Sci.* **14**, 866

THYROID ACTIVITY AND GROWTH

IAN R. FALCONER and S. A. DRAPER

University of Nottingham School of Agriculture, Sutton Bonington, Loughborough

THE RELATIONSHIP between the thyroid gland and growth has been known for nearly a century. Fagge (1871) described the absence or atrophy of the thyroid gland in a case of human cretinism and concluded that the lack of the thyroid was responsible for this condition. Later, experimental removal of the thyroid gland from young monkeys was shown by Horsley in 1886 to produce symptoms resembling human cretinism. The characteristics of cretinism can be summarized as much reduced skeletal and somatic growth, arrested sexual development and frequently mental deficiency. In cretins the basal metabolic rate may be down to 55 per cent of normal and the susceptibility to infection increased.

These symptoms of defective thyroid secretion illustrate the extreme case, where the gland is almost completely destroyed or absent. They apply to domestic animals as well as man, as illustrated in *Figure 1* which shows two lambs born in the same week, the smaller of which was surgically thyroidectomized at one month of age. The photograph was taken three months after thyroidectomy and clearly shows the reduced growth subsequent to the operation in the lamb lacking a thyroid gland. This cretin lamb only survived one month after the photograph was taken and died of pneumonia. Studies by Rac, Hill, Pain and Mulhearn (1968) have indicated that sheep with biochemical defects in the synthesis of thyroid hormone, which lead to a low hormone output, do not become cretins although the growth rate and, particularly, the viability of these animals are adversely affected. The author's (I.R.F.) own studies on these animals have shown that the goitre which is present develops as a result of excessive stimulation by thyroid stimulating hormone leading to substantial enlargement of the defective gland. This enlarged gland does, however, manage to produce sufficient thyroid hormone for the continued life of the animal, though measurements have shown it to be significantly lower in circulating thyroid hormone than normal sheep (Falconer, 1965; 1966a, b).

109

Figure 1. Normal (left) and thyroidectomized (right) Merino lambs at 4 months of age. Both animals were kept under similar dietary and environmental conditions. Thyroidectomy was at 1 month of age

In the adult animal thyroidectomy has less dramatic effects, though wool growth in the sheep is greatly reduced when the level of thyroid hormone in the circulation is appreciably lower than normal. *Figure 2* illustrates this point by showing two animals, both shorn ten months previously, one of which was surgically thyroidectomized about 8 months before shearing. These animals are Australian Merinos and it is apparent that there is a substantial difference in wool growth between the two. The mortality of these adult thyroidectomized sheep was considerable, only a minority of the animals surviving for more than 2 years after surgery under normal field conditions.

Spielman and co-workers have investigated the effect of thyroidectomy on dairy cows which have shown a substantial reduction in milk production to only 30 per cent of normal in some cases and an increase in mortality of the animals (Spielman, Petersen and Fitch, 1944; Spielman, Petersen, Fitch and Pomeroy, 1945). The same authors also showed the cessation of growth after thyroidectomy of dairy calves. It is clear from these considerations that effective functioning of the thyroid gland is a prerequisite for growth and production in animals, but it must be remembered that a variety of

Figure 2. Normal (upper) and thyroidectomized (lower) Merino ewes at 4 years of age. Thyroidectomy was at 2·5 years, both animals were shorn 10 months prior to the date of photography

other factors, in particular nutrition and pituitary function, have equally potent effects.

Several studies have been carried out on hypophysectomized animals and hypophysectomized-thyroidectomized animals with replacement of the missing hormones by thyroxine, growth hormone and corticosteroids in various combinations. Scow (1959) carried out carcass analysis on hypophysectomized-thyroidectomized rats given replacement therapy with growth hormone and/or thyroxine. He concluded from part of his study that thyroxine specifically influenced the growth of muscle and some viscera and had a powerful influence on the length and maturation of bone. Other effects

111

were identified that were largely related to growth hormone dosage and not thyroxine. A recent study by Goodall and Gavin (1966) indicated that hypophysectomized rats which were administered adequate levels of L-thyroxine and cortisone acetate failed to grow, but if growth hormone and thyroxine were administered together growth was restored. De Groot (1963) showed that under conditions of adequate thyroxine administration in hypophysectomized rats, the rate of growth was dependent on the dosage of growth hormone. It is clear from this and other work that both thyroxine and growth hormone are essential for normal growth but that under adequate levels of thyroxine, growth is dependent upon the quantity of growth hormone.

In an animal with an intact pituitary, the relationship between rate of growth and thyroid output is not a simple one. This aspect will be dealt with at some length in a later section of this paper which includes some experimental work carried out at Nottingham University to try to correlate thyroid activity and growth rate.

As a direct consequence of the early realization that the thyroid was important in growth, there have been numerous attempts to improve growth by altering thyroid activity. In the case of human cretinism, detected early in infancy, thyroid therapy has met with great success (Wilkins, 1957). In the domestic animals the administration of thyroxine to growing lambs has been reported by Coop and Clarke (1958) to cause weight loss and by Matsushima and co-workers (1960) to increase the lamb's rate of growth. The feeding of protein containing thyroxine to growing lambs has similarly been reported by Dinusson, Andrews and Beeson (1950) to decrease growth rate and food conversion, and conversely reported by Desai and Patel (1962) to increase the growth rate of calves. Thyroid antagonists which decrease the rate of secretion of the gland have also been fed to cattle, and whereas Schultz, Hathaway and Loeffel (1950) report little effect of the feeding of these materials, Beeson, Andrews and Brown (1947) report an increased rate of gain during the maturation phase of growth.

It is not surprising in the face of this conflicting evidence, that no practicable and beneficial thyroid treatment is yet in use to improve growth in normal animals.

The approach of the authors of this paper to the problem of the relationship between thyroid activity and growth has been to investigate in normal sheep the changes in thyroid activity with respect to age, and to measure the thyroid activity of growing lambs in order to correlate the rate of growth with the activity of the thyroid gland. Only with this basic information can one hope to devise any methods or recommendations for the optimization of thyroid activity in growing animals, or to understand any limitations that the thyroid imposes upon growth rate.

THYROID ACTIVITY AND AGE

Since the application of radioactive iodine to thyroid studies, very many methods have been adopted to measure the function of the gland. The methods that provide the most information are those that allow the calculation of the rate of secretion of thyroid hormone as done, for example, by Post and Mixner (1961) and Robertson and Falconer (1961). Methods other than these are often used, however, and provide valuable information if the conditions of use are sufficiently closely defined. The most effective among these other methods are those utilizing the rates of uptake or release of radioactive iodine by the thyroid. Another method which has been much used is analysis of the concentration of protein bound iodine in the plasma or serum, which in normal animals reflects quite accurately the concentration of circulating thyroid hormone. Unfortunately, the concentration of circulating hormone is not related to the rate of secretion from the thyroid, since it represents a balance between secretion and tissue utilization. An example of the poor correlation between protein bound iodine and thyroid secretion rate is shown in *Figure 3*, which is largely data from Robertson and Falconer's paper of 1961 with the addition of some further points calculated since that time. It can be seen quite clearly from this figure that there is no correlation whatsoever between these two parameters of thyroid function. The line which has been drawn on this graph represents the linear regression equation for the data, which is not significantly different from parallel to the base line and, in addition, the fit of the points to the line is not significantly different from random.

Measurements of thyroid activity in children carried out by radioactive iodine uptake determination, have shown in a study by Ponchon, Beckers and De Visscher (1966) that the peak of thyroid uptake is reached at 2 years of age and declines thereafter. A similar study by Berezkov (1958) demonstrated a higher thyroid uptake in children of 8–9 years of age than in younger or older children. Using the measurement of actual thyroid secretion rate in children, a study by Croughs, Visser and Woldring (1965) failed to show any clear change in thyroid secretion in children from 1 year

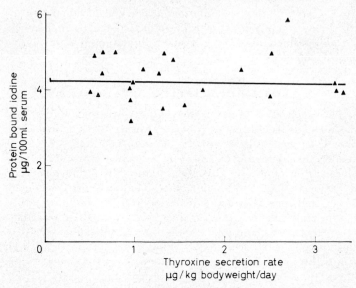

Figure 3. Correlation between serum protein bound iodine concentration and thyroid secretion rate in adult sheep during various reproductive states. No significant correlation; $R = 0.053$, Variance ratio $= 0.064$

to 15 years of age, though they did not have a very large number of children to investigate. Each of these papers clearly showed, however, that the thyroid activity of children was appreciably higher than that of adults. After childhood thyroid activity steadily declines, as has been shown by Fisher, Oddie and Wait (1964) comparing a group of children aged 5–9 years with a group of adults aged 20–24 years. In a study by McGavack and Hoch (1965) a comparison of a group of people of average age 51 years with a group of average age 87 years showed that during this period there is also a decline in thyroid secretion.

In the rat a similar pattern is observed with a steady decline in thyroid activity in adult animals with increasing age. Johnson, Kibler and Silsby (1964) and other studies by Johnson's group, using the measurements of radioactive iodine release rate from the thyroid, assessed the thyroid activity of animals kept at different environmental temperatures throughout their life-span. At 28°C the release rate of rats rose to a maximum at about 90–100 days of age and declined thereafter for the rest of the life of the animal. Several other studies using a variety of methods have shown a declining thyroid activity with ageing in rats kept under normal environmental conditions (Verzar and Freydberg, 1956; Wilansky, Newsham and Hoffman, 1957; Johnson, Ward and Kibler, 1966).

Measurements of basal metabolic rate in man have shown a decline parallel to thyroid activity during ageing (Boothby, 1924), as could be expected since thyroid activity is the major factor controlling the basal metabolic rate (*see* Tata, 1964, for review).

In domestic animals basal metabolic rate does not show a steady decline with age. As Brody (1945) demonstrated in the pig and

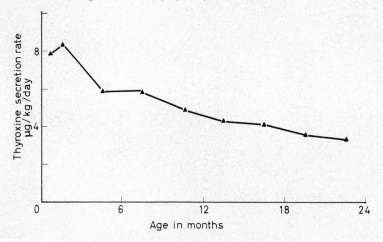

Figure 4. *Thyroid secretion rate/100 kg body weight in young cattle from* 3 weeks *to* 2 years *of age.* *Data plotted from Post and Mixner (1961) and Mixner, Szabo and Mather (1966), by kind permission of the authors*

cow, a peak in metabolic rate expressed as calories per square metre per day occurs at or near puberty and only after this is there a decline with increasing age. Brody (1945) also noted a similar pattern in the growth rate curve for these animals, which reaches a peak near to puberty.

In cattle, measurements of thyroid secretion rate per hundred kilogrammes body rate have been reported by Mixner, Szabo and Mather (1966) using a technique of considerable accuracy developed by Post and Mixner (1961). By courtesy of Dr. Mixner, we are able to show the results of part of this study in *Figure 4*, which illustrates the changes in thyroxine secretion rate in microgrammes per kilogramme per day with age of dairy calves from a few weeks to 2 years. As can be seen from the figure, there is a steady decrease in the secretion rate per unit body weight with age over this period in these experiments. If, however, one investigates the fractional turn-over rate of thyroxine, which measures the proportion of the body pool of thyroid hormone used each day, a peak in turn-over is evident in the data of Mixner in the 6–9 month old group of animals (*Figure 5*).

Flamboe and Reineke (1959) working with adult goats, Falconer and Robertson (1961) and Henneman, Reineke and Griffin (1955) with adult sheep, have shown a steady decline in thyroid secretion with age. In *Figure 6*, a graph which is drawn from the author's (I.R.F.) data, there is some evidence of an increase in thyroid

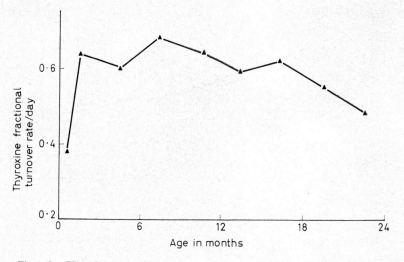

Figure 5. Thyroxine fractional turnover rate, calculated from the rate of decrease of injected thyroxine in the serum in young cattle from 3 weeks to 2 years of age. Data plotted from Post and Mixner (1961) and Mixner et al. (1966)

secretion rate per unit body weight up to 6–9 months of age in lambs, which is in agreement with the changes in basal metabolic rate in the domestic animals. If this data is expressed in terms of thyroxine secretion rate per animal per day instead of per unit body weight, one gets the type of curve that is shown in *Figure 7*. This shows evidence of a maximum secretion rate of thyroid hormone in sheep at around 6 months to a year of age with a decline thereafter. The changes in the activity of the thyroid during the period before puberty in domestic animals appear worthy of further investigation but considerable technical difficulties are involved, since the ability

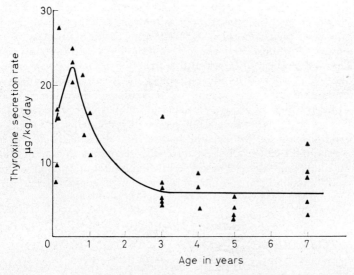

Figure 6. *Thyroid secretion rate in female sheep from 6 weeks to 7 years of age, calculated as µg thyroxine/kg body weight/day*

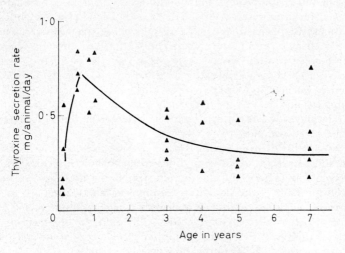

Figure 7. *Thyroid secretion rate, calculated as milligrammes thyroxine/animal/ day, in female sheep from 6 weeks to 7 years of age*

117

of the serum proteins to carry thyroid hormone changes markedly during early life. As this period, in the case of lambs at any rate, is of major importance with respect to the growth of animals to be slaughtered for meat, it would look as if the study of this period of a sheep's life might well be rewarding in terms of our understanding of the role of the thyroid during rapid growth and during puberty.

To sum up these investigations on the activity of the thyroid gland at different stages of growth of man and animals, it is clear that the young growing individual of all species investigated has a higher thyroid hormone output per unit body weight than the adult. It is also apparent that the adult has a steadily decreasing thyroid output with increasing age. The experimental studies of changes in thyroid function before sexual maturity show no consistent pattern, indicating the need for further investigation in this field.

THYROID ACTIVITY AND RATE OF GROWTH

Comparatively little work has been done on this aspect of the thyroid gland, probably because of the difficult experimental methods for the measurement of thyroid function, combined with the overriding effects of other factors upon the rate of growth. Studies by Kunkel and co-workers on the correlation between rate of gain and serum protein bound iodine concentration in beef cattle showed no clear results, though some indications of a curvilinear relationship appeared (Kunkel, Colby and Lyman, 1953; Kunkel, Green, Riggs, Smith and Myrtis, 1957). Since protein bound iodine concentration has been shown to have no correlation with thyroid secretion rate in cattle by Johnson and others (Johnson, Hindry, Burnett and Guidry, 1959), this method is not directly informative and it will be extremely difficult to show any valid correlation with growth because of its inherent failure to correlate with thyroid secretion.

A study by Armstrong and Hansel (1956), assessing the rate of gain in bodyweight of cattle on different diets, showed a correlation between rate of gain and the thyroid stimulating hormone content of the pituitary of the slaughtered animals. The cattle on a high plane of nutrition had a significantly higher pituitary T.S.H. content, and smaller thyroid glands. The difficulty of interpreting data of this type is relating pituitary T.S.H. content to thyroid activity; it could, for example, be inversely related, directly related or, like the protein-bound iodine concentration, merely represent a steady state level saying nothing about the rate of secretion of the hormone.

The work of the present authors on this problem was carried out in two parts. The first was a study of growth rate and thyroid weight at slaughter in a total of 138 lambs, and the second a study of growth rate and thyroid radio-iodine release rate in, up to the time of writing, 55 experiments on 30 lambs. In the experiments on the correlation between thyroid weight at slaughter and growth rate, the lambs were divided into four groups. Two groups were weaned at 6 weeks of age and then fed on pasture until slaughter at

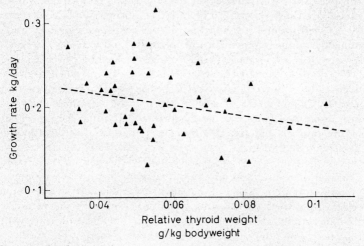

Figure 8. *Correlation between growth rate of lambs to* 36·4 kg (80 lb) *live weight and thyroid weight/kg body weight. The negative correlation is not significant;* $R = -0·26$

60 or 80 pounds live weight. The other two groups were fed (*ad lib.*) a pelleted ration indoors, and slaughtered at 80 or 100 pounds live weight. The thyroid glands were removed after death and weighed. In *Figure 8*, is shown a plot of the growth rate against relative thyroid weight (grammes of thyroid per kilogramme body weight) for the group that was slaughtered at 80 pounds live weight. As can be seen, there is some evidence of a correlation in a negative direction between thyroid weight and growth rate but it is not statistically significant. In the other three groups of animals there were no statistically significant correlations between growth rate and relative thyroid weight.

In order to improve our parameters for the measurement of

thyroid activity, of which thyroid weight is a relatively poor example, we decided to measure the rate of radioactive iodine released from the thyroid. This parameter had been shown earlier by Robertson and Falconer (1961) to be significantly correlated with thyroid secretion rate, and has the advantage that it is not influenced by variations in dietary iodine intake. It is a relatively simple parameter to measure compared to thyroid secretion rate, and therefore is more applicable to large studies of the type that we are undertaking. The lambs for these experiments were housed indoors, in a room with a 12 h day-length and at temperatures of $15 \pm 5°C$. All the groups of lambs were fed *ad lib.* on a ration of adequate but not excessive iodine content. The ration comprised 65 per cent cereal, 20 per cent ground straw or hay, 12 per cent soyabean meal and a mineral/vitamin supplement low in copper and iodine.

Growth rates were measured by weekly weighings over a period of 2–3 months, when the lambs were between 13 and 50 kilogrammes live weight. Thyroid activity was measured at intervals during this period. A dose of 20–40 μc ^{131}I was injected intramuscularly and directional scintillation counting used to determine the percentage of the dose in the thyroid, using the mathematical methods of Robertson and Falconer (1961) and Falconer (1963) for the calculation of the true rate of radioactive iodine release from the thyroid. Considerable variations in thyroid activity at any live weight or growth rate were observed, so that it was decided to attempt to relate the growth rate during a short period with the measured thyroid activity at that time and to carry out these measurements repetitively on lambs over the whole range of ages. The results of these experiments to the time of writing, and we would like to make it clear that we are still busily engaged in this work, are shown in *Figure 9*. The data has been plotted in terms of growth rate in kilogrammes per week against the logarithm to the base e of the true rate constant (K_4) of the equation defining the release of radioactive iodine from the thyroid. On this scale the activity of the thyroid is increasing exponentially from left to right. Statistical analysis of this data has demonstrated a curvilinear regression, the line of which is drawn on the figure and the equation of which is quoted in the figure. As can be seen by the analysis of variance for this curvilinear regression, the variance ratio is very highly significant with a probability of 0·001. This highly significant correlation shows both the essentially curvilinear relationship between thyroid activity and growth and also the range of thyroid activities over which the precise thyroid activity does not appear to influence growth. The data agree well with a straightforward interpretation of the role of the thyroid, especially if one considers the location on *Figure 9* which would be occupied by thyroidectomized, or partially thyroidectomized animals. These animals, as illustrated in the very first figure of this paper, grow extremely slowly and, of course, have a very low thyroid secretion rate, zero if one has completely removed the thyroid tissue. These animals if plotted as in *Figure 9* would therefore be grouped in the bottom left corner of the figure and show even more powerfully the curvilinear nature of the relationship between thyroid activity and growth rate.

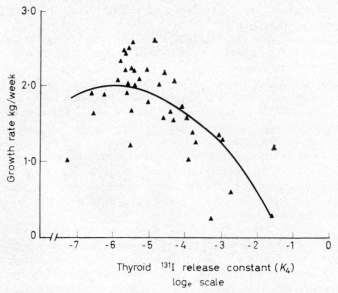

Figure 9. Correlation between the growth rate and thyroid activity of lambs kept in a controlled environment and fed ad lib a standard diet. Highly significant curvilinear correlation; $P < 0.001$, Variance ratio $= 17.3$

It is clear from this relationship that thyroid activities considerably above or below the normal rate lead to a reduced rate of growth. The reduction in metabolic rate, body protein synthesis and bone growth during hypothyroidism or cretinism account for the adverse effects on growth rate of the low end of the range of thyroid activities.

In hyperthyroidism the utilization of dietary energy is less efficient, due to the increased basal metabolic rate. This could be matched by an increased food consumption to maintain a normal growth rate, but in these experiments a lowered growth rate has been observed in the animals with the highest thyroid activities.

It is apparent from *Figure 9* that the lambs with the higher rates of growth show a considerable range in thyroid activity. This is understandable if one considers that the optimal thyroid activity may change between different stages of growth and maturation, and that the animals investigated were at different stages of growth. From the work of Brody (1945) on the metabolic rate of domestic animals, and the studies of thyroid activity discussed earlier in this paper, it would appear that the pre-pubertal rapidly growing animal requires an appreciably higher thyroid activity than the mature fattening animal.

These concepts may explain the variable and contradictory results of therapy designed to increase or decrease thyroid hormone availability to growing animals. Any method designed to elevate thyroid activity in a population of animals will only increase the growth of those having an endogenous secretion below the level that is optimal for that stage of development. A low endogenous secretion at later stages of development may itself be adequate, and any elevation further may have growth depressant effects. It may reasonably be concluded that a knowledge of the adequacy of an animal's own supply of thyroid hormone is needed before any beneficial treatment can be applied, and hence the inadequacy of general thyroid hormone or anti-thyroid drug treatment as a means of improving animal growth.

It appears, however, that the acquisition of optimal thyroid activity would increase the rate of live weight gain of a population of growing animals. A selective breeding programme which incorporated this factor among others, could therefore be expected to give rise to beneficial results.

REFERENCES

Armstrong, D. T. and Hansel, W. (1956). *J. Anim. Sci.* **15**, 640

Beeson, W. M., Andrews, F. N. and Brown, P. T. (1947). *J. Anim. Sci.* **6**, 16

Berezkov, L. F. (1958). *Pediatriya* **36**, 32

Boothby, W. M. (1924). *Physiol. Rev.* **4**, 69

Brody, S. (1945). *Bioenergetics and Growth.* New York; Reinhold

Coop, I. E. and Clark, V. R. (1958). *N.Z. Jl. agric. Res.* **1**, 365

Croughs, W., Visser, H. K. A. and Woldring, M. G. (1965). *J. Pediat.* **67**, 343

De Groot, C. A. (1963). *Acta endocr. Copenh.* **42**, 423

Desai, I. D. and Patel, B. N. (1962). *Indian Jl. Dairy Sci.* **15**, 15

Dinusson, W. E., Andrews, F. N. and Beeson, W. M. (1950). *J. Anim. Sci.* **9**, 321

Fagge, C. H. (1871). *Tr. med. chir. Soc. Lond.* **54**, 155

Falconer, I. R. (1963). *J. Endocr.* **25**, 533

— (1965). *Nature, Lond.* **205**, 978

— (1966a). *Biochem. J.* **100**, 190

— (1966b). *Biochem. J.* **100**, 197

— and Robertson, H. A. (1961). *J. Endocr.* **22**, 23

Fisher, D. A., Oddie, T. H. and Wait, J. C. (1964). *Am. J. Dis. Child.* **107**, 282

Flamboe, E. E. and Reineke, E. P. (1959). *J. Anim. Sci.* **14**, 419

Goodall, C. M. and Gavin, J. B. (1966). *Acta endocr. Copenh.* **51**, 315

Henneman, H. A., Reineke, E. P. and Griffin, S. A. (1955). *J. Anim. Sci.* **14**, 419

Horsley, V. (1886). *Proc. R. Soc.* **40**, 6

Johnson, H. D., Kibler, H. H. and Silsby, H. (1964). *Gerontologia* **9**, 18

— Ward, M. W. and Kibler, H. H. (1966). *J. appl. Physiol.* **21**, 689

Johnson, J. E., Hindry, G. A., Burnett, W. T. and Guidry, A. (1959). *J. Dairy Sci.* **42**, 927

Kunkel, H. O., Colby, R. W. and Lyman, C. M. (1953). *J. Anim. Sci.* **12**, 3

— Green, C. G., Riggs, J. K., Smith, R. L. and Myrtis, C. (1957). *J. Anim. Sci.* **16**, 1030

Matsushima, J. K., Mukhtar, A. S., Clanton, D. C. and Payne, L. C. (1960). *J. Anim. Sci.* **19**, 1274

McGavack, T. H. and Hoch, H. (1965). *J. Geront.* **20**, 383

Mixner, J. P., Szabo, K. T. and Mather, R. E. (1966). *J. Dairy Sci.* **49**, 199

Ponchon, G., Beckers, C. and De Visscher, M. (1966). *J. clin. Endocr. Metab.* **26**, 1392

Post, T. B. and Mixner, J. P. (1961). *J. Dairy Sci.* **44**, 2265

Race, R., Hill, G. N., Pain, R. W. and Mulhearn, C. (1968), *Res. Vet. Sci.* In press.

Robertson, H. A. and Falconer, I. R. (1961). *J. Endocr.* **21**, 411

Schultze, A. B., Hathaway, I. L. and Loeffel, W. J. (1950). *J. Anim. Sci.* **9**, 571

Scow, R. O. (1959). *Am. J. Physiol.* **196**, 859

Spielman, A. A., Petersen, W. E. and Fitch, J. B. (1944). *J. Dairy Sci.* **27**, 441

— — — and Pomeroy, B. S. (1945). *J. Dairy Sci.* **28**, 329

Tata, J. R. (1964). *Adv. metab. Dis.* **1**, 153

Verzar, F. and Freydberg, V. (1956). *J. Geront.* **11**, 53

Wilansky, D. L., Newsham, L. G. S. and Hoffman, M. M. (1957). *Endocrinology* **61**, 327

Wilkins, L. (1957). In: *The Diagnosis and Treatment of Endocrine Disorders in Childhood and Adolescence.* 2nd edn., Ch. VI. Springfield, Ill.; Thomas.

DISCUSSIONS

DR. GREENWOOD (London) observed that work with growth hormone was at a more advanced stage with humans than with animals, to which PROFESSOR LAMMING (Nottingham) replied that available supplies of animal trophins had been much less pure than those of human growth hormone and that this lack of purified materials had inhibited development in the animal field.

DR. FALCONER (Nottingham) expressed surprise that there seemed to be some doubt about the influence of GH on protein synthesis, as Korner's work appeared to have established this as a recognized effect. DR. HUNTER replied that there was no dispute about whether or not growth hormone influenced protein synthesis but only about the mechanism. What Dr. Korner had in fact shown was that ribosomes isolated from the liver of hypophysectomized rats showed only minimal ability to synthesize protein; this ability could be partly restored by *in vivo* administration of growth hormone but not by addition of the hormone to the ribosomes *in vitro*. This was not evidence for direct action as it may be simply that the ribosomes from hypophysectomized animals represented 'rusty parts' from a disused machine. DR. GREENWOOD observed that metabolism occurred in the cells and not in the blood, so plasma levels were only an index of activity.

DR. SPEAKE (I.C.I., Ltd.) had understood that GH was relatively ineffective as a lipolytic agent compared with other pituitary factors, yet Dr. Hunter had ascribed lipolysis as one of its main functions. Was the evidence that GH was a lipolytic agent any stronger than Korner's view of it as being a protein synthesizer? DR. HUNTER thought that it was much stronger and that there were three very good indications that this was so. First, the *in vitro* lipolytic action on isolated fat cells, second the similar *in vivo* action in forearm perfusion studies (both actions had been demonstrated with physiological concentrations of growth hormone), and third, the intricate association between NEFA mobilization and HGH secretion which had been accumulated now over 4 years application of the radio-immunoassay.

DR. GREENWOOD emphasized that GH was undoubtedly required as a factor to maximize growth potential, and went on to comment on the role of GH in situations of stress, where there were no changes in blood sugar but a rise in plasma cortisols *and* in ACTH; in the monkey, for example, this was a highly developed mechanism. There were two mechanisms which controlled GH; 'feed-back', or energy homeostasis, and stress.

DR. FALCONER asked for comment on the claim by Bornstein to have separated the lipolytic from the growth promoting activity of GH, to which DR. HUNTER replied that people were continually producing new lipolytic peptides and new lipolytic activities for established hormones. The radio-immunoassay data did provide, however, an interesting link between the whole HGH molecule, which they were estimating immunochemically, and NEFA regulation which was so well associated with changes in HGH. DR. GREENWOOD observed that work was currently in progress on this aspect at Leeds.

DR. ADLER (Jerusalem) inquired whether the results of castration had been related to efficiency of feed utilization, but MR. BRADFIELD replied that they had not, as the sheep concerned had been at pasture throughout.

Further to this, DR. PRESCOTT (Newcastle upon Tyne) outlined an experiment at Newcastle in which lambs slaughtered at 100 lb live weight were kept either on a high plane throughout or a low plane to 70 lb followed by the high plane; no differences were found between entire and castrate animals on the high plane throughout, but following restriction entire animals grew faster and were leaner.

DR. WARNER (Cornell) inquired whether the sheep had been fasted before final live weight measurements were taken, to equate killing-out percentages. MR. BRADFIELD replied that they were starved for 24 h to give a final empty live weight figure from which calculations were made.

DR. BRAUDE (Reading) commented on the statistical dubiety of the results because of the considerable individual variation between animals and questioned the use of groups of only 5 sheep. MR. BRADFIELD agreed that there was considerable variation but that the effects were nevertheless real. PROFESSOR LAMMING (Nottingham) considered that in this experiment five animals per subgroup had been sufficient and that the purpose of the statistical analysis was to assess the validity of the data.

DR. GREENWOOD inquired on the economic value of the statistically significant differences, to which MR. BRADFIELD replied that the entire male would be economically worth while but that the ovariectomized animal was only of academic interest; the results contradicted Hammond's original statement that the inhibitory effect of oestrogen was greater than the influence of testosterone.

DR. J. MCCARTHY (Dublin) asked how useful was the measure of RNA/DNA ratio, to which DR. HUNTER (Edinburgh) replied that while it was known that testosterone would suppress LH secretion, it was not known what would suppress FSH secretion; the amount of LH in the pituitary was about one per cent of the amount of GH, so it would be extremely difficult to show a depression in RNA/DNA ratio when so little LH was being produced. DR. GREENWOOD commented that, in fact, there had been a good correlation between treatment and change in RNA/DNA ratio, but it was not clear what the changes meant.

DR. EVERETT (Ruakura) commented on the importance of the economic aspects of the carcass changes and asked whether the results might have been influenced by the fact that castration had been at 3–4 days whereas ovariectomy had been at 40 days. MR. BRADFIELD replied that there were considerable differences in carcass development of the four 'sexes'. At 80 lb live weight entire males had inadequate fat cover, castrates and entire females were acceptable, while ovariectomized females were overfat. At higher weights castrates and entire females were overfat, whereas entire males had a desirable fat cover. Any value of oestrogen was only apparent in the entire animal where it accelerated 'finish' at 80 lb or above, or in the castrate male where it increased nitrogen deposition and fat content and allowed the animal to be taken to a greater weight. Testosterone had no influence on growth rate and reduced rate of fat deposition in both entire and castrate males.

On the question of difference between animals in age at castration or ovariectomy, there was no indication that either of these operations caused any check in growth.

PROFESSOR HOWELL (Saskatchewan) asked for clarification on whether DR. FALCONER had said that the more rapidly growing animal was less efficient, in that the hyperthyroid animal had a higher BMR which resulted in a lower growth efficiency, but also that the fast growing animal had higher thyroid activity. DR. FALCONER replied that there was a curvilinear relationship between thyroid activity and growth rate, so that both extremes of thyroid activity depressed growth. DR. TAYLOR (Edinburgh) inquired whether the BMR of animals with high thyroid activity was elevated to the same degree as that of thyroidectomized animals was depressed, i.e. 55 per cent, so that in animals with high thyroid secretion rate and low growth rate the metabolic rate might be abnormally high. DR. FALCONER replied that this could perhaps be the case but BMR had not been measured in these animals.

Dr. Greenwood observed that the hair growth in the cretinous sheep indicated a lack of protein synthesis and that the effect could be measured quantitatively by injecting S-labelled amino acids. Dr. Falconer replied that it was well established that hair growth was very dependent upon thyroid status and that thyroxine therapy had been used in New Zealand to increase fleece production. Mr. Draper commented that it was noticeable in the sheep which were hyperthyroid and growing slowly that they had heavy fleeces; levels of thyroid activity detrimental to growth could give a high wool growth.

Dr. Hunter asked whether, because thyroid hormone not only speeded up growth but also caused more rapid maturity, the giving of thyroid hormone to a slightly hypothyroid animal would fail to produce the commercially desirable optimum. Dr. Jagger (I.C.I. Ltd.) then commented, at the request of Dr. Falconer, on the effect of thyroid hormone administration in mature animals. In mature mice it was possible to induce growth by giving about 50 µg thyroxine; they then ate 25 per cent more feed and lost about 50 per cent of depot fat in 8 days. Work on neonatal treatment with thyroxine indicated that a very mature small animal was produced.

Dr. Taylor inquired why *relative* thyroid weight had been plotted against *absolute* growth rate, to which Dr. Falconer replied that there was no need to justify the use of absolute growth rate but the reason for relative thyroid weight was because a bigger animal had a bigger thyroid and produced more hormone for normal function, although actual data relating thyroid weight to body weight was lacking. Mr. Draper commented that it was doubtful whether thyroid weight was of any significance, since in sheep with congenital goitre there was a considerable increase in size of thyroid (60–100 g v 2–4 g) with no increase in its efficiency. This suggested that in a normal population there could be wide variation in thyroid weight with no relationship to growth rate.

Dr. Greenwood then went on to inquire whether, under any circumstances, the thyroid gland could be at maximum output but still limiting in terms of growth, i.e. thyroxine the limiting factor to growth, to which Dr. Falconer replied that the capacity of the thyroid to secrete hormone was always vastly in excess of its actual secretion.

Mr. Carroll (Dublin) asked Mr. Bradfield whether he had calculated the moisture content of the meat of implanted animals on a fat-free basis, because it was widely felt that implanted animals merely produced meat of higher water content. As the fat content of a carcass determined to a large extent its water content, the only valid comparison was on a fat-free basis. Mr. Bradfield replied that carcass analysis had been done only on the growth trial and not on the stilboestrol or testosterone treated animals. There were significant differences in moisture content on a fat-free basis between the four 'sexes'.

Dr. Thurley (Weybridge) inquired of the endocrinologists present the truth of the belief that when one bred for rapid growth one was selecting for diminution of thyroid activity and increase in secretion of GH; also, what was the effect of GH activity on muscle *strength* as distinct from growth? Dr. Greenwood thought that little was known about the genetics of hormone activity but Dr. Timon (Dublin) mentioned work done with pigs by Nalbandov which showed that those selected for high growth rate had larger pituitaries but not larger 'unit potency' of somatotrophin.

Dr. Hunter stressed that GH was necessary for growth; administration of GH to dwarfs proved to be GH deficient was very effective. The mechanism of this was essentially through redistribution; it allowed consumption of excess food during the day and the mobilization of this during the night. On the question of muscle strength and acromegalics, they were more muscular and less fat than the normal male, and DR. GREENWOOD confirmed that their muscles were weaker. DR. HUNTER thought that the reason for this weakness was lack of fat for mobilization in close association with the muscle. An alternative explanation was suggested by MR. CARROLL, namely, that when a muscle grew and if the units in length were not additive the inter-digitation of the actin and myocin would be more spread out and lack cross linking. DR. MESSAGE thought that this was an unlikely explanation of muscle weakness and went on to observe that very little was known about the hormonal background to changes in muscle strength.

DR. ADLER asked whether the resting length of a muscle influenced the tension which it could develop, to which DR. MESSAGE replied that it was extremely difficult to measure accurately the length of a muscle fibre and so reliable data were not available.

MR. HOUSTON (Leeds) commented that under the electron microscope there was a substance present immediately post-slaughter in muscle tissue which disappeared rapidly but at different rates with different animals. It was necessary, therefore, to be very careful in relating findings on *post rigor* muscle to the *in vivo* status.

III. PRENATAL DEVELOPMENT

PRENATAL DEVELOPMENT OF UNIPAROUS ANIMALS, WITH PARTICULAR REFERENCE TO THE INFLUENCE OF MATERNAL NUTRITION IN SHEEP

G. C. EVERITT

Ruakura Agricultural Research Centre,
Hamilton, New Zealand

INTRODUCTION

OVER 300 years ago, Sir Thomas Browne (1642) exclaimed, '. . . and surely we are all out of the computation of our age, and every man is some months older than he bethinks him. For we live, move, have a being and are subject to the actions of the elements and malice of diseases in that other world, the truest microcosm, the womb . . .'

This quotation summarizes the major themes running through this review. In the first place it is hardly possible to remain unimpressed with the beauty of the uterine microcosm and the coordination of activity inherent in the formation of new life. Secondly, the sensitivity of the foetal body mass to nutritional regimes imposed upon the maternal organism is notable. At some risk of pedantry it may be emphasized that undernutrition bordering upon starvation actively threatens a large proportion of the world human population, the burden resting most heavily upon the reproducing female. Barcroft (1946), recognizing the dangers, commented, 'The prospect of semi-starvation which hangs over much of the world at the present time suggests the possibility of other researches on the effect of maternal malnutrition on foetal development'. Much dissension exists at present, however, as to the manner in which nutritional effects on development processes are manifested. This conflict of views is of great interest to students of growth and is of material importance in assessing productive efficiency of domesticated animals. Thirdly, however, recent studies have revealed an ability of the mammalian foetus to sustain imposed stresses with remarkable preservation of the 'interieur mileau' and 'anatomical harmony' (Boccard and Dumont, 1960). Foetal growth proceeds along its course, protected by a resilient maternal buffer, attaining developmental horizons with a biological flexibility essential to survival.

Apart from the academic accumulation of knowledge for its own sake, critical studies of prenatal growth are of interest in the agricultural context, for growth itself is the product of diverse and powerful forces, blending with production on the one hand and reproduction on the other.

Among uniparous species multiple births occur to a greater extent in sheep and goats than in other members of the class, and this has an important bearing on the economy of sheep production. Although the sheep is essentially a uniparous animal, the quest for increased productivity demands an increasing degree of multiparity. Observations on multiparous animals thus assume considerable importance from an agricultural viewpoint.

The present knowledge of prenatal development in uniparous animals is based primarily on the sheep (*Ovis aries* L.), although the contribution of Weinbach (1941) to prenatal growth of humans needs noting. Maternal factors, such as maternal hormones, embryo endometrial interaction, size and age of dam, litter size, length of gestation, placental efficiency and maternal nutrition, are held responsible for 50–75 per cent of the variability in foetal size (Hafez, 1963); and of these factors the last named, maternal nutrition, is clearly of paramount importance for uniparous farm animals (Moustgaard, 1959) and humans (Hytten, 1964).

PRENATAL INFLUENCES ON POSTNATAL PRODUCTIVITY

Prenatal growth of domesticated farm animals has received scant attention as compared with postnatal studies. It is important that this should be rectified in spite of greater financial costs and difficulties of prenatal investigations.

An increasing proportion of the life-span of meat producing animals is being spent *in utero* due to a general policy of reducing slaughter age through application of new and improved husbandry techniques. Calves intensively housed and fed for veal production, for example, spend over twice as much time in incubation as they do in their equally cramped independent existence.

Less obvious is the fact that the potential for later production of wool and milk, as well as meat, may be defined during gestation.

Cells, tissues, organs and systems are all initiated here in a sequential developmental pattern, events of later life merely permitting or restricting their productive possibilities. The extent to which the latter may be modified by factors operating during the intra-uterine formative stages appears insufficiently appreciated.

Apart from ewe and lamb mortalities arising from impaired maternal metabolism (Reid, 1963), the reduced weight and associated neonatal incompetence of lambs at birth (Alexander, 1964a), coupled with restrained behaviour of dam and offspring, can be held responsible for a large proportion of the massive neonatal mortalities inflicted upon the world sheep industry (Thomson and Aitken, 1959). Restricted postnatal performance of offspring born to ewes undernourished in pregnancy is of equal concern as, if less striking than, reduction of birth weight. On average, any deviation below the mean in lamb birth weight may be magnified 3–4 fold by weaning time (Thomson and McDonald, 1956). Single lambs born to ewes severely ill-fed throughout the whole of pregnancy in the experiment of Schinckel and Short (1961) were about 34 per cent lighter at birth, 9 per cent smaller at maturity, with about 15 per cent fewer wool follicles per sheep, and they produced about 9 per cent less wool than lambs born to well-fed ewes.

Wool production of twin lambs, and of singles born to immature ewes, may suffer from a maternal handicap (Turner, 1961; Dun and Grewal, 1963; Wiener and Slee, 1965), although adequate nutrition in early postnatal life tends to compensate for prenatal penalties (Doney and Smith, 1964; Taplin and Everitt, 1964; Everitt, 1965, 1967). Prenatal and postnatal nutritional treatments contributed equally and additively to the permanent effects on Merino sheep recorded by Schinckel and Short (1961).

Donald (1958) cites evidence that although cattle twins are manifestly smaller than singles at birth and for some time thereafter, by the time they are 2 years old they are, on average, very similar to singles in body weight and dimensions, fertility and milk production.

The magnitude, additive nature and degree of permanency of undernutrition effects in early pregnancy (before 90 days) and late pregnancy (between 91 days and lambing) in sheep have been reported recently (Everitt, 1967). *Figure 1* records mean differences of main treatment effects on the post-weaning body weights of Merino sheep over a 3-year period in the Mediterranean-type environment of South Australia. The size and cyclical nature of the sex difference in body weight can be related to seasonal fluctuations in pasture nutrition. Undernutrition in late pregnancy substantially affected postnatal growth, persisting throughout the experimental period. A new finding was that severe undernutrition in early pregnancy retarded postnatal growth. This relatively small effect, cumulative with the undernutrition effect in late gestation, progressively diminished with time through compensatory growth to become negligible and non-significant at about 7 months of age.

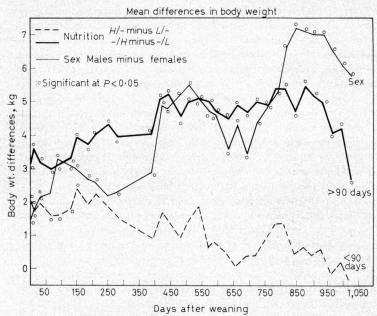

Figure 1. Mean differences of main effects on the post-weaning body weights of Merino lambs subjected to undernutrition in early pregnancy (conception—90 days) and/or late pregnancy (91 days—lambing). H/− represents high 0–90 days; L/− low 0–90 days; −/H represents high 91 days to lambing and −/L low 91 days to lambing. Differences attaining statistical significance are shown by (P < 0·05) (From Everitt, 1967)

Nevertheless, both these penalties could be of considerable importance when lambs are killed for meat production at, say, 4–5 months of age.

No information is yet available on the reproductive performance of offspring derived from ewes ill-fed during pregnancy. An effect of this nature could prove, in fact, of greater practical importance than a restriction of lifetime wool productivity. Retarded early postnatal growth may delay the onset of first oestrus in maiden ewes (Joubert, 1963), extending the generation interval, with economic repercussions. Ovulation rate and lambing percentage are also positively related to the body weight of the ewe when she is mated (Wallace, 1961; Coop, 1962, 1966). A small maiden ewe may exert a maternal effect on foetal growth, producing a small lamb at birth with concomitant temporary and permanent disadvantages (Turner, 1961). Moreover, the lighter the body weight of the ewe at mating time the greater is the risk of damage to the conceptus through undernutrition in early pregnancy (Everitt, 1966a).

Nutritional influences during prenatal life on the postnatal performance of uniparous animals reared primarily for meat production have not been examined in detail and represent a rich field for research activity. Hight (1966a), at Ruakura, has studied the effects of a high plane in contrast to a very low plane of grazing nutrition prior to calving on the performance of hill country beef cows and their calves. Low plane calves were 12 lb, or 20 per cent, lighter than high plane calves at birth. This difference increased to 36 lb at weaning, the latter possibly being influenced by impaired lactational performance of the ill-fed cows (Foote, 1964), and remained at this level for some time after weaning (Joblin, 1966). Studies of residual effects on growth and reproductive performance are continuing (Hight, 1966b) but it is currently believed that the total effect of the precalving level of nutrition on calf weaning weight is similar to that of the postcalving plane of nutrition. The shorter the time interval between birth and slaughter the more important does an adverse effect on birth weight and weaning weight become, for not only is the weight of product involved but, equally important, the season of marketing (Everitt, 1966b).

Meat production must be closely related to characteristics of skeletal muscles (Everitt, 1963) but morphological and histological studies of the prenatal development of skeletal muscle are sadly lacking. Wallace (1948) observed the prenatal development of ovine musculature as a whole, but the present knowledge rests largely upon the contribution of Joubert (1956). He inferred that the number of skeletal muscle cells is a genetic characteristic determined by 90 days of foetal age in the sheep, subsequent growth of muscular tissue occurring by increase in diameter and length of this constant number of fibres. Recent evidence (Everitt, 1965) strongly suggests that hyperplasia continues past 90 days of age in foetuses carried by well-fed ewes, but foetuses born to severely underfed ewes have fewer fibres per unit area of muscle (Table 1).

Table 1

Mean weight of the semitendinosus muscle in ovine foetuses and the estimated number of muscle fibres in the cross-section (adjusted for the sex effect)

Character	Maternal nutrition		S.E.	Significance of difference
	High plane	Low plane		
At 90 days post conception				
Muscle weight (g)	1·0643	0·9989	0·0220	P < 0·001
No. fibres/cross-section (10³)	63·81	55·27	3·167	P < 0·05
At 140 days post conception				
Muscle weight (g)	7·9651	4·0526	0·3138	P < 0·001
No. fibres/cross-section (10³)	85·63	59·10	3·587	P < 0·001

Note: *Muscle fibre numbers*

Logarithmic regression coefficients for foetal sex as a discontinuous covariate where female = 0, male = 1;

at 90 days: $b = 0·1830 \pm 0·0125 : P < 0·001$

at 140 days: $b = 0·1489 \pm 0·0021 : P < 0·001$

Muscle cell division in such restricted animals may be prolonged into postnatal life given adequate feeding after birth. Additional study of this topic is required for it may assist explanation of the retarded postnatal growth and restricted mature size of animals ill-fed in early life (Schinckel and Short, 1961; Taplin and Everitt, 1964; Everitt, 1967).

Table 1 also shows that male foetuses had approximately 15–18 per cent more fibres in the semitendinosus muscle cross-section than females of the same age and (statistically) the same weight. This interesting result helps to explain the plump, blocky shape of males described for the ovine foetus by Stephenson (1962), recorded in *Figure 2*, and the bovine foetus by Keller (1920) and Beer (1925). To what extent these findings relate to the recorded superior growth rate of males over females in postnatal life (Everitt and Jury, 1966a; Everitt, 1967; *see Figure 1*) and body compositional differences due to sex (Everitt and Jury, 1966b) is not known. It could be significant that the condition of muscular *hypertrophy* in cattle (Butterfield, 1966) may reflect, in fact, *hyperplasia*.

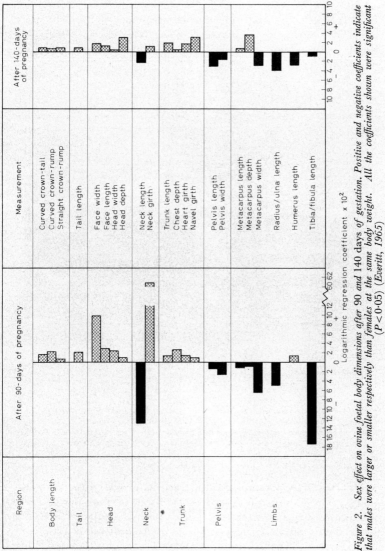

Figure 2. Sex effect on ovine foetal body dimensions after 90 and 140 days of gestation. Positive and negative coefficients indicate that males were larger or smaller respectively than females at the same body weight. All the coefficients shown were significant (P<0·05) (Everitt, 1965)

PRENATAL DEVELOPMENT AND MATERNAL NUTRITION

Effects of defective nutrition during pregnancy in farm livestock have been reviewed by Blaxter (1957), Moustgaard (1959) and Reid (1960), while Giroud (1959) deals with embryonic nutrition in broader terms.

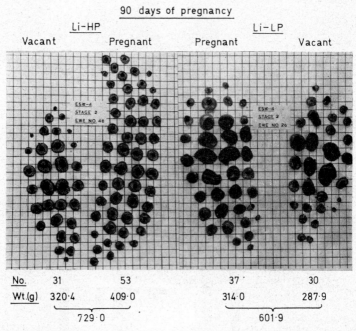

Figure 3. Semi-schematic representation of the functional cotyledon array after 90 days of pregnancy from selected well-fed (Li–HP) and ill-fed (Li–LP) ewes. The distorted cotyledons of the ill-fed ewe tend to lie more towards the centre of each uterine horn. Reduction in weight and number of cotyledons is also shown. 2 × 2 cm grid (Everitt, 1965)

It is clear that gestation tests the integrity of every structure in the body. The foetus, however, is not entirely at the mercy of the microcosm in which it exists. Under severe nutritional stress the ovine foetus attempts augmentation of its nutritional supply by compensatory development of placental components (Bellows, Pope, Chapman and Casida, 1963), for example, as shown in *Figures 3* and *4*.

Vertical sections of cotyledons

Li–HP   Li–LP

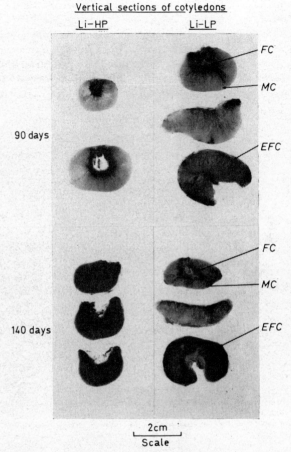

Figure 4.   Vertical sections of cotyledons from well-fed (Li–HP) and ill-fed (Li–LP) ewes after 90 and 140 days of pregnancy.   Cotyledons at 140 days are generally smaller and darker than those at 90 days.   Three stages in compensatory development of the foetal component under nutritional stress are shown.   FC = foetal component;   MC = maternal component;   EFC = everted foetal component.   2 × 2 cm grid (Everitt, 1965)

Similar morphological changes in ovine cotyledons associated with the level of progesterone (Alexander and Williams, 1966), and due to surgical reduction in the number of caruncles (Alexander, 1964c) have been reported.   Given the opportunity to do so, the nutritionally restricted foetus possesses powers of compensatory growth and development (Taplin and Everitt, 1964; Everitt, 1966a, 1967). Time, however, dictates the issue. Alexander (1956) clearly showed that maternal undernutrition can abbreviate gestation in sheep; premature senility of the placenta due to severe maternal undernutrition (*Figure 5*) may assist in advancing parturition, with the early expulsion of young into the hazards of an independent existence. Dawes and Parry (1965) comment that lambs of

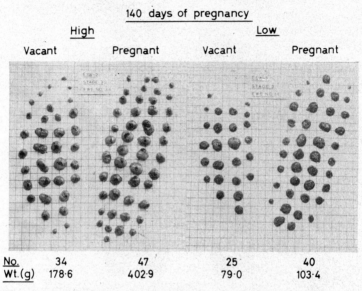

140 days of pregnancy

	High		Low	
	Vacant	Pregnant	Vacant	Pregnant
No.	34	47	25	40
Wt.(g)	178·6	402·9	79·0	103·4

Figure 5. Semi-schematic representation of the functional cotyledon array after 140 days of pregnancy from selected well-fed (High) and ill-fed (Low) ewes. The firm, spherical appearance of cotyledons from the High ewe may be compared with the relaxed, flat, atrophied appearance of those from the Low ewe (2 × 2 cm grid photographed about 2 h after slaughter of ewe) (Everitt, 1965)

gestational age less than 95 per cent of the normal are not of normal viability; the lamb differing from the monkey, cattle and man in this respect. Time also limits the degree of compensatory growth possible before birth. Furthermore, a restriction in placental efficiency through, for example, maternal undernutrition in early pregnancy (*Figure 4*) may restrict the foetal food supply in late pregnancy, even though the mother is herself feeding *ad lib*. There is now considerable evidence (Alexander, 1964b, c) to show that foetal and placental size in late pregnancy of the sheep, and other species, are closely correlated.

Table 2

Mean single foetus and functional cotyledon weights (g) after 90 and 140 days of pregnancy under high and low planes of maternal nutrition (Everitt, 1964, 1965, by courtesy of the Editor of *Nature*)

Character	Maternal nutrition		Low as percentage of High
	High	Low	
After 90 days			
Foetal wt.	547·8	490·2	89·5
Cotyledon wt.	489·7	340·7	69·6
After 140 days			
Foetal wt.	4,036	2,163	53·6
Cotyledon wt.	489·9	284·0	58·0
140 days as % of 90 days			
Foetal wt.	736·8	441·2	—
Cotyledon wt.	100·1	83·4	—

Alexander (1964b) concluded that very little placental tissue (cotyledons only) was necessary to permit gestation to reach 100 days in the sheep, but the size which the placenta had then attained limited subsequent foetal growth. A sensitive balance between embryonic loss and retardation of foetal growth is suggested, for Table 2 shows that placental development is restricted by maternal undernutrition to a greater extent than foetal growth, in agreement with earlier reports (Wallace, 1948; El-Sheikh, Hulet, Pope and Casida, 1955; Bellows *et al.*, 1963). Placental size must be restricted by undernutrition in early pregnancy below a critical threshold level before subsequent foetal growth is affected, for abundant feeding in late pregnancy of ewes which were severely ill-fed earlier can result in lambs of average birth weight (Taplin and Everitt, 1964; Everitt, 1966, 1967). Drastic curtailment of feed intake after conception and prior to completion of implantation may lead to embryonic loss (Edey, 1965, 1966); incipient undernutrition of a less precipitous nature may permit completion of implantation but restrict full placental development and thus foetal growth. Restricted placental development may also be associated with the abortion in late pregnancy of ewes severely undernourished in early pregnancy but abundantly fed thereafter (Taplin and Everitt, 1964; Everitt, 1967). The rapidly increasing weight of the foetus in late

141

gestation (*Figure 6*) may overload a tenuous maternal:foetal placental adherence at a time when placental weight is declining (McLaren, 1965a), especially under poor nutrition (Table 2).

A differential pattern of placental retardation due to under-nutrition is indicated by the weights and numbers of cotyledons in

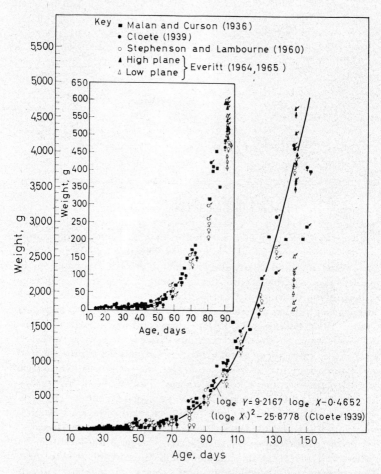

Figure 6. Prenatal growth of Merino sheep. Constructed from the data of Curson and Malan (1935), Cloete (1939), Stephenson and Lambourne (1960) and Everitt (1964, 1965). Males and females are shown where the separate sexes were recorded in the original data. The curve relates only to the data of Cloete (1939)

142

each uterine horn, and the weight and morphology of individual cotyledons according to their position within the horn (*Figures 3 and 4*). Further study of these characteristics is required for this may lead to appreciation of a *local* haemodynamic control of pre-natal growth (McLaren and Michie, 1960; McLaren, 1965b) in uniparous species, as distinct from the *general* systemic principle enunciated by Hammond (1944); and may also be of importance in understanding the placental insufficiency syndrome reported in man (Gershon and Straus, 1961).

The ratio of *local* to *general* effects on foetal growth varies with the species. In the ox and sheep, with small litters, only a general effect is at present recognized. When the litter becomes numerically large, as in the pig, rabbit or rat, a local effect of contiguity of foetuses within the same uterine horn also operates. McLaren and Michie (1960) concluded, however, 'It remains to be seen whether such differences in weight at birth as that between Shire-Shetlands and Shetland-Shires of Hammond and Walton can be accounted for in terms of vascular architecture and the mechanics of fluid flow'. The same comment might be offered for nutritional in-fluences on uniparous foetal growth.

Development sequelae of foetal undernutrition have received little attention. Although numerous studies on the nutrition of the pregnant ewe have been undertaken (Thomson and Aitken, 1959; Schinckel, 1963), the contribution of Wallace (1948) remains as the classical analysis of nutritional effects on ewe and foetus. Birth weight, as a comprehensive index of nutritional stress in gestation, as an easily measured parameter of the intense physiological activity expended *in utero*, and also because of its relationship to postnatal performance, has understandably received the greatest attention. Poor feeding of the ewe in late pregnancy markedly reduces this parameter in multiple births (Wallace, 1948). The degree of nutritional stress needed to reduce the birth weight of single lambs is substantially greater than for twins (Schinckel and Short, 1961; Taplin and Everitt, 1964; Everitt, 1964, 1966a, 1967). In fact, many of the reports of undernutrition effects on lamb birth weights, collated by Schinckel (1963), fail to demonstrate an appreciable reduction for single lambs, especially when group feeding was prac-tised. Nutrient requirements for maintenance of grazing sheep appear substantially greater than those of pen-fed animals (Coop, 1961). Foetal weight after 140 days of pregnancy in Merino sheep was found to increase by 3·5 g for every 1 g of digestible organic matter (D.O.M.) intake per day, while birth weight increased by

143

4·8 g for each 1 g D.O.M. intake per day during pregnancy (Everitt, 1965). Russel (1966) estimated nutrient requirements of Scottish Blackface ewes by studying the amount of D.O.M. required to maintain a prescribed physiological state (plasma-free fatty acid and ketone levels) and foetal weight at term. This elegant procedure by-passes the often misleading parameter of gross body weight as a measure of undernutrition in animals (Reid, 1963; Everitt, 1966a).

When reviewing the experimental evidence for the influence of prenatal nutrition on lamb birth weight, Thomson and Aitken (1959) and Schinckel (1963) deplore that most workers have presented nutritional response of ewe body weight in terms of absolute weight gain or loss and not relative to ewe body weight at the commencement of the feeding period. Response of lamb birth weight appears, nevertheless, to vary with the magnitude of the difference in ewe body weight between relatively 'high' and 'low' planes of nutrition but depends, too, on how inadequate the 'low' plane may have been. In some experiments 'low' plane animals show, in fact, an increase in body weight over a critical period (Coop, 1950) while 'high' plane animals may exhibit weight loss, as in the cattle experiment of Hight (1966a).

Few studies of pregnancy nutrition in uniparous farm animals have incorporated a serial slaughter programme, so that neither maternal nor embryo tissue changes have been studied in detail. The work of Wallace (1948) is an exception in this regard, but the degree of nutritional stress applied in his trials was not particularly great and only a few foetuses were examined at any one stage. Furthermore, in field grazing trials death may claim those ewes and/or their lambs subjected to the greatest stress, providing a bias from an academic viewpoint.

In contrast to the many studies of the effects of undernutrition in late pregnancy, the influence of the level of maternal feeding in early pregnancy on foetal development has received little attention. This is, in part, understandable for the additional requirements for nutrients even in late pregnancy of the cow are small (10 kcal energy retained per kg body weight $^{0.73}$) as compared with lactation (Blaxter, 1962).

Hammond (1944) postulated that foetal metabolic rate was highest in early pregnancy and only slightly less than that of the maternal nervous system. He drew support for the foetal component of his *Theory of the Partition of Nutrients According to Metabolic Rate* from the research of Wallace (1948), who was unable to demonstrate any significant effect of maternal nutrition on the

weight of twin sheep foetuses at 91 days of gestation. El-Sheikh
et al. (1955) found no significant differences in ovine foetal development at 40 days of gestation attributable to different levels of
maternal feeding. Likewise, no appreciable effect on foetal weight
at 25, 40 or 140 days of gestation was found by Foote, Pope, Chapman and Casida (1959) who imposed high and low planes of
nutrition on pregnant ewes of two breeds. Schinckel (1963), however, made brief reference to unpublished work where he found a
significant difference in foetal weight at 90 days of age between
single and multiple Merino foetuses, and concluded, 'This might
well imply that nutrient supply to the foetuses can be a limiting
factor before 90 days'.

Recent investigations have revealed that ovine foetal growth in
early gestation is not inviolate and can depend upon the level of
nutrition supplied to the ewe after mating (Everitt, 1964, 1965,
1966a). Table 2 shows that continuation of maternal undernutrition through late pregnancy dramatically increased the difference in weight between High Plane and Low Plane single foetuses
from approximately 12 per cent at 90 days of age, to 46 per cent at
140 days. This comparison, together with the 16 per cent mortality
of ill-fed ewes in late pregnancy, emphasizes the harsh stress imposed
in these trials, as in others reporting large reductions in birth weight
of single lambs (Schinckel and Short, 1961; Taplin and Everitt,
1964; Everitt, 1967). Poor nutrition clearly manifests itself most
on prenatal growth when the foetal mass is increasing rapidly in
late gestation (*Figure 6*).

Unlike the animal in postnatal life the foetus has little opportunity
to accumulate nutrient reserves and, if retarded, is limited by time
in making up lost ground before birth. These interactions of food,
growth and time in foetal life place a different emphasis on the
phenomenon of compensatory growth. This biological mechanism,
inherent in the growth of all animals (Tanner, 1963), is especially
important in the growth and production of domesticated species
(Wilson and Osbourn, 1960; Schinckel, 1963).

In postnatal life the duration, severity, and particularly the age of
the animal at the time of feed restriction play important roles, with
marked differences between species in their reaction (McCance,
1964). Work with laboratory animals (McCance and Widdowson,
1962; Widdowson and McCance, 1963), poultry (Wilson, 1954),
pigs (Robinson, 1964), goats (Wilson, 1958) and cattle (Crichton,
Aitken and Boyne, 1960; Dickinson, 1960; Everitt, 1961; Hight,
1966a, b; Joblin, 1966), as well as with sheep (Hammond, 1932;

Coop and Clark, 1955; Donald and Allden, 1959; Bradford, Weir and Torell, 1960; Schinckel and Short, 1961; Gunn, 1964; Purser and Roberts, 1964; Meyer and Clawson, 1964; Dun, Alexander and Smith, 1964) all points to the fact that the earlier in developmental life the stress is applied, the greater and more persistent will be the effect.

The sheep approximates the ox more closely than the rat in respect of developmental maturity at birth; in sheep the influence of nutritional stress after birth on ultimate size is not so marked as in rats but more so than in cattle. On the other hand, growth retardation of the ovine foetus in late gestation has a much greater and lasting effect than restriction in early gestation (Everitt, 1967; *see Figure 1*).

These differences between species in their ability to compensate for growth restriction appear related to differences in physiological age (Brody, 1945). Completion of organogenesis, in a metrical sense, may mark the period of life when a temporary retardation of growth ceases to exert a permanent effect on ultimate size, development and production. Prior to this time is a period of increasing sensitivity to stress and after it a period of diminishing sensitivity. Thus, the rat may cease organogenesis at, say, 6 weeks after birth as compared with the sheep which appears most susceptible in late prenatal-neonatal life. These suggestions are embodied in Dickinson's (1960) attractive hypothesis for the control of mammalian growth to which Schinckel (1963) added an additional component of sensitivity to stress in relation to physiological age of sheep. Extension of Schinckel's model to include other species is presented in *Figure 7*. Evidence to test the basic hypothesis needs accumulating, especially for the ox.

FOETAL GROWTH GRADIENTS

Thomson (1917) showed that by employing Cartesian co-ordinates the shape of one species could be derived from the shape of a closely related species by controlled distortions of the grid. This gave rise to the concept of growth gradients which Huxley (1932) subsequently examined in a wide variety of species through application of the allometric equation $(y = bx^k)$. Differential development of body parts has been known for centuries. Xenophon (400 B.C.), for instance, knew that 'The colt that is largest in the shanks at the time he is foaled makes the biggest horse. For in all quadrupeds the shanks increase but little in size as time goes on, whereas the rest of

146

the body grows to them, so as to be in the right proportion' (Marchant, 1925). Many other examples are discussed by Needham (1964) in his recent classical treatise on animal growth.

The studies of Scammon and Calkins (1929) on the prenatal growth of humans provided the first major contribution to the description of foetal growth gradients. Later significant contributions to knowledge of prenatal development in farm animals have

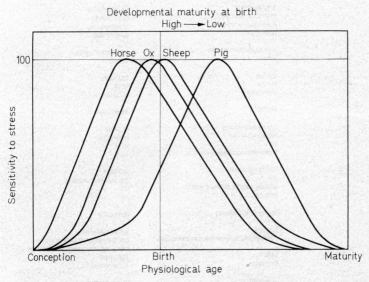

Figure 7. Generalization of the 'sensitivity to permanent stunting' proposal of Schinckel (1963) as applied to the mammalian growth model of Dickinson (1960). Considerable evidence exists for the sheep curve but relatively little for the curves drawn for other species

been those of Hammond (1932), Wallace (1948), Pálsson and Vergés (1952), Joubert (1956), Stephenson and Lambourne (1960) and Stephenson (1959, 1962, 1964). These studies reveal gradients proceeding antero-posteriorly along the main axis and centripetally along the limb axes, with similar gradients for internal organ systems, as well as the external form. Special adaptive growth of particular organs or parts has been noted. The various foetal organs grow at markedly different rates (Wallace, 1948), as shown in Figure 8, and with change in external body form (Curson and Malan, 1935; Cloete, 1939; Joubert, 1956; Stephenson and Lambourne, 1960; Stephenson, 1959, 1962, 1964), as shown in Figure 9, resulting in

147

continuous changes in conformation. Late gestation, especially in the sheep from 90 days onwards, is a stage of '. . . rapid absolute increment rather than of striking differentiation, and the changes in external body form take place quite slowly through differences in the relative growth rates of the various segments and parts of the body'. (Scammon and Calkins, 1929.)

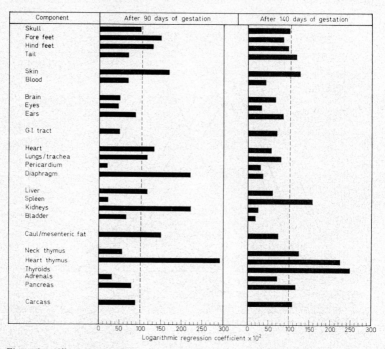

Figure 8. Allometric growth patterns of foetal body components in the sheep. Logarithmic regression coefficients in the relation between the weight of body components and foetal empty body weight at 90 and 140 days of gestation, after adjustment for the effect of sex, are shown. All the coefficients shown were significant $(P < 0.05)$ (Everitt, 1965)

Hammond (1921) first theorized that one of the chief advances made by man in creating improved breeds of meat-producing animals has been simply to extend and steepen growth gradients which already operate during the life of ancestral forms. Many of the concepts developed by the traditional Cambridge school under Hammond and since strongly advocated (for example Hammond, 1940, 1944, 1961; Pálsson, 1955), rest upon this proposed ability of man to exert genetical and nutritional control over the growth gradients. Wallace (1948), however, from his studies of nutritional influences on prenatal and postnatal growth of sheep, opposed the view presented by his colleagues Hammond (1932, 1940) and Pálsson and Vergés (1952) working with sheep, and of McMeekan (1940–41) and Pomeroy (1941) with pigs. Even rigorous nutritional stress, Wallace contended, did not alter in great degree the

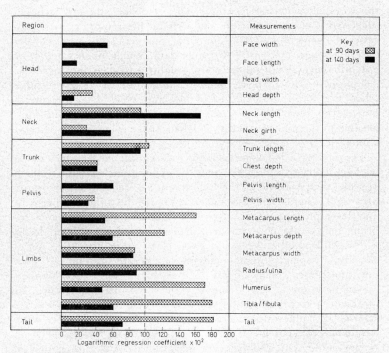

Figure 9. Relative growth of foetal body dimensions. Logarithmic regression coefficients between body measurements and curved crown-rump length at 90 and 140 days of gestation, after adjustment for the sex effect, are shown. All the coefficients shown were significant (P < 0·05) Everitt, 1965)

allometric relationships of different parts of a tissue or body component. This suggested either that the effect of such stress was a general one on the part concerned and not differential in nature, or that a control mechanism within the individual maintained the normal allometric relationships under such stress conditions.

Recent prenatal research (Everitt, 1965) supports the allometric growth principle and makes untenable the concept of a marked *differential* effect of undernutrition on ovine development. Although effects of prenatal nutrition on numerous parameters of foetal development were recorded, these were strikingly proportional to the effects recorded on foetal body mass. Allometric growth patterns of foetal body components (*Figure 8*) and external body dimensions (*Figure 9*) agreed with earlier expositions (Wallace, 1948, Joubert, 1956, Stephenson, 1962), and appeared little disturbed by the level of maternal feeding. The evidence adduced strongly suggested that undernutrition in prenatal life retards growth processes in a general, and not differential, manner. The two basically different interpretations of the same data depend upon whether or not differences incurred in body weight are taken into account. Adoption of the methods used by the Hammond or Cambridge school necessitates presentation of data as in *Figure 10*. Interpretation of this type of analysis leads inevitably to conclusions comparable to those forwarded by Pálsson (1955) in his review of farm animal development phenomena.

The data recorded on foetal stages substantiate the view forwarded by Taplin and Everitt (1964) that the undernourished lamb at birth appears malproportioned in relation to a well-fed lamb simply because it is lighter and at an earlier stage of the differential growth and development process. Recent re-analyses of the original Cambridge data on the postnatal growth of sheep and pigs (Tulloh, 1963; Elsley, McDonald and Fowler, 1964) clearly show that the form of an animal depends almost solely on its absolute size and restricted nutrition causes a more or less uniform retardation of development except in so far as fatty tissue is concerned. In foetal stages fat represents a very small proportion of body weight and thus the latter assumes the same importance in this context as does fat-free body weight in postnatal life (Everitt and Jury, 1966a, b). Stephenson (1964) found that 99·8 per cent of variation in skeletal muscle of ovine foetuses could be accounted for by differences in body weight.

Application of allometry to the effects of undernutrition in foetal life may, nevertheless, have to take cognizance of the difference between size and physiological maturity. Alexander (1964a) has shown that, to a large extent, the size of an animal at birth determines its functional maturity and ability to survive. Although a high plane of nutrition may promote rapid growth and development, and a small animal of the same age is generally less mature than a bigger one, if one animal attains a given size more slowly than another it will in some respects be more mature by the time it does so (McCance, 1962). Ossification of the foetal skeleton, for example, is more closely allied to body weight than age (*see Figure 11*) in the ovine (Wallace, 1948; Everitt, 1965), the pig (Pomeroy, 1960) and the rabbit (Appleton, 1929). A newborn animal of small birth weight after a normal gestation period will be more mature than one

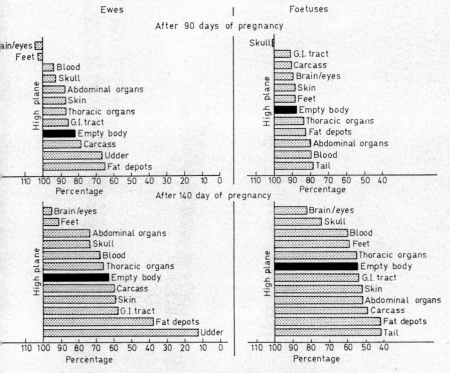

Figure 10. *Relative effects of plane of nutrition on components of the empty body of ewes and foetuses after 90 and 140 days of pregnancy. Mean weights of components of low plane animals are expressed as percentages of mean weights of high plane animals. Abdominal organs = liver, gall bladder, spleen, kidneys, urinary bladder. Thoracic organs = heart, lungs and trachea, diaphragm. Feet = fore + hind-feet (Everitt, 1965)*

of the same weight born prematurely. In other words, a distinction between size, developmental and physiological maturity with age may need to be made.

Opposition to Hammond's (1921) hypothesis that man exerts control over evolutionary genetical growth gradients has also been

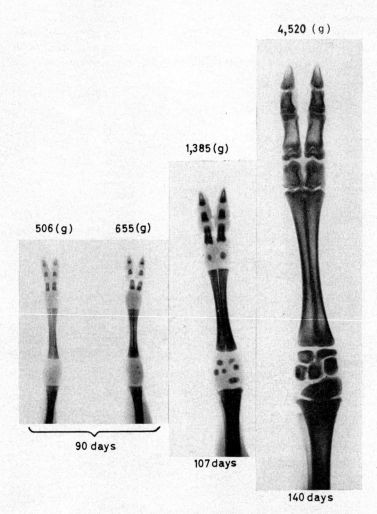

Figure 11. X-ray positives to illustrate skeletal development of the fore-leg during foetal life of the sheep. At 506 g body weight (Low plane: 90 days of age) nuclei of the carpals and proximal sesamoids are absent; at 655 g body weight (High plane: 90 days of age) carpal nuclei can be seen; at 1,385 g body weight (107 days of age) proximal sesamoid nuclei are clearly seen. The unfused state of the metacarpal bone at the lighter weights, the general increase in bone density with increasing body weight and age, and the appearance of the distal end of the ulna at heavier weights may also be noted (Everitt, 1965)

forwarded recently by Stephenson (1962). He compared prenatal growth of the Australian Merino and Romney × Southdown sheep foetuses at several ages. Breed differences in several dimensions were clearly shown, for which two explanations were offered. Firstly, that the allometric growth gradients differed between the two breed groups in those measurements showing most variation. The gradients, Stephenson contends, were altered only for certain specific types of measurements and the altered rate must take effect for almost the whole growth period. Secondly, morphogenesis may produce different dimensions of the parts at the start of the growth period, with similar subsequent growth rates, leading to the conformation differences characteristic of the two types. The second of these two premises appears a priori most plausible. A marked positive relationship in blastocyst diameter and racial size of parents in rabbits was established by Castle and Gregory (1929) and the results of Beatty (1958) also indicate genetical control of blastocyst size. These size differences may result from different amounts of glutathione (Gregory and Goss, 1933), a co-enzyme important in cell division, endocrine and metabolic activity (Crook, 1958). Alternatively, they may reflect differences between eggs in time of ovulation, or the time of fertilization, or maternal effects, as well as genetical factors (Hafez and Rajakoski, 1964).

Sheep and cattle embryos develop at the same rate during the first week after fertilization (Green and Winters, 1945) but with subsequent divergence in the rate of development. Clark (1934) found that ova from the same ewe corresponded more closely in their measurements than those from other individuals of the same breed. The recent finding of Tulloh and Romberg (1963) of a marked gravity effect on the shape of the long bones in sheep, overwhelming nutritional effects, also negates the plausibility of Hammond's (1921) views on genetical and nutritional control of growth gradients. These recent nutritional and genetical studies all contribute to a needed revision of developmental theories. They illustrate, too, the foetal ability to strive for the 'anatomical harmony' (Boccard and Dumont, 1960) recorded in postnatal life, although there are limits to the amount of distortion which can be tolerated.

CONCLUSIONS

Control of prenatal growth in the sheep and other uniparous species is substantially more complex than in the animal after birth. The latter is influenced directly by manifold factors; the foetus indirectly through circuitous pathways. Further studies of these factors and their pathways are needed. Maternal nutrition, in particular, can profoundly affect prenatal development with important residual effects on postnatal performance. Clearly, however, the association is by no means a simple one, but represents the action and inter-action of a large number of diverse factors.

There is a prime necessity in the science of animal production for integration of growth studies with investigations of the prenatal existence. Losses of substantial magnitude in the world animal industries impinge upon the breeding unit and it is to problems of the reproducing animal and the *in utero* progeny that considerable research resources need devoting, for the potential rewards are great.

REFERENCES

Alexander, G. (1956). *Nature, Lond.* **178**, 1058
— (1964a). *Proc. Aust. Soc. Anim. Prod.* **5**, 113
— (1964b). *J. Reprod. Fert.* **7**, 289
— (1964c). *J. Reprod. Fert.* **7**, 307
— and Williams, D. (1966). *J. Endocr.* **34**, 241
Appleton, A. B. (1929). *C.r. Ass. Anat.*, 24th Meet., Bordeaux, 3–25
Barcroft, J. (1946). *Researches on Prenatal Life.* Oxford; Blackwell
Beatty, R. A. (1958). *J. Genet.* **55**, 325
Beer, G. (1925). *Jber. VetMed.* **45**, 237
Bellows, R. A., Pope, A. L., Chapman, A. B. and Casida, L. E. (1963). *J. Anim. Sci.* **22**, 101
Bennett, D., Axelson, A. and Chapman, H. W. (1964). *Proc. Aust. Soc. Anim. Prod.* **5**, 70
Blaxter, K. L. (1957). *Proc. Nutr. Soc.* **16**, 52
— (1962). *The Energy Metabolism of Ruminants.* London; Hutchinson
Boccard, R. and Dumont, B. L. (1960). *Annls. Zootech.* **9**, 355
Bradford, G. E., Weir, W. C. and Torrell, D. T. (1960). *J. Anim. Sci.* **19**, 1216
Brody, S. (1945). *Bioenergetics and Growth.* New York; Reinhold
Browne, Sir T. (1642). *Religio Medici.* (Ed. and annot. by J. Winny.) Cambridge (1963); Cambridge Univ. Press
Butterfield, R. M. (1966). *Aust. Vet. J.* **42**, 37
Castle, W. E. and Gregory, P. W. (1929). *J. Morph.* **48**, 81
Clark, R. T. (1934). *Anat. Rec.* **60**, 135
Cloete, J. H. L. (1939). *Onderstepoort. J. vet. Sci. Anim. Ind.* **13**, 417
Coop, I. E. (1950). *J. agric. Sci., Camb.* **40**, 311
— (1961). *Proc. N.Z. Soc. Anim. Prod.* **21**, 79
— (1962). *N.Z. Jl. agric. Res.* **5**, 249

Coop, I. E. (1966). *J. agric. Sci., Camb.* **67**, 305
— and Clark, V. R. (1955). *N.Z. Jl. Sci. Technol.* **A**, **37**, 214
Crichton, J. R., Aitken, J. N. and Boyne, A. W. (1960). *Anim. Prod.* **2**, 45
Crook, E. M. (1958). *Nature, Lond.* **181**, 887
Curson, H. H. and Malan, A. P. (1935). *Onderstepoort. J. vet. Sci. Anim. Ind.* **4**, 481
Dawes, G. S. and Parry, H. B. (1965). *Nature, Lond.* **207**, 330
Dickinson, A. G. (1960). *J. agric. Sci., Camb.* **54**, 378
Donald, C. M. and Allden, W. G. (1959). *Aust. J. agric. Res.* **10**, 199
Donald, H. P. (1958). *Proc. 10th Int. Conf. Genet.* **1**, 225
Doney, J. M. and Smith, W. F. (1964). *Anim. Prod.* **6**, 155
Dun, R. B. and Grewal, R. S. (1963). *Aust. J. exp. Agric. Anim. Husb.* **3**, 235
— Alexander, R. and Smith, M. D. (1964). *Proc. Aust. Soc. Anim. Prod.* **5**, 141
El-Sheikh, A. S., Hulet, C. V., Pope, A. L. and Casida, L. E. (1955). *J. Anim. Sci.* **14**, 919
Elsley, F. W. H., McDonald, I. and Fowler, V. R. (1964). *Anim. Prod.* **6**, 141
Edey, T. N. (1965). *Nature, Lond.* **208**, 1232
— (1966). *J. agric. Sci., Camb.* **67**, 287
Everitt, G. C. (1961). *Proc. Ruakura Fmrs' Conf. Week*, pp. 85–102
— (1963). *Symp. Carcass Composition and Appraisal of Meat Animals.* Paper 10-1. (Ed. by D. E. Tribe.) Melbourne; C.S.I.R.O.
— (1964). *Nature, Lond.* **201**, 1341
— (1965). *Ph.D. Thesis*, Univ. of Adelaide
— (1966a). *Proc. Aust. Soc. Anim. Prod.* **6**, 91
— (1966b). *N.Z. agric. Sci.,* **1** (9), 20
— (1967). *Proc. N.Z. Soc. Anim. Prod.* **27**, 52
— and Jury, K. E. (1966a). *J. agric. Sci., Camb.* **66**, 1
— — (1966b). *J. agric. Sci. Camb.* **66**, 15
Foote, A. S. (1964). *Jl. R. agric. Soc.* **125**, 170
Foote, W. C., Pope, A. L., Chapman, A. B. and Casida, L. E. (1959). *J. Anim. Sci.* **18**, 463
Gershon, R. and Straus, Lotte (1961). *Am. J. Dis. Child.* **102**, 645
Giroud, A. (1959). *The nutritional requirements of embryos and the repercussions of deficiencies in World Review of Nutrition and Dietetics* **1**, p. 233. (Ed. by G. H. Bourne.) London; Pitman
Green, W. W. and Winters, L. M. (1945). *Tech. Bull. Minn. agric. Exp. Stn.* **169**, 1–36
Gregory, P. W. and Goss, H. (1933). *J. exp. Zool.* **66**, 335
Gunn, R. G. (1964). *J. agric. Sci., Camb.* **62**, 123
Hafez, E. S. E. (1963). *J. Anim. Sci.* **22**, 779
— and Rajakoski, E. (1964). *J. Reprod. Fert.* **7**, 229
Hammond, J. (1921). *J. agric. Sci., Camb.* **11**, 267
— (1932). *Growth and Development of Mutton Qualities in the Sheep.* Edinburgh; Oliver and Boyd

Hammond, J. (1940). *Farm Animals: Their Breeding, Growth and Inheritance.* 2nd edn. 1952. London; Arnold
— (1944). *Proc. Nutr. Soc.* **2**, 8
— (1961). In: *'Somatic Stability in the Newly Born'.* p. 5; *Symp. Ciba Fdn.* (Ed. by G. E. W. Wolstenholme and M. O'Connor.) London; Churchill
Hight, G. K. (1966a). *N.Z. Jl. Agric. Res.* **9**, 479
— (1966b). *Proc. Ruakura Fmrs' Conf. Week,* p. 19
Huxley, J. S. (1932). *Problems of Relative Growth.* London; Methuen
Hytten, F. E. (1964). In: *Proc. 6th Int. Congr., Nutr.,* p. 59. (Ed. by C. F. Mills and R. Passmore.) Edinburgh; Livingstone
Joblin, A. D. H. (1966). *Proc. Ruakura Fmrs' Conf. Week,* p. 33
Joubert, D. M. (1956). *J. agric. Sci., Camb.* **47**, 382
— (1963). *Anim. Breed. Abstr.* **31**, 295
Keller, K. (1920). *Wien. tierarztl. Mschr.* **7**, 137
McCance, R. A. (1962). *Lancet,* 7257, 621
— (1964). *Proc. 6th Int. Congr. Nutr.,* p. 74. (Ed. by C. F. Mills and R. Passmore.) Edinburgh; Livingstone
— and Widdowson, Elsie (1962). *Proc. R. Soc.* **B, 156**, 326
McClymont, G. L. and Lambourne, L. J. (1958). *Proc. Aust. Soc. Anim. Prod.* **2**, 135
McLaren, Anne (1965a). *J. Reprod. Fert.* **9**, 343
— (1965b). *J. Reprod. Fert.* **9**, 79
— and Michie, D. (1960). *Nature, Lond.* **187**, 363
McMeekan, C. P. (1940–41). *J. agric. Sci., Camb.* **30**, 276; **30**, 387; **30**, 511; **31**, 1
Marchant, E. C. (1925). *Xenophon: scripta minora.* English translation: Loeb Classical library. London; Heineman
Meyer, J. H. and Clawson, W. J. (1964). *J. Anim. Sci.* **23**, 214
Moustgaard, J. (1959). In: *Reproduction in Domestic Animals* **2**, p. 70. (Ed. by H. H. Cole and P. T. Cupps.) New York; Academic Press
Needham, A. E. (1964). *The Growth Process in Animals.* London; Pitman
Pálsson, H. (1955). In: *Progress in the Physiology of Farm Animals* **2**, p. 430. (Ed. by J. Hammond.) London; Butterworths
— and Vergés, J. B. (1952). *J. agric. Sci., Camb.* **42**, 1
Pomeroy, R. W. (1941). *J. agric. Sci., Camb.* **31**, 50
— (1960). *J. agric. Sci., Camb.* **54**, 31
Purser, A. F. and Roberts, R. C. (1964). *Anim. Prod.* **6**, 273
Reid, J. T. (1960). *J. Dairy Sci.* **43**, 103
Reid, R. L. (1963). *J. Aust. Inst. agric. Sci.* **29**, 215
Robinson, D. W. (1964). *Proc. Nutr. Soc.* **23**, II–III, A
Russel, A. J. F. (1966). *Anim. Prod.* **8**, 350
Scammon, R. E. and Calkins, L. A. (1929). *The Development and Growth of the External Dimensions of the Human Body in the Foetal Period.* Minneapolis; Minneapolis Univ. Press
Schinckel, P. G. (1963). *Proc. 1st Wld Conf. Anim. Prod.* **1**, 199

Schinckel, P. G. and Short, B. V. (1961). *Aust. J. agric. Res.* **12**, 176

Stephenson, S. K. (1959). *Aust. J. agric. Res.* **10**, 433

— (1962). *Aust. J. agric. Res.* **13**, 733

— (1964). *Anim. Prod.* **6**, 309

— and Lambourne, L. J. (1960). *Aust. J. agric. Res.* **12**, 1044

Tanner, J. M. (1963). *Nature, Lond.* **199**, 845

Taplin, D. E. and Everitt, G. C. (1964). *Proc. Aust. Soc. Anim. Prod.* **5**, 72

Thomson, D'Arcy W. (1917). *On Growth and Form.* 2nd edn. (1942).
Cambridge; Cambridge Univ. Press

Thomson, W. and Aitken, F. C. (1959). *Tech. Comm. Commonw. Bur. Anim. Nutr.* No. **20**

— and McDonald, I. (1956). *Proc. Br. Soc. Anim. Prod.*, p. 38

Tulloh, N. M. (1963). Symp. *Carcase Composition and Appraisal of Meat Animals, C.S.I.R.O.* Paper 5-1. (Ed. by D. E. Tribe.) Melbourne

— and Romberg, Barbara (1963). *Nature, Lond.* **200**, 438

Turner, Helen N. (1961). *Aust. J. agric. Res.* **12**, 974

Wallace, L. R. (1948). *J. agric. Sci., Camb.* **38**, 93; **38**, 243; **38**, 367

— (1961). *Proc. Ruakura Fmrs' Conf. Week*, p. 14

Weinbach, A. P. (1941). *Growth* **5**, 217

Wiener, G. and Slee, J. (1965). *Anim. Prod.* **7**, 333

Widdowson, Elsie M. and McCance, R. A. (1963). *Proc. R. Soc.* **B**, **158**, 329

Wilson, P. N. (1954). *J. agric. Sci., Camb.* **44**, 67

— (1958). *J. agric. Sci., Camb.* **50**, 198

— and Osbourn, D. F. (1960). *Biol. Rev.* **35**, 324

PRENATAL DEVELOPMENT IN THE PIG AND SOME OTHER MULTIPAROUS ANIMALS

E. SALMON-LEGAGNEUR

Centre national de Recherches Zootechniques, Jouy-en-Josas, France

INTRODUCTION: THE MEANING OF MULTIPARITY

A FEMALE is multiparous, or polytoccous, when she gives birth to several young at a single parturition, and the pig is generally considered to be one of the best examples of a multiparous animal. The pig, among the more common species of farm animal, is the most prolific in terms of number of young per litter, averaging ten born with numbers ranging as high as 20–22. However, among other multiparous species we find animals as physiologically different as the mouse, the lemur and the brown bear, which raises the question, especially in relation to prenatal growth, of whether any character other than multiovulation is common to all these species. Moreover, most species, Man included, can at least occasionally carry more than two young simultaneously, instanced by a recent report of a Mexican woman who gave birth to eight babies at one delivery. Selection can transform breeds of ostensibly uniparous animals, like sheep, into multiparous ones. For instance, the Romanov ewe produces an average of 3 lambs per crop, while, on the other hand, the bat, which is not far removed from the prolific rodents and insectivora, is remarkable for its relative infertility, ordinarily producing only one young at a time.

These variations from the usual situation raise the question of what features are common to multiparous species. Broadly speaking, the average number of young produced in a litter in any species of mammal is inversely proportional to the average mature size of that species. Thus, the large elephant never produces twins, whereas the small mouse bears up to 20 young. One exception is the bat and another the sow which, on this basis, would be less prolific than the ewe. Spencer (1899) elaborated a theory of fertility which can explain these exceptions: he suggested that where the energy expenditure for life is high or the food supply restricted, the rate of reproduction was lowered correspondingly.   Thus, in the bat the habit of flying results in a higher rate of energy expenditure than that of a mouse of similar size on the ground, and the pig is less likely to be deprived of food than the sheep.   It is also evident that in polytoccous species the rate of growth after birth is generally higher than in other species.

Among morphological characteristics, the number of teats offers an approximate, but not reliable, indication of average litter size in a species.   This is probably a physiological correlation which results from natural or induced selection.   On the other hand, the shape of the uterus, the size of the ova and the type of placentation could not be related to the number of young born.   Multiparity more probably results in a particular endocrine equilibrium in those species the ovaries of which release many ova at one time (Hammond, 1952).

Two other important trends should be mentioned.   First, is the relationship between the mean weight of litter (W) and the number of young (N) shown by the equation of Crozier (1940):

$$W = aN^{0.83}$$

The higher the number born, the higher is the weight of the litter in every species, and the higher is the weight of the litter relative to that of the mother between species (with the exception of the sow). Second, when the number of young increases, the duration of pregnancy decreases (Biggers, Curnon and Finn, 1963).   Thus, in the less prolific wild pig the duration of pregnancy is 130 days, compared with 115 days in the domestic forms and the length of pregnancy varies from 57 to 72 days with decreasing number of young in guinea-pigs.   This variation in duration of pregnancy seems to be the best criterion to compare species when related to the size of the mother or the rate of growth.   An approximate estimation of prolificity may be obtained by dividing the 0·7 logarithm of the dam's weight, in grammes, by the duration of her pregnancy, in days (Table 1).   The deviation between computed and observed values could indicate the tendency to multiparity among different species. By this reasoning, the pig is probably the most prolific, followed by the mouse.

With regard to prenatal growth, attention should be drawn to the strong negative correlation existing between number and weight of individual young in a litter, which has been observed for many species (Hammond, 1940; Vengé, 1950; Bredeck and Mayer, 1955;

Hafez, 1963; Salmon-Legagneur, Legault and Aumaitre, 1966). Therefore, every analysis of variation in prenatal development in the multiparous animal has to consider at least three points:

(*1*) Variation in the number of foetuses, weight being taken as a constant.

(*2*) Variation in the individual weight, number of foetuses being constant.

(*3*) Physiological or chemical variations, regardless of number and weight.

Table 1

Some quantitative aspects of pregnancy in those mammalian species where multiparity can occur

	Offspring wt. / *Maternal wt.* (W) (%)	*Pregnancy duration* (D) (days)	$\frac{0.7 \times \log W}{D}$	*Litter size*
Ox	5·8	284	1·4	1·0
Man	7·0	281	1·2	1·0
Bat	8·5	72	1·5	1·5
Sheep (Romanov)	20·0	145	2·4	3·0
Guinea-pig	28·7	68	3·0	3·5
Pig	7·0	115	3·3	10·0
Mouse	35·0	20	4·5	7·4
Rabbit	20·0	31	7·4	8·0
Rat	25·0	21	7·6	9·0

1. CAUSES OF VARIATION IN NUMBER OF EMBRYOS

One of the differences between uniparous and multiparous females is that the former can only be either pregnant with one embryo or not pregnant at all, whereas the latter when pregnant carry an undetermined number of embryos. In fact, the number of embryos is seldom maximal, owing to the considerable loss of ova, blastocysts and embryos which occurs in most polytoccous species.

The percentage of total losses during ovulation, fertilization and prenatal growth varies between species but is always surprisingly high; 80 per cent in the oppossum (Hartman, 1929), 60 per cent in the wild rabbit (Brambell, 1942), 30–50 per cent in the pig (Squiers, Dickerson and Mayer, 1952; Pomeroy, 1960; Rathnasabapathy, Lasley and Mayer, 1956; Majerciak, 1965), 30 per cent in the rat and the mouse (Fahim, Mayer and Lasley, 1961; Runner, 1951), and 29 per cent in the guinea-pig (Nicol, 1933).

The greater part of these losses occurs between ovulation and implantation (Bredeck and Mayer, 1955; Reddy, Mayer and Lasley, 1958; Corner, 1923; Pomeroy, 1952). The remainder can occur at any stage of embryonic growth with, however, two critical periods in the pig; the first during the first month of pregnancy (Squiers *et al.*, 1952; Pomeroy, 1952; Lerner, Mayer and Lasley, 1957) which could be due to failures in implantation, and the second late in pregnancy (Pomeroy, 1960; Hafez, 1963; Majerciak, 1965) resulting probably from over-crowding in the uterus.

The specific causes of embryonic death are not well known. Embryologists suggest that defective internal environment is the preliminary cause, geneticists believe that the embryo itself may possess hereditary defects, while nutritionists provide evidence for the influence of some external factors.

(a) *Influence of genetic factors*

The influences of breed, strain and individual are known in every species and can be demonstrated in any experiment with inbreeding or cross-breeding, but the low heritability ($h^2 < 0.1$) of litter size observed in pigs, mice, rabbits and rats has to be compared with the higher heritability of ovulation rate ($h^2 > 0.3$) in the same species (Newman, 1963; Falconer, Edwards, Fowler and Roberts, 1961). This arises from the high incidence of mortality giving a low correlation between ovulation and litter size ($r = 0.2$).

Falconer, Edwards, Fowler and Roberts (1961) have shown that selection in mice for both high and low litter size results in more corpora lutea in the first case and no change in the second. This means that a high reproductive performance depends on potential capacity of the ovaries but that the reciprocal is not necessarily true. The same applies to pigs, breeders of which have observed that high litter size (more than 12) is hereditary and small litter size is not. Other evidence was given by Pomeroy (1960) when he observed that most sows culled for low fertility could produce, however, a normal number of fertilized ova.

Furthermore, genetic influence is not simple and sometimes is more closely associated with heterosis or lethal factors (Hammond, 1928; Boyd and Hamilton, 1952; Squiers, Dickerson and Mayer, 1952; Taketomi, 1956; Hazel, 1963; Ino, 1966) than with egg number (Table 2). This is the reason why, despite the assumption

6+   161

that the male has a lesser influence on litter size than the female, the choice of male can greatly influence the efficiency of reproduction in multiparous animals. For instance, it has been found that average prolificity between boars on A.I. can vary from 7 to 12 piglets from one ejaculation (Legault and Ollivier, 1965). Another example may be drawn from blood groups and serum studies; Gruhn (1966) has shown that such homozygous combinations as AA(Tf) in serum

Table 2

Effect of cross-breeding upon fertility in rabbits (Hammond, 1928)

Breed	Number of ova shed	Number of embryos	Embryonic mortality (%)
H	11·1	4·8	45·6
F	5·8	3·2	13·5
H × F	10·4	6·8	15·0

transferrins may result in increased mortality. Haemolytic disease could also result from the same mechanism and several other lethal genetic characters have been described in many species (Hammond, 1952).

As regards the maternal influence, several factors may be involved. First, genetic differences in the endocrine activity of the hypophysis (Mauleon and Pelletier, 1964) and the reactivity of follicle populations to FSH (Mauleon and Rao, 1963) have been noted. Second, the embryo capacity of the uterus depends upon its length, which is likely to be representative of the family character (Rathnasabapathy et al., 1956). At least in some species, like rabbits, rats and pigs where foetal atrophy is inherited as a maternal character, there is some support for the belief that a lack of some substance, probably progesterone, in the blood of the mother could act as a limiting factor (Hammond, 1952). The same could be deduced from experiments with ova transplantation (Polge, 1965) in which, whatever the number of implanted ova, the number of embryos is fairly constant.

(b) Intra-uterine environment

Since an excess of ova is generally produced in the multiparous female, the primary problem of embryo survival concerns the uterus itself. Uterine size, determining space per embryo, is important in providing an optimum intra-uterine environment. Moreover, physiological, anatomical and biochemical changes induced in uterine tissues by hormones or chemical agents may prevent or stimulate embryonic life. Therefore, many aspects are involved; in species with a high ovulation rate, the number of ova shed by each

Table 3

Relationship between the distribution of young in the right and left uterine horns

	Mouse*		Pig†	
	Left	Right	Left	Right
	1·71	4·07	3·33	3·83
	3·38	3·53	5·42	5·36
	5·69	2·97	7·75	7·00
Average	3·62	3·41	5·97	5·56

* Runner, 1951
† Salmon-Legagneur, 1967

ovary, right and left, during one ovulation can vary widely. Although nearly the same, there are, on average, slightly more ova from the left side than from the right (Majerciak, 1965; Salmon-Legagneur, 1967). The greatest difference we have encountered in a sow was one ovum on the right side and 21 ova on the left. Nevertheless, the number of young lying in each horn is nearly the same (Majerciak, 1965; Amoroso, 1952). It follows, therefore, that either a migration of eggs occurs in those species where the shape of uterus allows it, or an equilibration process takes place in the other species (such as rodents).

In pigs, intra-uterine migration and mixing occur in 40–60 per cent of the cases (Corner, 1923; Majerciak, 1965; Dziuk, Polge and Rowson, 1964) resulting in identical partition of embryos between both sides whatever the total number may be (Table 3). In the mouse and rodents which possess a duplex uterus, migration is impossible except in the case of an external migration (Boyd and Hamilton, 1952). Here, the distribution of young demonstrated a tendency for an inverse relationship between numbers in the opposite sides (Runner, 1951; McCarthy, 1965). The same was found in guinea-pigs by Eckstein, McKeown and Record (1955). This, however, failed when extra ova were implanted experimentally in

163

one horn, in which case, the number of embryos was the same for both horns, as in the pig. This seems to indicate that factors reducing the number of young were operating maternally and active in both uterine horns.

Thus, whatever the species and process the uterus acts as a whole allowing predetermined numbers of blastocysts to implant. The mechanism of implantation is rather obscure. The eggs, distributed

Table 4

Influence of some hormonal treatments on pregnancy in swine (Reddy *et al.*, 1958)

	Ovulation rate	Litter size	Mortality (%)	Linear capacity of uterus (cm)
Controls	10·3	7·9	23·3	2·60
25 mg progesterone + 25 μg oestrone per day (14th–24th day)	11·5	9·0	21·7	2·81
25 mg progesterone + 12 μg oestrone per day (14th–24th day)	12·3	10·7	13·5	3·38

by irregular myometral contractions, are induced to settle when favourable uterine mucose is achieved. There is general agreement that the presence of the luteal hormone is essential for successful implantation (Boyd, Hamilton and Hammond, 1944). Moreover, the eggs seem to exert some influence, both chemical and physical, on the endometrium (Wislocki and Streeter, 1938). So, spacing can be explained in part by an absence of receptivity in the region of the uterine mucosa adjacent to an attached blastocyst; also, in part, by the stimulation in uterine growth exerted by the embryo itself (Huggett and Hammond, 1952). Thus it is not surprising that any change in area, length, or hormonal or chemical condition of the uterus can induce modification in the number of implanted blastocysts or the well-being of embryos. This was clearly shown by Reddy, Mayer and Lasley (1958) and Romack, Zoellner and Day (1960) in pigs, and by Fahim *et al.* (1961) in rats, where minute doses of exogenous progesterone and oestrone administered at certain critical phases of pregnancy contributed to increased number of young (Table 4).

This occurrence and influence of different relationships between uterus, hormones and viability of embryos persists throughout pregnancy and seems quite complex. In the sow, abortion is induced if the number of corpora lutea is lowered to five or less (du Mesnil du Buisson and Dauzier, 1959). The same result is achieved, at least during the first month of gestation, if a portion of uterine horn is emptied by removing the embryos (du Mesnil du Buisson and Rombauts, 1963). As a general rule, and in contrast to the rat, unilateral pregnancy is not maintained in the sow (Rathmacher and Anderson, 1963). In this case, the corpora lutea regress due to a uterine luteolytic effect which may be prevented by injection of oestrone (du Mesnil du Buisson, 1964). In normal pregnancy, provided their number is higher than four, embryos prevent the production of the luteolyzing substance and contribute in this way to their own safety (du Mesnil du Buisson, 1966). On the other hand, pregnancy can develop normally, even with only one foetus, if the empty part of the uterus is removed surgically.

Of the chemical influences, mention may be made of acid phosphatase, the uterine concentration of which is proportional to the number of embryos in rats (Bredeck and Mayer, 1955) and in pigs (Goode, 1961), though it is not clear whether its accumulation is a cause or a consequence of the development of embryos.

All these facts illustrate one aspect of mutual relationship between embryos and uterus and may provide valuable information in the future on the possibility of modifying uterine environment to allow fuller expression of the animal's reproductive potentialities.

(c) General environment

Many factors can affect, in varying degrees, the number of young in a litter before birth. Some are characteristics of the mother (age or weight), others concern the external environment (climate), while yet others are management factors, including the major one of nutrition.

Hammond (1952) described the effect of age as a wave which, starting from sterility, rises rapidly to its maximum and then falls again to sterility. Litter size has been observed to increase up to the third litter in guinea-pigs (Eckstein et al., 1955), the second litter (7 months) in rats, and to the fourth or fifth litter in pigs (Krizenecky, 1942; Salmon-Legagneur, Legault and Aumaitre, 1966). The age at first service also has an influence; Stewart (1945) found that gilts farrowing for the first time at 320 days of age averaged one pig less and those farrowing at 410 days one pig more than those farrowing at 365 days. In this field some differences appear among species and breeds (Hammond, 1952). The weight of dam is said to have a great influence in some species, like rabbits (Venge, 1950) mice and rats (Bateman, 1954), but in the pig, the difference in litter size between heavy and light sows is small (Hetzer and Brier, 1940; Stewart, 1945). Since age and weight have a positive correlation, the true influence of the latter is negligible and none of the body characteristics are of importance except backfat thickness (Rathnasabapathy, Lasley and Mayer, 1956; Goode, 1961). On the other hand, although a matter of controversy (Salmon-Legagneur and Jacquot, 1961), excessive gain in weight or adiposity seem to exert an unfavourable effect on survival of embryos, but it is possible that sometimes sterility may be the cause rather than the result of fatness (Moustgaard, 1958).

Climate has a definite influence on reproduction, in that a hot climate reduces fertility in males and ovulation rate in females (Hammond, 1952) and that risks of abortion are increased (Heitman, Hughes and Kelly, 1951). Moreover, there exists a seasonal influence which is probably due to variation in hours of daylight at different times of the year. In the sow, spring or summer litters are often larger than winter ones (Allen and Lasley, 1954). On the other hand, Majerciak (1965), found a higher embryonic mortality (26–37 per cent, against 20–22 per cent) from January to May.

A considerable number of papers have been devoted to the influence of nutrition on fertility in multiparous animals. Those results reviewed by several authors (Russel, 1948; Moustgaard, 1958; Mitchell, 1962; Duncan and Lodge, 1960) are a little conflicting, particularly the between species comparisons. It is necessary here to differentiate between effects on ovulation rate, where serious failures in diet (lack of an essential nutrient) may impair fertility, and the effects on survival of embryos which are generally inconsistent. For instance, it has been shown that a diet containing suboptimal levels of protein or minerals does not disturb unduly the pregnancy of pigs or rats (Davidson, 1930) and there is no convincing data to prove that increasing levels of nutrients beyond the minimum levels required for maintenance allow litter size to increase in many multiparous animals. That this appears to be true for pigs is indicated by recent data of Dean and Tribble (1961), Goode (1961), Boaz (1962), Clawson, Richards, Matrone and Barrick (1963), Salmon-Legagneur (1965), Lodge, Elsley and MacPherson (1966a) and could be explained either by the small requirements of embryos or by their obtaining supplies from maternal stores (Salmon-

Legagneur, 1965). On the other hand, it is true that a temporary increase in energy intake at mating can improve ovulation and fertilization rate, as observed in pigs by many workers (Zimmerman, Spies, Self and Casida, 1960; Goode, 1961; Heap and Lodge, 1966), but generally speaking, this effect is transitory and not always followed by an increased number of young at term (Table 5).

Table 5

Influence of energy in the diet before and after mating upon reproduction in gilts (Salmon-Legagneur and du Mesnil, 1965)

Diet	Fertiliza-tion rate percentage	Ovulation rate	Number of embryos		per-centage losses	Wt. of corpora lutea (mg)	Wt. of embryos (g)
			living	dead			
H–H*	75	16·9	11·3	5·0	33	455	80·8
H–L	75	15·2	10·2	3·3	26	401	77·8
L–H	100	13·1	10·1	2·3	23	434	78·1
L–L	92	12·4	9·7	2·2	21	388	77·4

* Planes of feeding before and after mating.

2. SIGNIFICANCE OF EMBRYO WEIGHT

As the weight of foetuses and their growth throughout pregnancy is assumed to be indicative of prenatal development, this weight, or the mathematical law of weight increment, is often used as a basis for determination of nutritional requirements of pregnant females (Mitchell, Carroll, Hamilton and Hunt, 1931; Moustgaard, 1958), but this involves a series of discrepancies:

(a) Foetuses themselves only constitute a small part of the elaborate tissue structure built up during pregnancy; placenta, uterine growth and body gain of the mother being the other factors.

(b) Due to variations in litter size, the average foetus has no real existence and its true growth does not necessarily follow a smooth computed curve.

(c) Differential growth of organs impairs any general significance of total body weight at any one stage.

(d) Many factors other than nutrition have variable effects on prenatal growth.

(e) The nutritional efficiency of prenatal growth is as yet unknown.

Most of these aspects, therefore, must be kept in mind when considering weight increase of the embryos, particularly in multiparous species.

(a) *Foetuses as a part of the conceptus*

Provided that feed level is sufficiently high, the first external sign of pregnancy, before any enlargement of the uterus occurs, is a con-

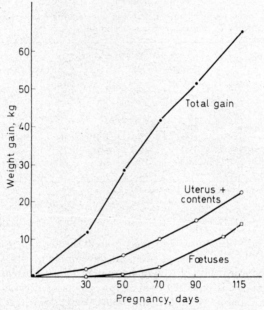

Figure 1a. Distribution of gain in weight in the sow during pregnancy (From Salmon-Legagneur and Jacquot, 1961)
By courtesy of Gauthier Villars

siderable gain in weight by the mother. Whether this 'pregnancy anabolism' results from increased food ingestion, as proposed by Blaxter (1964), or from a specifically better feed efficiency for both nitrogen and energy (Bourdel, 1960; Lenkeit, Gutte, Warnecke and Kirchoff, 1955; Salmon-Legagneur, 1965) is immaterial, the fact is that this 'mother-benefit' may continue extensively throughout pregnancy and may constitute an important proportion of the prenatal nutrient stream. It obviously has to be considered in so far as our estimation of pregnancy requirement is concerned. In sows, the maternal gain itself may constitute up to 75 per cent of the total live weight gain during pregnancy (or 90 per cent of the energy increase). This value seems considerably higher than that for uni-parous females (*Figure 1a* and *b*); there is no reason to believe that it depends on the number of young born but rather on species differ-ences and level of nutrition (Salmon-Legagneur, 1965).

Another important trend is the change during pregnancy in relative proportions by weight of foetal membranes, foetal fluids and foetuses (*Figure 2a* and *b*). During the first stages of foetal develop-ment, the proportion of fluids is greatest, followed by the placenta, then the foetuses. By the end of pregnancy the relationship is reversed. The physiological functions of the placenta (both maternal and foetal) are known and have been described com-prehensively by Amoroso (1952) and by Huggett and Hammond

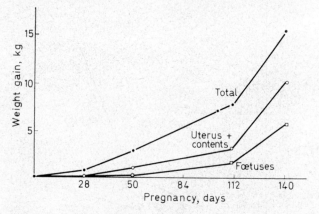

Figure 1b. Distribution of gain in weight in the sheep during pregnancy
(From Wallace, 1948)
By courtesy of Cambridge University Press

(1952). In early pregnancy there is no definite relationship between the weight of individual foetuses and the weight of the corresponding placenta. This position changes completely by late pregnancy when a close relationship exists between them, as demonstrated in guinea-pigs (Eckstein *et al.*, 1955; Ibsen, 1928), in sheep (Barcroft, 1946) and in pigs (Waldorf, Foote, Self, Chapman and Casida, 1957; Majerciak, 1965; Salmon-Legagneur, *et al.*, 1966). Furthermore, placenta weight is generally affected by the same factors as are the foetuses themselves. Thus, since the growth of placenta is limited mainly to early pregnancy, it would appear that the size to which the foetal placenta grows at this stage may determine the level of nutrition which is available to the foetus later on. In this respect, piglets and rabbits have a considerable advantage over lambs.

The function of the fluids (both amniotic and allantoic) is less well known and seems essentially to be physical. By distending foetal membranes and uterine wall they allow regular spacing and protection of embryos. This is particularly useful when there are many embryos and implantation in the upper part of the uterine horns is required. This may explain the relatively higher weight of fluids

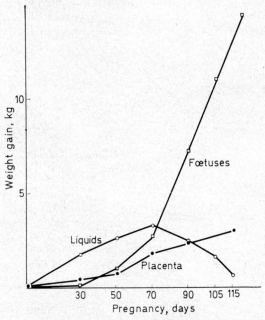

Figure 2a. Growth of the uterine contents during pregnancy of the pig

in multiparous animals. Moreover, these fluids are involved in exchanges of water and excreta from the foetuses, but from mid pregnancy to the last third in pigs and rodents, instead of increasing further, these fluids are absorbed at a rapid rate until by parturition only a 'few drops' are left (Hammond, 1937; Pomeroy, 1960; Salmon-Legagneur and Jacquot, 1961). This absorption allows space for the further growth of foetuses, but not in all species, nor to the same extent and stage, are the fluids replaced by growth. In the ewe and in the mare, allantoic fluid increases until the end of pregnancy and then occupies 25–35 per cent of the uterine content (Wallace, 1948). In women there are 20 per cent foetal fluids at parturition, compared with 3 per cent in the sow and none in the rabbit (Leitch, 1957; Hammond, 1937). The significance of these species differences is unknown as yet but they may be related to the relative physiological stage and weight of the young at birth.

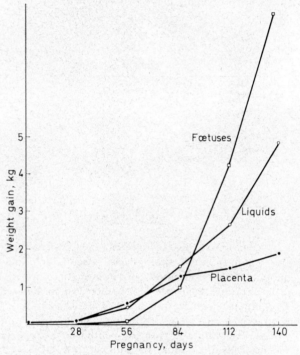

Figure 2b. Growth of the uterine contents during pregnancy of the sheep
(*From Wallace 1948*)

Finally, the net weight of foetuses as a part of the total conceptus differs slightly within and between species when compared with total live weight increment of pregnancy. It accounts for 50 per cent of the total in the rat (Salmon-Legagneur, Perissee and Jacquot, 1960), 20–30 per cent in the sow (Salmon-Legagneur, 1965), 25 per cent in the woman (Leitch, 1957) and 30 per cent in the ewe (Wallace, 1948) and could in some way characterize the

171

reproductive efficiency of these species, but gives little indication of nutrient needs.

(b) *Weight increase*

The variations in weight of the foetus afford no direct information on nutrient requirements but have a bearing on that problem since they emphasize stages where growth rate is highest.   Moreover they allow comparisons to be made between species.

Table 6

Summary of the constants of Weinbach's equation* as related to prenatal growth of various species

Species	K (%)	A (g)	t' (days)
Rat	45·4	0·1	12·9
Mouse	37·6	0·05	9·7
Rabbit	16·5	14·1	20·0
Cow	1·9	1775·0	98·1
Pig	1·5	521·3	48·0
Man	1·1	471·5	94·0

$$* \; W = Ae^{k(t-t')} - A$$

Much has been written on the mathematical expression of the prenatal time–weight relationship in several species, including pigs. Several equations have been used to define embryo weight and tentative generalizations have been made between species.   However, such equations describe much more realistically the ideal course that growth should follow, rather than the irregular picture secured by individual values.   Most frequently these equations are in the exponential form, as found by Mitchell *et al.* (1931), Brody (1945), Weinbach (1941), Moustgaard (1958):

$$W = Ae^{k(t - t')} - A$$

The constants A and k are representative of each given species and a correction term (t') is included for delayed implantation and the first stages when growth occurs without any sensible change in weight (Table 6).   As a practical consequence, embryo weight is not important for the first 30–40 per cent of pregnancy in the pig, and 40–50 per cent in the mouse, the rat or the guinea-pig, as compared with 20–30 per cent in Man, cattle or sheep.   On the other hand, rate of growth is accelerated correspondingly during the remaining period. One might well ask whether these species differences are related to multiparity.

From the comprehensive study of Pomeroy (1960) on prenatal growth in the pig, it appears that the best adjustment would be to the third power in that species:

$$W = \frac{1}{10}(0\cdot24t - 4\cdot06)^3$$

This could also be deduced from the placenta weight evolution (*see Figure 2*), in so far as both placenta and embryos are correlated. It can be concluded that in pigs and some other species maximal rate of growth is achieved before birth and therefore that growth proceeds by degrees.

However, the equation given above can be transformed into a linear one, as proposed by Hugget and Widdas (1951), when expressed as the cube root of weight:

$$W^{1/3} = 0\cdot113(t - 16\cdot59)$$

The same was found to be true for the mouse, rabbit and many other species, including both uniparous and multiparous (Spencer and Coulombe, 1965). Since the cube root of weight is the expression of body length or size, it is relevant that size is a simpler, more reliable and less variable characteristic of growth than weight (Barcroft, 1946). This assumption has been confirmed recently in pigs by Ullrey, Sprague, Becker and Miller (1965), who found a closer relationship between embryonic age and length of bones than weight of the foetal body or organs (Table 7). One can also mention, for multiparous animals, the positive relationship between stage of pregnancy and variability in weight of foetuses within litters, as observed by Ibsen (1928) in guinea-pigs, Hammond (1934) in rabbits and Pomeroy (1960) and Majerciak (1965) in the pig, which means that variation in weight increases with litter size and age. This could be due to crowding in the uterus and competition between foetuses in late pregnancy but Pomeroy (1960) suggested that this effect is more likely to be related to the maximum growth of the placenta.

(c) Causes of variation

In most multiparous species causes of variation in foetal weight are thought to be essentially of maternal origin. Thus, Wright

(1922) estimated that about 75 per cent of the variability in birth weight of guinea-pigs was due to maternal influence; Venge (1950) gave a figure of 25–70 per cent in rabbits for the same factors, whereas in pigs, figures of 43 per cent were estimated by Legault and Aumaitre (1966) and 47 per cent by Lush, Hetzer and Culbertson

Table 7

Correlations of foetal measurements with age of concepta (Ullrey *et al.*, 1965)

Measurement	Correlation coefficient (r)
Weight of body	0·83
Weight of liver	0·75
Weight of left adrenal	0·68
Weight of thyroid	0·91
Length of body	0·98
Length of humerus	0·98
Length of femoral diaphysis	0·98

(1934). On the other hand, according to the latter authors, only 1–7 per cent of the variability was accounted for by sire differences in pigs.

These litter effects may have been divided according to Runner (1951), Eckstein *et al.* (1955), Hafez (1963), into local, or intrinsic, and general, or extrinsic, effects. The former correspond roughly to intra-uterine factors, such as position, space, litter size and the foetuses themselves; the latter effects are concerned mainly with environment, including maternal characteristics and nutrition.

Genetic factors, both individual and parental, despite the well known influence of breed or strain on birth weight within species, seem of little importance as judged by their low heritability values (Lush *et al.*, 1934; Baker, Hazel and Reinmiller, 1943; Craig, Norton and Terrill, 1956; Legault and Aumaitre, 1966).

(*d*) *Intra-uterine factors*

It is well known that the birth weights of individuals in large litters tend to be less than those in small litters. This can be seen from the strong negative correlation which appears between number and weight of foetuses in most species (Wishart and Hammond, 1933; Waldorf *et al.*, 1957; Salmon-Legagneur *et al.*, 1966; Mc-Carthy, 1965). This relationship is determined partly by variations in placental function and partly by the duration of pregnancy (Hafez, 1963). As shown by Pomeroy (1960) in pigs, the inhibition of the growth rate in large litters increases with litter size and becomes more marked as pregnancy proceeds. By comparing weights of foetuses in uterine horns containing different numbers of embryos, Eckstein *et al.* (1955) with guinea-pigs and Waldorf *et al.* (1957) with pigs concluded that factors local to the horn were of more importance than were general factors, such as litter size or general nutrition. Thus, the amounts of incoming nutrient or growth substances which reach embryos in each horn are the limiting factor rather than the overcrowding itself. Since the foetus is dependent on the placenta for its nutrition, foetal growth is likely to be conditioned by the ultimate size or area of the placenta which in turn is influenced by the number of embryos (Pomeroy, 1960). In other words, in multiparous animals uterine accommodation exerts its effect early in pregnancy by limiting the growth of the placenta and this is reflected by variations in foetal weight in relation to litter size in the later stages of pregnancy.

An important question is whether intrahorn position and spacing of the embryos have an effect upon growth. According to Majerciak (1965) large and small piglets are found throughout the uterine horn and privileged positions do not exist. This is contrary to the observation of Waldorf *et al.* (1957) who found foetuses and membranes to be larger at the extremities of a uterine horn than in the middle. Our own observations are more complex, since we have found that privilege of position depends upon litter size and the stage of pregnancy (Table 8). When the foetal number averages five, a local effect appears towards the end of pregnancy, the largest embryos being at the extremities. When the number is higher than five, then the extremities are better than the middle whatever the stage of pregnancy, so the higher the number, the earlier is the effect of position. Furthermore, when a small or dead foetus is encountered in a horn, its neighbours are heavier than the average of litter mates. Also, according to Majerciak (1965), stillborn are mostly found in the middle parts of the horns. This emphasizes some aspects of nutritional competition *in utero* and leads to the conclusion that foetuses having the best uterine vascular contact and the least space restriction will be the largest.

175

(e) Environmental factors

Among maternal characteristics, large size and weight are thought to be positively correlated with faster prenatal growth. This phenomenon has been demonstrated in rabbits (Venge, 1950), mice (Bateman, 1954), pigs (Waldorf *et al.*, 1957) and also in many uniparous species but it does not seem to be of great importance.

Table 8

Influence of intrahorn position upon foetal weight (g) according to stage of pregnancy and litter size in pigs (Salmon-Legagneur, 1967)

Size of half litter	Intra-horn position	70 days	90 days	110 days
	Upper	167	849	950
3	Middle	168	728	1013
	Lower	177	732	1153
	Upper	168	591	1,109
5	Middle	165	789	882*
	Lower	187	759	1,214
	Upper	178	584	1,271
7	Middle	140*	472*	967*
	Lower	172	622	1,132

* Significant difference

Neither age of dam nor weight changes during pregnancy seem to exert an effect on birth weight in pigs (Salmon-Legagneur, *et al.*, 1966).

Among other factors, maternal nutrition in pigs and rats has given rise to considerable data which have been reviewed by Russell (1948), Duncan and Lodge (1960), Mitchell (1962) and Salmon-Legagneur (1965). The general conclusions are in agreement with those on sheep presented by Everitt in the present volume (page 131). There is, however, much variation and controversy when one considers each species, the level of nutrition, the nutrient involved, the stage of pregnancy and litter size. Conclusions, therefore, are rather ambiguous but may be classified as follows:

(i) Only those species in which litter weight is high in relation to the mother's (*see* Table 1) and where the maternal gain is rather low, will be susceptible to large variations of foetal weight due to diet. This is the reason why sheep or mouse embryos are more likely to be affected than those of Man or the pig. Personally, so far as can be judged from my past 10 years of work on nutrition of sows, I have never obtained any convincing evidence of an important effect of maternal diet on birth weight in pigs (Salmon-Legagneur, 1965). This is probably due to the natural facility with which sows can deposit body stores, even with restricted feeding, and so act as a

Table 9

Effect of maternal feed restriction on the birth weight of progeny in rats and pigs

Diet	Level feed	Litter weight	Average individual weight	Dam's weight gain
Rats (g)*				
High	ad lib.	54	5·4	+55
Restricted	50%	41	4·6	−18
Pigs (kg)				
High*	ad lib. (3·7 kg per day)	14·2	1·24	54
Restricted	50%	15·3	1·20	20
High†	2·7 kg per day	14·9	1·25	44
Restricted	50%	12·5	1·09	4

* Salmon-Legagneur, Perisse and Jacquot, 1960
† Lodge, Elsley and MacPherson, 1966a and b

buffer to protect, if necessary, the foetuses' growth against any acute deficiency.

(ii) As regards the level of nutrition, when the mother is in reasonable condition, the birth weight, because of the small requirements of the foetuses, is probably at its upper limit and cannot be improved upon by extra feeding (Salmon-Legagneur and Rerat, 1962). The only nutritional effect which could appear, therefore, is generally as a result of under-feeding. The critical level of restriction depends upon the susceptibility of the normal diet, or the feeding behaviour of species, to be borderline. In those species which may be fed ad lib., like pigs, the restriction has to be very severe, say 30–50 per cent, to induce any detrimental effect on birth weight (Lodge et al., 1966; Salmon-Legagneur, 1965). In this case, the effect is more pronounced if the mother suffers herself, as in the rat, than if she is still allowed to gain weight as is usual in the pig (Table 9).

(*iii*) The specific influence of nutrients may be seen when comparing the relative effects of energy and protein in the diet. It appears from experiments with rats (Champigny, 1963) and with pigs (Clawson, *et al.*, 1963; Salmon-Legagneur and Lodge, 1967) that energy restriction in the diet results in a greater depletion of foetal weight than does protein restriction, i.e., energy but not protein may be a limiting factor to prenatal growth (Table 10). For instance, less than 10 per cent protein in the diet allowed normal prenatal growth in pigs, provided that energy was liberally supplied (Salmon-Legagneur, 1963). Despite a 20 per cent weight reduction, rabbit embryos survived on a protein-free diet allowed for more than half of pregnancy but the same result was achieved with only 2 days of energy starvation (Seegers, 1938).

Table 10

Relative influence of energy and protein in the diet on reproductive performance of sows (Salmon-Legagneur, 1965)

Daily diet		Uterine content				Dam's weight gain
Energy (Mcal)	Protein (g)	Piglets (kg)	Placenta (kg)	Average piglet weight (kg)	Litter size	(kg)
8·9	425	14·4	3·2	1·19	12·7	46
13·7	425	13·3	2·6	1·22	11·2	68*
8·9	275	14·3	3·0	1·32	11·3	41

* Significant difference

(*iv*) It is generally assumed, regarding stages of pregnancy, that food restriction is more deeply felt by the embryos towards the end than earlier. This was observed in rats by Runner and Miller (1956) and in pigs by Salmon-Legagneur (1962) and Lodge *et al.* (1966a) (Table 11). However, the difference is small and it may be argued that limitation of growth through inadequate nutrition in early pregnancy is more dangerous, since it generally results in teratologic effects as seen from experiments inducing vitamin deficiencies by Giroud (1959) and Kalter and Warkany (1959).

(*v*) Influence of litter size is that the greater the size of litter, the higher is the possible influence of maternal nutrition on progeny.

In this respect Schultze (1954) was unable to find any detrimental effects of food restriction in rats when the litter size was less than six.

It may be concluded that the overall influence of maternal nutrition is generally lower in multiparous animals, like pigs, than in uniparous, like sheep. However, this assumption can be widely moderated according to practical conditions.

Table 11

Effect of chronological food restriction during pregnancy
(Salmon-Legagneur, 1962)

Diet	Uterine contents			Dam's weight gain (kg)
	Litter weight (kg)	Placenta weight (kg)	Average piglet weight (kg)	
High–low	13·4	2·4	1·26	21
Low–high	15·6	2·9	1·38	19

(Low = 30 per cent restriction)

An important question is whether these influences of prenatal nutrition influence the subsequent growth of neonates, as is stressed in sheep. Though controversial, the answer to this is probably 'yes', as indicated by the mortality rate and late effects of protein malnutrition in piglets (Stevenson and Ellis, 1957) and rats (Slonaker, 1938) but this effect could be variable, since the percentage of variation in weight of weaning accounted for by variation in birth weight may vary from 5 to 26 (Lodge, McDonald and MacPherson, 1961; Lynch, 1965; Aumaitre, Legault and Salmon-Legagneur, 1966).

III. BIOCHEMICAL ASPECTS OF PRENATAL DEVELOPMENT

Growth may be considered as a combination of physical processes, such as hyperplasy and hypertrophy, which induce volumetric development and chemical changes which are responsible for physiological activity and ageing. For these latter, mammalian species differ greatly among themselves according to the time and the site where each change takes place and it would be dangerous to

179

generalize and draw comparisons between multiparous and uniparous species. Moreover, although there are many studies dealing with body composition and placental transfusion mechanisms at particular stages of pregnancy in some species, such as the pig, information is lacking for others. Furthermore, the periods studied are insufficient to allow worth-while comparisons between species. This is a consequence of experimental difficulties which are encountered, partly in providing representative samples of embryos and partly in the accuracy of the techniques used.

Table 12

Chemical composition of the pig foetus (Salmon-Legagneur, 1965)

Stage of pregnancy (days)	Percentage fresh weight				Percentage dry matter		
	Water	Lipid	Protein	Ash	Lipid	Protein	Ash
30	94·7	0·5	3·6	0·9	9·4	67·9	16·9
60	89·5	0·9	6·2	1·9	8·5	59·0	18·0
100	85·3	1·3	9·1	3·1	8·8	61·9	21·1
107	83·6	1·4	9·7	3·2	8·5	59·1	19·5

For this reason, from many points of interest, the following two have been selected as being related to interspecies differences:

(a) changes in proportions of lipids and protein in the foetal body, and

(b) kinetics of glycogen formation during pregnancy.

(a) Changes in proportions of lipids and protein

Several workers have given comprehensive figures on changes in the gross chemical composition of foetuses in the pig (Mitchell et al., 1931; Gortner, 1945; Urbanyi, 1950; de Villiers et al., 1958; Pomeroy, 1960; Majerciak, 1965), in the sheep (Wallace, 1948), in Man (Widdowson, 1950; Widdowson and Dickerson, 1960; Jirasek, Uher and Uhrova, 1966) and in laboratory animals (Boyd, 1935; Widdowson, 1950; Villee and Hagerman, 1958; Lafon, 1960; Picon and Jost, 1963).

Most of these authors expressed results as percentages of fresh weight, but since water is by far the major constituent in the foetal body, this results essentially in the expression of constituents as percentages of water. As the water content decreases, which is assumed to be related to a loss of extracellular fluids (Widdowson,

1950; Lafon, 1960), percentages of other constituents increase correspondingly. Therefore, the observed variations reflect this phenomenon more than the true variation in composition of active substances. Results appear quite different when expressed as a percentage of the dry matter or relative to any other constituent (Table 12). It can be seen, for instance, that foetal dry matter maintains a relatively constant composition throughout pregnancy

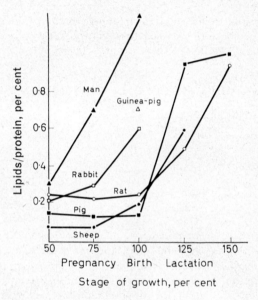

*Figure 3.   Body changes (lipids/protein) in different species
before and after birth*

in pigs and that there are only small changes during prenatal growth in the living substrate of this animal. There may be species differences, however, in this respect.

Fat content has been observed to be the most widely varying constituent of the foetal body in various species (Widdowson, 1950), with the highest value for Man and the lowest for the pig. Since fat percentage in the body was proposed sometimes as a criterion of 'chemical maturity' (Hammond, 1940), it would be of interest to follow its variation relative to dry matter or to protein content during foetal life. Picon and Jost (1963) have compared pregnancy in rats and rabbits in this way; in the rat, the lipid/protein ratio remains remarkably constant at around 0·25 during the whole of pregnancy and then sharply increases (as in pigs according to Elsley, 1964) as soon as birth occurs. In the rabbit the L/P ratio increases during the last third of pregnancy and more than doubles by the time of birth. Similar variations have been observed in sheep and in Man (*Figure 3*).

Mammalian species differ greatly, therefore, in the time in which deposition of fat occurs during perinatal life. According to Widdowson (1950) these differences cannot be related to the stage of maturity at birth, for, of all the mammals studied, the pig and the guinea-pig are born the most mature, yet the former contains 1 per cent of fat and the latter 10 per cent. There may be, however,

Table 13

Body composition in newborn mammals as percentage of fresh weight (Widdowson, 1950)

Species	Water	Protein	Fat	Fat/Protein
Man	69·1	11·9	16·1	1·35
Guinea-pig	70·9	14·9	10·1	0·68
Rabbit	84·6	11·1	2·0	0·18
Rat	86·0	10·8	1·1	0·10
Pig	84·1	11·3	1·1	0·09

some connection between the large amount of fat in some species (Man, guinea-pig) and their comparatively long gestation periods. Furthermore, one might observe that, as multiparity is related to relative pregnancy length, the same relationship may exist between multiparity and the L/P ratio in the embryo (Table 13); the lower the L/P ratio, the higher is the degree of multiparity. So, it appears that in multiparous species, lipid synthesis during prenatal life could be a limited process which takes place late in pregnancy and which is more related to a maternal limiting factor, or to placental permeability, than to the foetuses themselves. This point seems confirmed in some experiments by the slow degree of diffusion of labelled lipids, where passage from mother to foetuses appears more qualitative than quantitative (Sternberg, 1962; Duyne, Havel and Felts, 1962) and by an absence of relationship between maternal and foetal lipaemia (Friedman and Byers, 1961; Brody and Carlson, 1962). On the other hand, differences between species in embryonic lipid content could be related to hormonal factors *in utero* (thyroxine, cortisone or prolactin), as shown by Jost and Picon (1958).

(*b*) *Glycogen formation during pregnancy*

Since the historical discovery by Bernard (1859), it is well known that the placenta is rich in glycogen and, also, that in some species (ungulate) there is a significant amount of circulating fructose in the foetuses (Bacon and Bell, 1946; Goodwin, 1956). Placental glycogen appears early in pregnancy and is said to increase up to a maximum towards mid-pregnancy and to decrease thereafter. Its

Table 14

Variation in uterine glycogen content during pregnancy in the sow (mg/100 g) (Salmon-Legagneur, 1967)

Stage of pregnancy	30 days	60 days	114 days
Organs			
Uterus myometrium	504	210	472
Placenta	108	140	149
Foetuses	343	840	2,821
Foetal liver	—	336	9,208

failure at term could provoke anoxia and delivery of foetuses. Histological staining and isotopic labelling studies have shown that glycogen is mainly synthesized in the maternal part of the placenta (Hugget and Hammond, 1952) but only a small amount of muco-polysaccharide is found in the foetal part (Becze, 1962). Since glycogen is a relatively non-diffusible substance, despite some lack of parallelism between the observations and chemical determinations, it is to be expected that glycogen will not appear in large amount in the foetal body before its synthesis *in situ* (Shelley, 1961). This occurs relatively late in pregnancy, so liver glycogen is not found before the fifty-sixth day in the guinea-pig foetus (Hard, Reynolds and Wimburg, 1944) or before the eighteenth day in rats (Jacquot, 1955), although Hertig (1960) described the presence of glycogen in early human embryos. It may be argued also that glycogen may occur in some other part of the embryo (muscle or heart) before and at a higher rate than in the liver (Table 14). At a later stage, glycogen synthesis increases rapidly in the foetal organism and reaches at birth the level of 3–5 per cent in muscles and 6–11 per cent in the liver (Shelley, 1961), so that it becomes an important constituent of the foetal dry matter. There is no doubt that these considerable stores of glycogen are essential to neonates for the maintenance of body temperature, since they are nearly depleted within a few hours after birth.

Therefore, it could be tempting to suggest that species which are born capable of vigorous activity need larger amounts of glycogen than those born in less mature condition. Thus, the extent of stores and the rate of fall of glycogen are higher in the pig and the sheep than in the rabbit and the rat. It can be observed that this

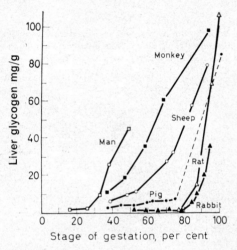

Figure 4. Liver glycogen in different species before birth

locomotor activity occurs markedly in some species where subcutaneous fat is lacking (pig), whereas in others, homeothermy and reduction in heat losses are achieved by immobility of the new-born lying close together under the mother. This is also the reason why, if deprived of food, the neonate pig is more exposed to hypoglycaemia and death than the other better protected species (Goodwin, 1956).

The rate and kinetics of glycogen synthesis during prenatal life differ greatly between species. In those with a relatively long gestation period (Man and sheep), liver glycogen begins to accumulate fairly early at a steady rate of about 2 mg/g per day, but where the gestation period is short (pig and rat) the rise occurs in the last fifth of gestation and proceeds at the considerable rate of 10–40 mg/g per day. Thus, species may be classified according to the kinetics of their liver glycogenesis, which corresponds roughly again to multiparity (*Figure 4*); the later the increase in liver glycogen, the higher is the prolificity. This trend parallels that of lipid deposition in the same species, so we find here another proof of a biochemical relationship between reproduction and body composition.

Little is known about the factors responsible for the first appearance of glycogen in foetal tissues and the subsequent changes in its concentration (Shelley, 1961). First, and contrary to the maternal glycogenesis, there is no definite relationship between the foetal

Table 15

Influence of energy in the maternal diet on glycogen and phosphatase synthesis during pregnancy in sows (Salmon-Legagneur, du Mesnil du Buisson and Mauleon, 1965)

Energy intake	Glycogen (mg/100 g)		Phosphatase (U/100 g)*	
	Low	High	Low	High
Liver	3,123	6,267	96	186
Uterus	216	212	2,483	3,494
Placenta	134	138	285	270
Embryo	709	692	5,676	5,373

* One phosphatase unit hydrolyzes 1 m. mole p-nitrophenol in 60 min at 37°C

glycogenesis and the maternal diet, as proved by experiments where increasing energy level failed to improve the glycogen deposition in the foetal pig (Table 15). Thus, foetal glycogen appears to be an endogenous metabolic product independent both of the transfusion of sugar and the level of sugar in the mother and foetus. Wislocki and Dempsey (1946) supposed that formation of glycogen is preceded by dephosphorylation of the hexose-phosphate which requires a splitting enzyme, such as alkaline-phosphatase. Phosphatase, therefore, should be associated with glycogen, which was proved in Man and the cat but is doubtful in the sow. On the other hand, foetal glycogenesis appeared to be dependent on foetal pituitary and adrenal function (Jacquot, 1955). Furthermore, its increase was shown to be related to development of PGM and UDPG activities (Jacquot and Kretchemer, 1964). It is probable, therefore, that the appearance of glycogen in large amounts in the foetal body coincides with the occurrence of both endocrine and enzymatic factors. This could be achieved early or late in pregnancy, according to species and the stage of development of embryos. That would mean that 'physiological maturity' occurs later in multiparous species than in uniparous ones, probably in relation to the smaller size of individuals at birth in the former.

CONCLUSION

In this short review we found several answers to the question of what constitutes multiparity. These include short pregnancy length, high litter weight when related to the mother weight, sharp weight increase late in pregnancy, particular hormonal equilibrium, favourable uterine environment related to large sized placenta, and biochemical factors, like fat and glycogen content of foetal body.

The pig and the rabbit are probably the only normally multiparous species to have an economical importance, but multiple birth is sought also in sheep and probably soon in the cow, so that it may become the general aim. At the present time when productivity of farm animals has to be improved, there is no doubt that time and money may be greatly saved in this way.

It may also be emphasized that among multiparous species like the pig, where embryonic losses constitute a considerable wastage of potential life, there is probably more to save by reducing prenatal deaths than by promoting any hypothetical improvement in growth. In this field, feeding pregnant females is an important factor but not the main one, as for some other species. Many practical mistakes have been made and many gaps still remain, as shown from recent data on sow nutrition. However, we must pay more attention to physiological factors of prenatal development, thus concentrating upon those correlated with multiparity, such as hormonal equilibrium during pregnancy. We have much to learn in this field but solutions are already anticipated.

REFERENCES

Allen, A. D. and Lasley, J. F. (1954). *J. Anim. Sci.* **13**, 955

Amoroso, E. C. (1952). In: *Marshall's Physiology of reproduction*, 3rd Edn. London; Longmans

Aumaitre, A., Legault, C. and Salmon-Legagneur, E. (1966). *Annls Zootech.* **15**, 313

Bacon, J. S. and Bell, D. J. (1946). *Biochem. J.* **40**, XLII

Baker, M. C., Hazel, L. and Reinmiller, C. F. (1943). *J. Anim. Sci.* **2**, 3

Barcroft, J. (1946). *Researches on Prenatal Life*, Oxford

Bateman, N. (1954). *Physiol. Zool.* **27**, 163

Becze, J. (1962). *Vet. Med.* **35** (9), 705

Bernard, C. (1859). *C.r. hebd. Seanc. Acad. Sci., Paris* **48**, 77

Biggers, J. D., Curnon, R. N. and Finn, A. (1963). *J. Reprod. Fert.* **6**, 125

Blaxter, N. L. (1964). In: *Mammalian Protein Metabolism*, Vol. II (Ed. by H. N. Munro and J. B. Allison). New York; Academic Press

Boaz, T. G. (1962). *Vet. Rec.* **74**, 1432

Bourdel, G. (1960). *Thèse Doct.* Fac. Sci. Paris

Boyd, E. M. (1935). *Biochem. J.* **29**, 985

Boyd, J. D., Hamilton, W. J. and Hammond, J. (1944). *J. Anat.* **78**, 5

—— (1952). In: *Marshall's Physiology of Reproduction*, 3rd edn. London; Longmans

Brambell, F. W. (1942). *Proc. R. Soc.* **B**, **130**, 462

Bredeck, H. E. and Mayer, D. T. (1955). *Bull. Mo. agric. Exp. Stn.*, 591

Brody, S. (1945). *Bioenergetics and Growth.* New York; Rheinhold

— and Carlson, L. A. (1962). *Clinica chim. Acta* **7**, 694

Champigny, O. (1963). *C.r. hebd. Seanc. Acad. Sci., Paris* **256**, 4755

Clawson, A. J., Richards, H. L., Matrone, G. and Barrick, E. R. (1963). *J. Anim. Sci.* **22**, 662

Corner, E. W. (1923). *Am. J. Anat.* **31**, 523

Craig, J. V., Norton, A. W. and Terrill, S. W. (1956). *J. Anim. Sci.* **15**, 242

Crozier, W. J. (1940). *J. gen. Physiol.* **23**, 309

Davidson, H. R. (1930). *J. agric. Sci. Camb.* **14**, 507

Dean, B. T. and Trible, L. F. (1961). *Res. Bull. Mo. agr. Exp. Stn.*, 774

Duncan, D. L. and Lodge, G. A. (1960). *Tech. Commun. Commonw. Bur. Anim. Nutr.* No. 21

Duyne, C. M. Van, Havel, R. J. and Felts, J. M. (1962). *Am. J. Obstet. Gynec.* **84**, 1069

Dziuk, P. J., Polge, C. and Rowson, L. E. (1964). *J. Anim. Sci.* **23**, 37

Eckstein, P., McKeown, T. and Record, R. G. (1955). *J. Endocr.* **12**, 108

Elsley, F. W. (1964). *Annls Zootech.* **13**, suppl. 75

Fahim, M. S., Mayer, D. T. and Lasley, J. F. (1961). *Bull. Mo. agr. Exp. Stn.* 783

Falconer, D. S., Edwards, R. G., Fowler, R. E. and Roberts, R. C. (1961). *J. Reprod. Fert.* **2**, 418

Friedman, M. and Byers, S. O. (1961). *Am. J. Physiol.* **201**, 611

Giroud, A. (1959). *Wld Rev. Nutr. Diet.* **1**, 231

Goode, L. (1961). *Ph.D. Thesis.* Univ. Florida

Goodwin, R. F. (1956). *J. Physiol.* **134**, 88

Gortner, W. A. (1945). *J. biol. Chem.* **159**, 135

Gruhn, R. (1966). *Zuchtungskunde* **38**, 104

Hafez, E. S. (1963). *J. Anim. Sci.* **22**, 779

Hammond, J. (1928). *ZuchtungsKunde* **3**, 523

— (1934). *Harper Adams Util. Poult. J.* **19**, 557

— (1937). *Sch. Sci. Rev.* **72**, 548

— (1940). *Farm animals: their breeding, growth, inheritance.* London; Longmans

Hammond, J. (1952). In: *Marshall's Physiology of Reproduction*, 3rd edn. London; Longmans

Hartman, C. (1929). *J. Mammal*, **10**, 127

Hard, W. L., Reynolds, D. E. and Wimburg, M. (1944). *J. exp. Zool.* **96**, 189

Hazel, L. N. (1963). *Proc. Wld. Conf. Anim. Prod.* **2**, 189

Heap, F. C. and Lodge, G. A. (1966). *9th Int. Congr. Anim. Prod., Edinb., Scient. Progm.* Abstr. Eng. edn., p. 28

Heitman, H., Hughes, E. H. and Kelly, C. F. (1951). *J. Anim. Sci.* **10**, 907

Hertig, A. (1960). In: *The Placenta*. (Ed. by C. A. Villee.) Baltimore; William and Wilkins

Hetzer, H. O. and Brier, G. W. (1940). *Proc. Am. Soc. Anim. Prod.* 33rd Meet. 135

Huggett, A. S. and Hammond, J. (1952). In: *Marshall's Physiology of Reproduction*. 3rd edn. London; Longmans

— and Widdas, W. (1951). *J. Physiol.* **114**, 306

Ibsen, H. L. (1928). *J. exp. Biol.* **51**, 51

Ino, T. (1966). *Natn. Inst. Anim. Hlth., Q. Tokyo* **6**, 178

Jacquot, R. (1955). *J. Physiol. Path. gen.* **47**, 857

— and Kretchemer, N. (1964). *J. biol. Chem.* **239**, 1301

Jirasek, J. E., Uher, J. and Uhrova, M. (1966). *Am. J. Obstet. Gynec.* **96**, 868

Jost, A. and Picon, L. (1958). *C.r. hebd Séanc. Acad. Sci. Paris* **246**, 1281

Kalter, H. and Warkany, J. (1959). *Physiol. Rev.* **39**, 69

Krizenecky, J. (1942). *Z. Tierzücht. ZüchtBiol.* **52**, 209

Lafon, M. (1960). *Archs. int. Physiol. Biochim.* **68**, 460

Legault, C. and Aumaitre, A. (1966). *Annls. Zootech.* **15**, 333

— and Ollivier, L. (1965). *Annls. Zootech.* **14**, 40

Leitch, I. (1957). *Proc. Nutr. Soc.* **16**, 227

Lenkeit, W., Gutte, J. O., Warnecke, W. and Kirchoff, W. (1955). *Z. Tiereznahr. Futtermittelk* **70**, 551

Lerner, E. H., Mayer, D. T. and Lasley, J. F. (1957). *Bull. Res. Mo. agric. Exp. Stn.* 629

Lodge, G. A., Elsley, F. W. H. and MacPherson, R. M. (1966a). *Anim. Prod.* **8**, 29

— — — (1966b). *Anim. Prod.* **8**, 499

— McDonald, I., MacPherson, R. M. (1961). *Anim. Prod.* **3**, 261

Lush, J. L., Hetzer, H. O. and Culbertson, C. C. (1934). *Genetics* **19**, 329

Lynch, G. (1965). *Meld. Norg. Landbz.Hoisk.* **44**, 9

Majerciak, P. (1965). *Plodnost a reprodukcia prasnic.* Bratislava; Vyd. Slov. Akad.

Mauleon, P. and Pelletier, J. (1964). *Annls Biol. anim. Biochim. Biophys.* **4**, 105

— and Rao, H. (1963). *Annls Biol. anim. Biochim. Biophys.* **3**, 21

McCarthy, J. C. (1965). *Anim. Prod.* **7**, 347

Mesnil du Buisson, F. du (1964). *Proc. Int. Congr. Endocr.* p. 680

Mesnil du Buisson, F. du (1966).   *Thèse Doct.*, Fac. Sci., Paris
— and Dauzier, L. (1959).   *Annls Zootech.* **8**, suppl. 147
— and Rombauts, P. (1963).   *Annls Biol. anim. Bioch. Biophys.* **3**, 445
Mitchell, H. H. (1962).   *Comparative Nutrition of Man and Domestic Animals.*
   Vol. I.   New York; Academic Press
— Caroll, W. E., Hamilton, T. S. and Hunt, G. E. (1931).   *Bull. Ill. agric.*
   Exp. Stn. 375
Moustgaard, J. (1958).   In: *Nutrition and Reproduction in Domestic Animals.*
   (Ed. by H. H. Cole and P. T. Cupps.)   Vol. II.   New York; Academic
   Press
Newman, J. A. (1963).   *Can. J. Anim. Sci.* **43**, 285
Nicol, T. (1933).   *J. Anat.* **68**, 75
Picon, L. and Jost, A. (1963).   *C.r. Seanc. Soc. Biol.* **157**, 1368
Polge, C. P. (1965).   *Symp. Int. Artificial Insemination in the Pig*, Noordwijk
Pomeroy, R. W. (1952).   *Proc. 2nd Congr. int. Physiol., path. anim.*
— (1960).   *J. agric. Sci. Camb.* **54**, 1
— (1960).   *J. agric. Sci. Camb.* **54**, 31
— (1960).   *J. agric. Sci. Camb.* **54**, 57
Rathmacher, R. P. and Anderson, L. L. (1963).   *J. Anim. Sci.* **22**, 1139
Rathnasabapathy, V., Lasley, J. F. and Mayer, D. T. (1956).   *Res. Bull.*
   Mo. agr. Exp. Res. Stn. 615
Reddy, V. B., Mayer, D. T. and Lasley, J. F. (1958).   *Mo. Agr. Exp. Res.*
   Bull. 667
Romack, F. E., Zoellner, K. and Day, B. N. (1960).   4th Swine *Day Rep.*
   Mo. agr. Exp. Stn.
Runner, M. N. (1951).   *J. Exp. Zool.* **116**, 1
— and Miller, J. R. (1956).   *Anat. Rec.* **124**, 437
Russell, F. C. (1948).   *Tech. Commun. Commonw. Bur. Anim. Nutr.* No. 16
Salmon-Legagneur, E. (1962).   *Annls Zootech.* **11**, 173
— (1963).   *Arch. Sci. physiol.* **17**, 233
— (1965).   *Thèse Doct.*, Fac. Sci., Paris
— (1967).   Unpublished data
— and Jacquot, R. (1961).   *C.r. hebd Seanc. Acad. Sci. Paris* **253**, 544
— Legault, C. and Aumaitre, A. (1966).   *Annls Zootech.* **15**, 215
— and Lodge, G. A. (1967).   To be published in *Annls Zootech.*
— Mesnil du Buisson, F. du and Mauleon, P. (1965).   Unpublished data
— Perisse, J. and Jacquot, R. (1960).   *C.r. hebd Seanc. Acad. Sci. Paris* **250**,
   1921
— and Rerat, A. (1962).   In: *Nutrition of Pigs and Poultry* (Ed. by J. T.
   Morgan and D. Lewis).   p. 207.   London; Butterworths
Schultze, M. O. (1954).   *J. Nutr.* **54**, 453
Seegers, W. H. (1938).   *Am. J. Physiol.* **121**, 231
Shelley, H. J. (1961).   *Br. med. Bull.* **17**, 137
Slonaker, J. R. (1938).   *Am. J. Physiol.* **123**, 526
Spencer, H. (1899).   *Principles of Biology.*   2nd edn.   London
Spencer, R. P. and Coulombe, M. J. (1965).   *Growth* **29**, 165

Squiers, C., Dickerson, G. E. and Mayer, D. I. (1952). *Res. Bull. Mo. agr. Exp. Stn.* 494

Sternberg, J. (1962). *Am. J. Obstet. Gynec.* **84**, 1731

Stevenson, J. W., and Ellis, J. R. (1957). *J. Anim. Sci.* **16**, 877

Stewart, H. A. (1945). *J. Anim. Sci.* **3**, 455

Taketomi, M. (1956). *Agric. Bull. Saga Univ.* **4**, 119

Ullrey, D. E., Sprague, J. I., Becker, D. E. and Miller, E. R. (1965). *J. Anim. Sci.* **24**, 711

Urbanyi, L. (1950). *Magy. Allatorv. Lap.* **5**, 6

Venge, O. (1950). *Acta zool. Stockh.* **31**, 1

Villee, R. A. and Hagerman, D. D. (1958). *Am. J. Physiol.* **194**, 457

Villiers, V. (de), Sorensen, P. A., Jakobsen, P. and Moustgaard, J. (1958). a. *Rep. Sterility Res. Inst. Copenh.* 139

Waldorf, D. P., Foote, W. C., Self, H. L., Chapman, A. B. and Casida, L. E. (1957). *J. Anim. Sci.* **16**, 976

Wallace, L. R. (1948). *J. agric. Sci., Camb.* **38**, 244

Weinbach, A. P. (1941). *Growth* **5**, 217

Widdowson, E. M. (1950). *Nature, Lond.* **166**, 626

— and Dickerson, J. M. (1960). *Biochem. J.* **77**, 30

Wishart, S. and Hammond, J. (1933). *J. agric. Sci., Camb.* **17**, 922

Wislocki, G. B. and Dempsey, E. W. (1946). *Am. J. Anat.* **78**, 1

Wislocki, G. and Streeter, G. L. (1938). *Contr. Embryol.* **27**, 1

Wright, S. (1922). *Bull. U.S. Dep. Agric.* 1121

Zimmerman, D. R., Spies, H. G., Self, H. L. and Casida, L. E. (1960). *J. Anim. Sci.* **79**, 295

DISCUSSION

DR. SHAW (Brussels) asked for further information on the *degree* of undernutrition which had induced the effects reported, to which DR. EVERITT replied that this was best thought of perhaps in terms of weight loss, which had amounted to 15 per cent by mid-pregnancy and 30 per cent by term.

DR. McCARTHY (Dublin) asked how important were haemodynamic effects on prenatal growth in the sheep; DR. EVERITT did not think haemodynamic control could be as important when only one foetus was present as when there were several, but it was noticeable that in a state of undernutrition the cotyledons did have a differential effect which would accord with the directional blood flow.

MR. PIKE (Leeds) enquired to what extent extra-uterine tissues could be used to offset undernutrition at different stages of pregnancy. DR. EVERITT replied that there did appear to be some buffering effect, if an incomplete one, in animals which had considerable fat reserves when undernutrition was imposed; it was necessary, however, to measure these effects in more precise terms than mere loss in body weight by studying plasma free-ketone levels, as used by Russel at H.F.R.O.

DR. FREDERICK (Ottawa) commented that dietary changes produced only small effects on reproductive phenomena which were not statistically significant by normal analytical methods. It was necessary to examine within-treatment variation more closely and compare it with between-treatment variation. Within-litter variation in birth weight decreased as diet improved and there was also a recognized tendency for variation in birth weight to increase with increase in litter size, one effect could therefore obscure the other. A further point made by DR. FREDERICK was that an increase in stillbirths might occur and be incorrectly interpreted, in that if the total number of live births increased then the number of still births might increase also in proportion and not be due to any specific defect.

DR. SALMON-LEGAGNEUR commented on the influence of level of feeding during oestrus and early pregnancy by saying that the result varied with the particular conditions; if the sows had a high ovulation rate then the influence of nutrition was small because foetal atrophy was mainly due to overcrowding, but if the number of ova was small, as in gilts, then level of feeding might increase this and also the number of foetuses.

DR. LODGE (Nottingham) commented on the apparent species difference in the effect of feed restriction on reproductive performance, mentioned by both speakers, and wondered whether this may be due simply to the fact that work with different species had been pitched at different levels. In Wallace's work with sheep, both his medium and low planes of feeding resulted in net weight loss and were much lower than anything reported with sows.

PROFESSOR BRAMBELL (Bangor) raised the question of the relative maternal effects of nutrition during pregnancy and lactation. The shrew produced five young after about 18 days gestation, weighing 0·5 g each, so the total litter increment during pregnancy was some 2·5 g whereas during 23 days suckling the litter increased to 30 g. The relative outputs of the mother to the young were vastly different, therefore, in the two periods and was this true also of sheep and pigs? DR. SALMON-LEGAGNEUR agreed that the contribution of the mother to the young was much greater during lactation than during pregnancy but there was, in the sow, an inverse relationship between pregnancy body stores and lactation losses; the higher the pregnancy gain the greater the lactation loss, but utilization of pregnancy stores to produce postnatal growth appeared to be an inefficient way to utilize food. DR. EVERITT believed the same situation to be true for sheep but neither pig nor sheep could equal the performance of the shrew.

DR. TIMON (Dublin) enquired what was the extent of prenatal death in sheep and pigs, when did it occur and was the cause primarily physiological, nutritional or genetical? DR. EVERITT replied that for sheep it was about 30 per cent and mostly before implantation; DR. SALMON-LEGAGNEUR confirmed a similar situation in the pig with most losses between fertilization and implantation, due mainly to physiological factors, such as progesterone levels, but related to nutrition and other external influences.

PROFESSOR BUTTERFIELD (Sydney) asked whether the Merino, which was frequently subjected to very low nutritional levels during pregnancy, had adapted to this situation, but DR. EVERITT thought that there was no direct evidence of this; any apparent breed differences in ability to withstand low nutrition were most probably due to differences in body stores.

IV. POSTNATAL DEVELOPMENT

BODY DEVELOPMENT AND SOME PROBLEMS
OF ITS EVALUATION

V. R. FOWLER

Rowett Research Institute, Bucksburn, Aberdeen

THE PERIOD of growth considered here, between birth and puberty, is one during which the interaction of the growing animal with its environment is particularly intense. The great importance of this phase of growth to those concerned in the production of meat makes an understanding of the principles involved in this interaction of special value.

The growth process is generally considered from two aspects; firstly, the increase of body mass in time, usually described by the live weight growth curve for the whole animal, and secondly, changes in the form of the animal resulting from differences in the relative growth rates of the component parts of the body. Gross live weight changes are comparatively easily measured and data of this type have been the subject of extensive mathematical and philosophical treatment (Brody, 1945a; Bertalanffy, 1960; Laird, Tyler and Barton, 1965).

The measurement and evaluation of body development is a more complex problem than the measurement of changes in total body weight and satisfactory data are obtained less easily. Many experiments in this field have been confined necessarily to linear measurements between external or radiographic anatomical landmarks. Although such studies have yielded much valuable information they are limited in scope, firstly because the body components which can be measured in this way are in the main only indicative of skeletal size, and secondly, because any attempt to interpret the results in volumetric or gravimetric terms raises the errors to approximately the third power. Clearly for the most penetrating inquiry into the subject, data from animals which have been subjected to comprehensive anatomical dissection are required.

The formidable amount of painstaking and unglamorous work involved in such studies, combined with the expense of the carcasses, has daunted all but a few dedicated workers and it is greatly to the credit of Sir John Hammond and his associates, notably McMeekan, Pálsson and Vergés, that they achieved so much in this field with so few facilities. Although this work was initiated over three decades ago, the studies of these workers are still the best documented and most comprehensive available.

SOME PRELIMINARY CONSIDERATIONS

Before discussing the problems of evaluating body development it is necessary to consider some of the basic principles involved. D'Arcy Thompson was the first to focus within one treatise the flashes of insight which were published in this context towards the end of the last century. He demonstrated the wide application of the axiom 'form is related to function' (Thompson, 1917) and highlighted the biological significance of Galileo's principle of similitude —'nor can Nature grow a tree nor construct an animal beyond a certain size, while retaining the proportions and employing the materials which suffice in the case of a smaller structure.'

Some geometric principles which require a change in the form of land animals as they increase in size were summarized by Brody (1945b) as follows:

'(1). Weight which tends to crush the land animal's limbs and which has to be moved by muscles varies with the cube of linear size.

(2). Tensile strength of the muscles and bones which move and support the animal varies with the square of linear size.

(3). Surfaces through which diffusion, nutrition and excretion take place vary with the square of linear size.'

This concept Brody epitomized in the phrase 'the organism changes geometrically so as to remain the same physiologically'.

In addition to these purely geometrical requirements for changes in form during growth, the changing functional and physiological needs of the animal must also be considered. For example, the transference of the young animal from dependence to independence, particularly at weaning, makes its chances of survival progressively more dependent on its own physical abilities to find food, escape from predators and compete successfully with its contemporaries. The characteristic post-weaning peaks in bone and muscle growth rates, when compared with the development of nervous and fat tissue, is an example of the way in which functional priorities affect the relative proportions of the tissues of the animal.

If the theme of form being determined by functional requirements

196

is pursued, it is clearly reasonable to postulate that the stage at which form is most critical in the mammal is somewhere between puberty and early maturity. This is the stage at which the stress of reproduction, or competition for the right to reproduce, is superimposed on the animal's existing functional requirements, i.e. to survive itself, and is the stage at which functional competence must be at its maximum. It would be reasonable to expect that in the competitive environment of natural selection, only those genotypes capable of reaching puberty with the essential functional relationships between body components unimpaired by the hazards of the environment, will be represented in the next generation. This view was advanced by Darwin over a century ago when he suggested that the most important part of the selection process occurs when birds or other animals are 'nearly grown up' (Darwin, 1859). After maturity, form is to some extent less critical as factors such as experience tend to offset impaired physical efficiency.

The great importance of form from puberty to maturity in terms of functional competence and representation in the next generation, suggests that it will be under very close genetic control. By the same reasoning it is also apparent that the pattern of body development, which eventually culminates in this 'target' conformation of 'climax' form, should also be under close genetic control.

Huxley (1932) was the first to demonstrate experimentally the fixed nature of the relationships between body components during growth. He showed that within a species or genotype the weight of a particular tissue or organ was virtually determined by the total weight of the animal, the relationship being of the form

$$y = bx^\alpha$$

where y = weight of organ or part, x = weight of animal (or another appropriate independent variable), α = growth coefficient of the organ.

This equation, which is a simple power function, was shown to have a wide application in biological growth and is consistent with the foregoing argument that morphogenetic changes will be of a regular and fixed kind. There are several examples in the literature of the need for caution in applying this formula too extensively, for example Reeve and Huxley (1945). It should be noted, therefore, that although the allometric equation provides a valuable and simple mathematical approximation for describing differential growth, there are no intrinsic biological laws which make it apply exactly.

PATTERNS OF 'NORMAL' DEVELOPMENT

The changes which take place in the general conformation of animals from birth to maturity were investigated in farm animals by the Cambridge School of workers, the results being summarized by Pálsson (1955). The order of maturity which was demonstrated is well known, with nervous tissue, bone, muscle and fat developing in that order. Head and distal regions were regarded as early maturing and proximal regions, including ribs and loin, late maturing. It is interesting to note that the application of Huxley's allometry equation enables the maturity of a part or organ in relation to another determining part of the animal, or indeed the whole animal, to be expressed mathematically. This is illustrated in Table 1, where growth coefficients calculated by Elsley, McDonald and Fowler (1964) from the data of McMeekan (1940a, b and c) and Pálsson and Vergés (1952) are presented.

Table 1

Growth coefficients of bone + muscle in different body parts relative to total bone + muscle. (From Elsley, McDonald and Fowler, 1964, by courtesy of Oliver and Boyd)

	Pigs	Sheep	
Bone + muscle of:			
Head and neck	0·95	0·84	Early maturing
Fore-limbs	0·96	0·97 ⎫	Intermediate in relative maturity
Hind-limbs	1·00	0·99 ⎬	
Thorax	1·00	1·06 ⎭	
Pelvis and loin	1·11	1·14	Late maturing

Commenting on such systematic changes, Pálsson (1963) made the important point that the order in which the various parts and tissues develop is much the same in all species of farm animals for it is based on the functional importance of the parts or tissues for survival of the individual.

THE EVALUATION OF BODY DEVELOPMENT

There are wide differences of opinion on what should constitute the guide lines for assessing differences in body conformation. The remainder of this paper is concerned mainly with the problems of evaluating body development in nutritional rather than genetic studies. It should be noted that although the problems of geneticist and nutritionist in this context are similar, they are not identical. Thus, when the suggestion is made that a particular relationship between body components cannot be influenced by nutrition, it should not be construed to mean that it would be unalterable by breeding also.

Before considering in detail the problems of the nutritionist in evaluating body development, it will be helpful to introduce into the discussion a further biological principle which was first expressed towards the end of the last century by Bernard. He suggested that vital mechanisms have one object, which is preserving constant the internal environment (Bernard, 1878). This idea was adapted by Cannon (1929) and given the respectable title of 'physiological homeostasis'. Like all principles of such a general nature it needs applying with some care, but it provides a valuable starting point for describing the response of the animal to environmental insults, particularly nutritional ones.

If we were now to condense the principles mentioned earlier in this paper along with this latter one into a single basic proposition it would be this:

'That the animal tends to adjust to environmental (nutritional) changes in such a way that the vital functional relationships between essential body components are preserved, or modified to a form which gives the animal its best chance of survival and successful reproduction.'

It is in the light of this proposition that some major experiments on the effects of nutrition on body development will be considered.

The first example is that of McMeekan (1940a, b and c) who grew pigs along four different growth curves by adjusting feed intakes. The actual growth curves are shown in *Figure 1*. These patterns were designated High-High, High-Low, Low-High and Low-Low, the words describing the plane of nutrition from birth to 16 weeks and 16 weeks to slaughter at 200 lb respectively. The dissection data were assessed in various ways using live weight, age, or the treatments expressed as a percentage of the Low-Low treatment, as the basis for comparison. McMeekan showed clearly that when development was considered from any of these base-lines, then large differences in the relationships of body components had occurred in response to nutritional treatment. One of the main findings was that the principle of allometry relating the weight of an organ or part to the weight of the animal as a whole was not tenable for animals on different nutritional regimes.

This finding was confirmed in sheep by Pálsson and Vergés (1952) and, as a result of these two experiments, the Cambridge School advanced the hypothesis that:

(1) Under-nutrition penalized the growth of body components differentially in the reverse order of their maturity; late maturing tissues and parts were the most affected.

(2) That those body components having their maximum growth intensity at the time of restriction were the most retarded (Pálsson, 1955).

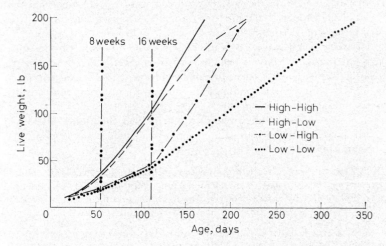

Figure 1. *Actual growth curves of pigs (From McMeekan, 1940a, b and c)*
By courtesy of Cambridge University Press

These generalizations went some way towards rationalizing the facts. For example, early maturing head parts are manifestly resistant to under-nutrition and become disproportionately large when it is prolonged. In the same way the 'later maturing' fat tissues are the first to be reduced when the level of feeding is lowered, a fact which receives recognition in the accepted method of producing bacon pigs by restricted feeding towards the end of the growing period.

However, several features of these experiments could not be explained in these terms. The ratios of bone to muscle were virtually unaltered by the treatments. When bone weight is expressed as a percentage of muscle weight the values for the sheep range between 25·7 per cent for the High-High and 23·4 per cent for the Low-Low, and for the pigs between 27·2 per cent for the High-High and 25·0 per cent for the High-Low. If any principle were to be deduced from these results as they stand, it would be that low levels of feeding in later life tend to increase the proportion of muscle relative to bone, which is the reverse of the hypothesis advanced by the Hammond school as muscle is later maturing than bone.

The clue to resolving such apparent anomalies in the data was provided by Wallace (1948) and Wilson (1954). Wallace suggested that had McMeekan's data been analysed on a within-tissue basis

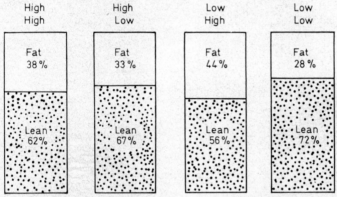

Figure 2. *Carcass proportions* (*Original data from McMeekan, 1940c*)
By courtesy of Cambridge University Press

then many of the treatment effects would disappear, and pointed to the fact that the total weights of tissues were vastly different between the treatment groups. Wilson showed that the main effect of the nutritional treatments was upon the fat tissues. The resulting disproportion between the total weight of the lean tissues for each treatment is illustrated in *Figure 2*.

A similar situation applies to some extent to the sheep data of Pálsson and Vergés (1952). Clearly, the functional significance of the majority of fat tissue is distinct from the function of the lean tissues. The lipid fraction of fatty tissue can be regarded to a large extent as metabolically inert, functioning largely as a concentrated energy store (Benedict, 1938). Some lipid undoubtedly has other functional significance, such as insulation in the pig (Pomeroy, 1955) and in metabolic turnover. It is, however, difficult to distinguish between essential lipid and lipid functioning as stored energy, and the best approximation to the concept of metabolically active mass

is usually taken as fat-free body mass. In turn, the best approximation to this base-line which is available from dissection data alone is
dissectable fat-free mass or lean carcass.

Wilson (1960) incorporated this idea into a plane of nutrition
experiment with Dwarf East African goats. These were killed when
it was estimated that they had equal fat-free carcass weights. The
striking results of this experiment were that, compared on this basis,
there were no very obvious differential effects of treatments on the
relationships between lean tissue components.

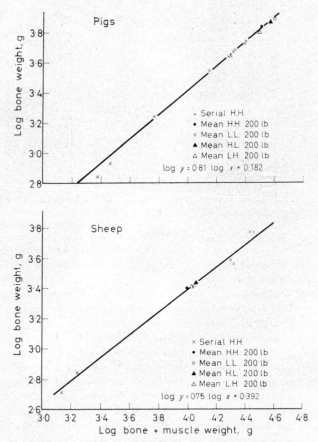

*Figure 3.   Regressions of log bone on log bone + muscle.   Pig data from McMeekan
(1940a, b and c); sheep data from Pálsson and Vergés (1952)*

By courtesy of Cambridge University Press

202

Bone + muscle was the independent variable chosen by Elsley *et al.* (1964) to test the allometric relationships of body tissues and parts using the original Cambridge School data. They showed that an almost perfect allometric relationship persisted between bone and muscle tissues in spite of the very different nutritional treatments; some of these results are shown graphically in *Figure 3*.

In the same way, with the exception of the head, the sheep data demonstrated that the allometric relationship between body parts considered on a dissectible fat-free basis was not disturbed by nutritional treatment.

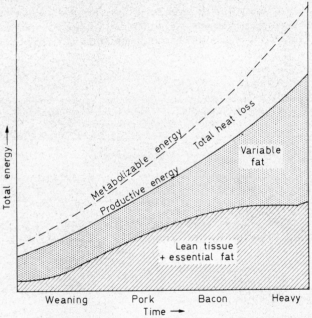

Figure 4. Partition of metabolizable energy in the growing pig on unrestricted feeding

This inquiry pointed to the fact that the animal has a tremendous capacity to buffer the functionally important relationships between tissues and parts from changes in dietary energy intake, by adjusting the amount of variable fat in the body. A simple model illustrating this point in the case of the pig is given in *Figure 4*. This shows qualitatively how the cumulative metabolizable energy is partitioned between heat loss and the energy retained in the two functionally distinct tissue groups. When the level of productive energy is reduced at any stage, the effect is mainly on the variable fat component with only a comparatively small reduction in the rate of energy retention in the essential 'lean' tissues, the intrinsic form and function of which are preserved.

This simplification, however, does not fit all the facts. Pálsson and Vergés analysed some of their data on the bones and showed that there were significant differences in the growth coefficients of head bones when related to the weight of the total skeleton, between lambs grown on a continuously high and those on a continuously low level of feeding. A more complete examination of their data by statistical analysis of the type used by Elsley *et al.* (1964), shows that even after adjustment for differences in total bone weight, significant effects of treatments still occur; some of these results are shown in Table 2.

Table 2

Skeletal parts of sheep, adjusted to the same *total bone weight* as, and expressed as percentages of, the High–High group. (Original data from Pálsson and Vergés, 1952, by courtesy of Cambridge University Press)

	Treatment group					
	High–High	Low–Low	High–Low	Low–High	Approximate S.E.	Sig.
Head	100	132	115	110	$\pm 2 \cdot 1$	$P < 0 \cdot 001$
Ribs + sternum	100	95	93	111	$\pm 3 \cdot 9$	$P < 0 \cdot 05$
Humerus	100	91	98	92	$\pm 2 \cdot 0$	$P < 0 \cdot 01$
Radius-ulna	100	91	95	92	$\pm 2 \cdot 0$	$P < 0 \cdot 01$
Femur	100	95	102	93	$\pm 3 \cdot 5$	N.S. ($P < 0 \cdot 1$)
Tibia-fibula	100	102	98	94	$\pm 3 \cdot 1$	N.S.
Total cannons	100	99	96	97	$\pm 3 \cdot 1$	N.S.

Such results show clearly that some relationships within the essential tissue component are altered by nutrition, but these changes are not adequately explained by either of the principles advanced by the Cambridge School.

The major effect is plainly upon the bones of the head which continue to grow even during periods of restricted nutrition. Once, however, an effect is established for the bones of the head, the independent variable, total skeletal weight, can no longer be regarded as completely independent. The significant results in the other bones may be merely a mirrored effect of the change in relative size of the head bones.

In our search for a unifying principle at this stage, we shall refer to our earlier proposition that changes in form are likely to be related to functional priorities.

Some parts of the animal operate as a functional unit, for example the legs for locomotion, or the head acting primarily as a protective container for the brain. *Figure 5* shows a preliminary breakdown of a representative farm animal into functional units. The functions of head and legs are to a large extent independent, but the functions of the components of the legs are very much dependent on each other for the limbs to function with maximum efficiency.

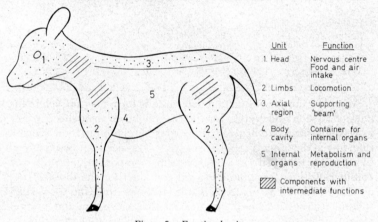

Unit	Function
1. Head	Nervous centre Food and air intake
2. Limbs	Locomotion
3. Axial region	Supporting 'beam'
4. Body cavity	Container for internal organs
5. Internal organs	Metabolism and reproduction
▨ Components with intermediate functions	

Figure 5. Functional units

Thus, during periods of under-nutrition the animal is not placed at a functional disadvantage by increases in the relative size of its head whilst the rest of its body remains in strict proportion. The increase in head size is associated with the growth of components with a priority survival function, particularly the brain.

In order to test whether or not it is feasible to consider the 'functional unit' as the major determinant of component size, some of the Pálsson and Vergés data have been re-analysed using as base-lines variables which represent a functional entity, e.g. total weight of limb bones. Table 3 shows how comparison of the proportions of the bones of the sheep limbs, based on total limb bone weight, virtually eliminates the treatment differences which appear in Table 2 when total skeleton was taken as the independent variable. This supports the idea that the concept of allometric growth is most useful when it is applied to anatomical components which are integral parts of a particular functional unit.

Table 3

Limb bone weights of sheep, adjusted to the same *total limb bone* weight as, and expressed as percentages of, the High–High group. (Original data from Pálsson and Vergés, 1952, by courtesy of Cambridge University Press)

| | Treatment group | | | | | |
	High–High	Low–Low	High–Low	Low–High	Approximate S.E.	Sig.
Humerus	100	99	101	100	± 1.9	N.S.
Radius-ulna	100	98	98	99	± 1.2	N.S.
Femur	100	103	105	100	± 2.3	N.S.
Tibia-fibula	100	104	100	102	± 2.0	N.S.
Total cannons	100	105	98	103	± 2.4	N.S.

Another example of function determining form may be seen in the data of Pálsson and Vergés for the rib weights of their sheep. The Low-High sheep had heavier rib-weights in relation to total skeleton than any of the other treatment groups. This need not be explained by reference to the ribs as a late maturing component, but rather by considering their function as a rigid container for the thoracic and anterior abdominal organs. Table 4 shows that the Low-High treatment produced sheep with heavier thoracic organs and fore-gut weights than the other treatments and it follows that these organs would require a larger container. The results in Table 4 confirm that this is the case, for when rib weights are expressed as a percentage of thoracic organ + fore-gut weight there are no differences between treatments.

A parallel example to this can be cited from the work of Lodge and Heap (1967), who showed that during pregnancy the muscle in the abdominal wall of sows increased relative to the rest of the muscles. The proper functional base-line for considering such a change as this is, of course, the size of the contents of the abdominal cavity and not the total muscle.

The theory advanced in this paper is not far removed from the idea of Guttman and Guttman (1965). They suggest that two

features that are closer to each other in terms of structural or pro-
cessual relations should also be closer in the sense of statistical
correlations. The theory is also analogous to the grouping of
muscles suggested by Butterfield and Berg (1966a and b) but which,
in their case, refers particularly to the economic significance of the
muscle group and to its 'relative growth impetus'.

Table 4

The relationship of rib + sternum weights to the weight of the organs of the thorax
and anterior alimentary tract. (Original data from Pálsson and Vergés, 1952)

| | Treatment group | | | | | |
	High–High	Low–Low	High–Low	Low–High	Approximate S.E.	Sig.
a Wt. of ribs + sternum (g)	345	335	350	398	—	—
b Wt. of organs of the thorax + anterior alimentary tract (g)	2,393	2,445	2,438	2,923	—	—
$\dfrac{a}{b} \times 100$	14·4	13·7	14·3	13·6	—	—
Values of a, adjusted to the same value for b as, and expressed as percentages of, the High–High group	100	96	102	95	$\pm 7\cdot4$	N.S.

THE CHOICE OF INDEPENDENT VARIABLE

The question which now arises is which independent variable should
be chosen for evaluating body development? To this there can be
no single answer, as the correct way in which to evaluate body
development depends on the question being asked.

If the question is economic, then, so also should be the base-line;
if the question is anatomical, physiological, or chemical, the base-
line should be selected accordingly. For example, if the measurements are linear ones taken on the growing animal, then an
appropriate base-line to take is that used by Lyne and Verhagen
(1957) of linear equivalence $\sqrt[3]{W}$. If the question is a detailed
anatomical one, then it is best answered by regression analysis which
allows a number of different base lines to be selected. To obtain

the maximum benefit from such an analysis the experiment should be designed with this objective in view and a serial slaughtering technique adopted.

The experimental 'cost' of using regression analyses rather than analyses of the difference between means is small and can be roughly assessed as the cost of the extra degree of freedom required for each treatment regression line; in other words, one further replication.

In recent investigations of my own (Fowler, 1965 and 1968) I compared three different independent variables when the slaughter weights of pigs were deliberately spread to facilitate regression analysis. The object of the experiment was to assess the effect of different feeding levels from 10 lb live weight to slaughter at a mean live

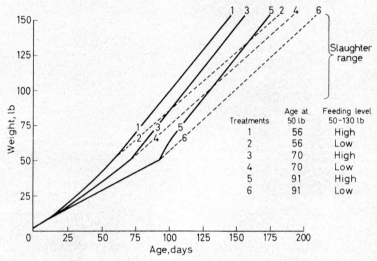

Figure 6. Growth curves of pigs showing effect of feeding level

weight of 130 lb. The mean growth curves produced by the six feeding patterns are illustrated in *Figure 6*. The treatments were the factorial combination of three rates of growth to 50 lb live weight which were 56, 70 and 91 days from birth respectively, and two levels of feeding from 50 lb to slaughter designated 'High' and 'Low'. Pigs were slaughtered at equal weight intervals between 105 and 155 lb live weight.

Regression lines were computed for three different independent variables, which were weight of total carcass, weight of muscle in the ham, and weight of the long bones of the hind limb, femur +

tibia-fibula. The treatment regression lines were used to calculate the predicted values of a number of dependent variables at a given value of each independent variable, this value being selected at, or close to, the overall mean. Two examples of the results are given in Table 5. The dependent variable in the first example is the

Table 5

Two measures of pig carcass composition (dissectable fat/muscle of ham × 100 [F/M × 100] and weight of femur) adjusted to equal values of three different independent variables

Independent variable	Feed level 50–130 lb	Age at 50 lb (days) 56	70	91	Approximate S.E.	Sig.
		F/M × 100				
Total carcass	High	51	41	37	1·7	$P < 0.001$
	Low	36	34	29		
Ham muscle	High	46	42	37	2·4	$P < 0.001$
	Low	37	34	29		
Tibia-fibula + femur	High	50	42	38	1·7	$P < 0.001$
	Low	37	34	28		
		Weight of femur (g)				
Total carcass	High	184	173	166	± 10·2	$P < 0.001$
	Low	190	179	200		
Ham muscle	High	178	184	167	± 9·1	N.S.
	Low	186	174	174		
Tibia-fibula + femur	High	189	191	180	± 8·3	N.S.
	Low	183	188	183		

dissectible fat in the ham expressed as a percentage of the dissectible muscle in the ham. The results show that the nutritional treatments greatly affected the ratio of fat to lean tissue in this joint and this is apparent whichever independent variable is used. In the second example a dependent variable which is a component of the dissectible fat-free body, namely the femur, has been selected. The results show that differences between treatments depend on which independent variable has been used in adjusting femur weights. When the independent variable includes fat, as is the case when carcass weight is used, then there is a profound effect of the treatments, fatter pigs obviously having smaller femurs. However, when femur weights are adjusted using independent variables which are themselves components of the dissectible fat-free body, i.e. ham muscle or tibia-fibula + femur weights, then it is clear that there are little or no differential effects of the treatments on the relationship of the femur weights to these components.

CONCLUSION

Many issues have unfortunately to be omitted from a paper of this length. Gravimetric measures of tissues and parts are more informative than linear measurements alone, but are still comparatively crude. They can give little information in depth of the chemical, physical, or mechanical properties of the body components and much remains to be done in this field. Even so, existing studies allow some first approximations to be made of the principles which regulate the form of the animal during growth. From the point of view of the animal nutritionist, it is useful to bear in mind that conditions of prolonged under-nutrition alternating with periods of good nutrition are characteristic environmental situations in the wild state, to which even domesticated stock have a long history of adaptation. Level of feeding experiments reveal the nature of these adaptive mechanisms in the animal and we can reasonably expect them to be based on functional priorities. It is apparent that we are greatly indebted to our chairman, Dr. Pálsson, and other members of the Cambridge School, for their outstanding work in the field of growth and development, and I hope that this paper in some small measure is an extension to the great wealth of knowledge brought to light by the genius of the late Sir John Hammond and of his colleagues.

ACKNOWLEDGEMENTS

I should like to express my thanks to Dr. F. W. H. Elsley and Mr. I. McDonald for their generous help in the preparation of this paper.

REFERENCES

Bernard, C. (1878). *Leçons sur les phenomenes de la vie.* Vol. 2, p. 7. Paris
Bertalanffy, L. von (1960). In: *Fundamental Aspects of Normal and Malignant Growth.* p. 137. (Ed. by W. W. Nowinski.) Amsterdam; Elsevier
Benedict, F. G. (1938). *Vital Energetics.* Publ. No. 503. Carnegie Institution of Washington
Brody, S. (1945a). *Bioenergetics and Growth.* p. 484. New York; Reinhold

Brody, S (1945b). *Bioenergetics and Growth.* p. 575. New York; Reinhold

Butterfield, R. M. and Berg, R. T. (1966a). *Res. vet. Sci.* **7**, 326

—— (1966b). *Res. vet. Sci.* **7**, 389

Cannon, W. B. (1929). *Physiol. Rev.* **9**, 399

Darwin, C. (1859). *The Origin of Species.* London; John Murray

Elsley, F. W. H., McDonald, I. and Fowler, V. R. (1964). *Anim. Prod.* **6**, 141

Fowler, V. R. (1965). *Anim. Prod.* **7**, 284 (abstr.)

— (1968). 'Effects of different rates of growth to 50 lb. followed by high or low levels of feeding for pigs killed in the range 105–155 lb live-weight.' (In preparation)

Guttman, R. and Guttman, E. (1965). *Growth* **29**, 219

Huxley, J. S. (1932). *Problems of Relative Growth.* London; Methuen

Laird, A. K., Tyler, S. A. and Barton, A. D. (1965). *Growth* **29**, 233

Lodge, G. A. and Heap, F. C. (1967). *Anim. Prod.* **9**, 237

Lyne, A. G. and Verhagen, A. M. W. (1957). *Growth* **21**, 167

McMeekan, C. P. (1940a). *J. agric. Sci. Camb.* **30**, 276

— (1940b). *J. agric. Sci. Camb.* **30**, 387

— (1940c). *J. agric. Sci. Camb.* **30**, 511

Pálsson, H. (1955). In: *Progress in the Physiology of Farm Animals.* Vol. 2, p. 430. (Ed. by J. Hammond.) London; Butterworths

— (1963). *Proc. 38th Meet. Br. Soc. Anim. Prod. Mimeo*

— and Vergés, J. B. (1952). *J. agric. Sci. Camb.* **42**, 1

Pomeroy, R. W. (1955). In: *Progress in the Physiology of Farm Animals.* Vol. 2, p. 395. (Ed. by J. Hammond.) London; Butterworths

Reeve, E. C. R. and Huxley, J. S. (1945). In: *Essays on Growth and Form.* p. 121. (Ed. by W. E. Le Gross Clark and P. B. Medawar.) Oxford; Clarendon Press

Thompson, D'Arcy W. (1917). *On Growth and Form.* Cambridge; University Press

Wallace, L. R. (1948). *J. agric. Sci. Camb.* **38**, 367

Wilson, P. N. (1954). *J. agric. Sci. Camb.* **45**, 67

— (1960). *J. agric. Sci. Camb.* **54**, 105

211

THE EFFECT OF GROWTH RATE OF MUSCLE IN CATTLE ON CONFORMATION AS INFLUENCED BY MUSCLE-WEIGHT DISTRIBUTION

R. M. BUTTERFIELD* and E. R. JOHNSON

Department of Veterinary Anatomy, University of Queensland, Australia

IF THE distribution of muscle over the bodies of cattle varied this could affect significantly the size of some parts of their bodies and the relative weights of the cuts of meat from their carcasses. Thus it is important that the effect of any factor, such as rate of growth, which could influence the relative size of muscles (muscle-weight distribution) should be understood.

Although Hammond (1935) showed that external form or conformation of animals of 'improved' meat breeds could be made to approach that of unimproved breeds by poor nutrition, the structural changes involved were not clearly demonstrated. Changes in the external form of meat animals have been claimed to result, at least in part, from differences in the muscle-weight distribution (Hammond, 1960), although Wallace (1948) showed that it was unlikely that levels of nutrition would affect significantly the relative size of muscles at any given weight of total muscle, a view subsequently supported by Elsley, McDonald and Fowler (1964).

Dissection of individual muscles has provided a means to study the growth of the musculature in considerable detail. From the growth patterns of individual muscles, Butterfield and Berg (1966a) showed that most differential growth of those muscles with diphasic growth patterns occurs in early postnatal life. Most of these muscles change their rate of growth to the same rate as that of total muscle soon after birth but, as shown by Butterfield and Berg (1966c), the time taken to attain this uniform rate is considerably modified by changes in the plane of nutrition.

McMeekan (1940) showed with pigs that when considered over

---

* Present address: Department of Veterinary Anatomy, University of Sydney, Sydney, New South Wales.

an age interval or at equal body weight, tissues with the highest growth impetus were those likely to be retarded most by inadequate nutrition. However, Wallace (1948) contended that to determine whether or not there had been a differential effect within a tissue, it was necessary to compare proportionate development at equal tissue weights rather than at equal body weight or at the same age.

AN ANATOMICAL APPROACH TO STUDIES OF MUSCLE-WEIGHT DISTRIBUTION

It is only by weighing dissected muscles that the muscle-weight distribution of individual animals can be shown. From data accumulated by the dissection of many animals slaughtered at different points along the growth curve of the species, the approximate growth patterns of muscles can be derived.

To illustrate the influence that different levels of nutrition may have on conformation, through their effects on the relative growth of muscles, grouping of muscles according to anatomical location is desirable. However, the demonstration of nutritional effects on such groups is difficult because muscles of different and possibly compensating growth patterns are included in the same anatomical groups, so that nutritional or other effects may be confounded. The effect of nutrition can, therefore, be better illustrated by the use of individual muscles or, better still, groups of muscles with similar growth patterns. The muscles grouped according to their growth patterns are termed 'impetus groups' and these are used in this paper to enunciate the effect of growth rate on relative growth of muscles and hence on muscle-weight distribution. Further, 'standard muscle groups' (Butterfield, 1963; Butterfield and Berg, 1966b), in which the anatomical location determines the grouping, are used to relate the principles involved to changes observed in conformation under different planes of nutrition.

Butterfield and Berg (1966a) described growth patterns of the individual muscles by the use of growth coefficients (Huxley, 1932) over five age phases. These growth patterns were classified into six groups, of which three were monophasic (low, average and high impetus) and three were diphasic (low-average, average-high and high-average impetus). If the rate of growth of a muscle was uniformly less than the rate of growth of total muscle (that is 'b' was less than 1·0) it was classified as 'low impetus', if 'b' was greater than 1·0 the muscle was classified as 'high impetus' and if not significantly different from 1·0, as 'average impetus'.

A muscle was allotted a diphasic classification if the 'b' values were significantly different for the various phases. For example, a muscle with its 'b' value significantly greater than 1 in the first phase and not different from 1 in subsequent phases was classified as 'high-average impetus'. Table 1 shows that for most muscles the change of impetus occurred early in life.

Table 1

The number of muscles and the percentage of total muscle with diphasic growth patterns whose 'b' values are significantly different from 1, in five age phases

Phase		1	2	3	4	5
Range (days)		0–84	85–365	366–730	731–1,460	1,460+
Muscles with 'b' values significantly different from 1	Number of muscles	21	8	6	6	3
	Percentage of total muscle weight (adult)	41·5	5·5	8·9	3·9	9·0

IMPETUS GROUPS

When all the muscles with similar growth patterns are grouped together, so-called impetus groups are formed. These groups are ideal to show the influence of growth rate of the entire musculature on the relative growth rates of muscles and hence on muscle-weight distribution. However, as muscles in any one growth impetus group may be anatomically remote from each other, it is difficult to relate these findings to changes in conformation. Generally, however, the muscles in the low and low-average impetus groups are small and located either distally on the limbs or deep in the carcass; those which are in the high-average group are either large muscles of the hind quarters and back or muscles of the abdominal wall. The average impetus muscles and those which are unclassified in any of the growth patterns are scattered throughout the body.

THE EFFECT OF GROWTH RATE ON THE RELATIVE GROWTH OF THE IMPETUS GROUPS

Pálsson (1963) stated that the tissues making most proportionate gain in weight over any age interval are likely to be those most influenced by rate of growth. Thus the high-average impetus group, which has the highest 'b' value in the immediate postnatal period, is ideal to demonstrate nutritional effects. This group contains many of the biggest and economically most important muscles of the body.

Figure 1, based on the findings of Butterfield and Berg (1966c), shows that rate of growth of the musculature of calves had no effect on the proportionate changes of the group of muscles with a high-average impetus when compared at the same muscle weight. Wider

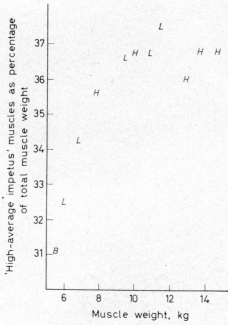

Figure 1. *Percentage of total muscle-weight represented by the high-average impetus group of muscles related to half-carcass muscle weight in fast ('H') and slow ('L') grown calves. Each symbol is the mean value for 2 calves. 'B' = mean value for 7 calves slaughtered when 1 day old (After Butterfield and Berg, 1966c)*

differences in rate of growth, in an unpublished experiment of Butterfield and Johnson, produced a similar result even though calves of one group were held at their birth weight for 70 days.

Wallace (1948) suggested that when comparisons are made at equal tissue weight, the normal weight relationships will be shown to exist within the tissue irrespective of rate of growth. These findings support that claim.

When comparison of the relative rate of growth of this group of muscles was made on an age basis (Butterfield and Berg, 1966c) it was clear that in calves which grew slowly, muscles of this group took a longer time to achieve their adult proportion of total muscle than in calves which grew more rapidly (*Figure 2*).

These results show that within the range from birth until muscle weight had been approximately doubled (the high impetus phase),

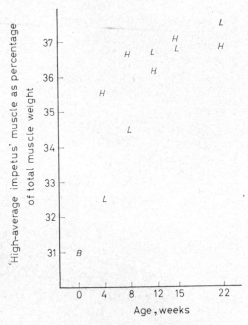

Figure 2. Percentage of total muscle weight represented by the high-average impetus group of muscles related to age in fast ('H') and slow ('L') grown calves. Each symbol is the mean value for 2 calves. 'B' = mean value for 7 calves slaughtered when 1 day old (After Butterfield and Berg, 1966c)

rate of growth of the musculature had no effect on the percentage of total muscle represented by the high-average group when animals of the same total muscle weight were compared. However, during the high impetus phase of growth of this group of muscles, comparisons made at the same age showed that these muscles had been more retarded than the whole musculature by slow growth rate.

Muscles with a low growth impetus pattern which constitute a higher proportion of the total musculature at birth than in the adult,

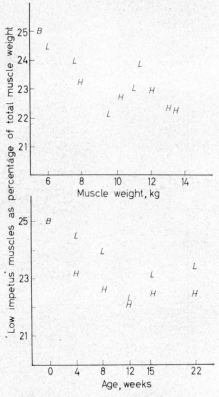

Figure 3. Percentage of total muscle weight represented by the 'low' impetus group of muscles related to half-carcass muscle weight and to age in fast ('H') and slow ('L') grown calves. Each symbol is the mean value for 2 calves. 'B' = mean value for 7 calves slaughtered when 1 day old (Based on data from experiments of Butterfield, Pryor and Berg (1966))

take longer to decline to their adult values in animals retarded by poor nutrition. Again, at any particular weight such muscles constitute a constant proportion of total muscle (*Figure 3*).

The group of muscles which has an average growth impetus is unaffected in its relationship to total muscle by rate of growth on the basis of either weight or age (*Figure 4*).

If the rate of growth of total muscle is increased, then there is a similar increase in the growth rate of this group. If it is slowed this

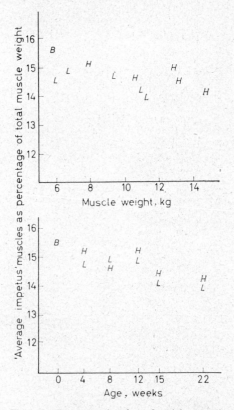

Figure 4. Percentage of total muscle weight represented by the 'average' impetus group of muscles related to half-carcass muscle weight and to age in fast ('H') and slow ('L') grown calves. Each symbol is the mean value for 2 calves. 'B' = mean value for 7 calves slaughtered when 1 day old (Based on data from experiment of Butterfield, Pryor and Berg (1966))

group is similarly slowed and therefore it is not possible to show any differential effect of nutrition on this group of muscles at any age or weight. Similarly, muscles with a diphasic pattern are not differentially affected during the average impetus phase of their growth.

From the findings described the following principles emerge:

(1) Sub-optimal nutrition during a high impetus growth phase of muscles retards these rapidly-growing muscles more than total muscle when considered on an age scale, but the high impetus muscles and total muscle are retarded uniformly, when considered on a total muscle weight basis.

(2) Sub-optimal nutrition during a low impetus phase retards these slow-growing muscles less than total muscle when considered on an age scale but the low impetus muscles and total muscle are retarded uniformly when considered on a total muscle weight basis.

(3) Sub-optimal nutrition during an average impetus phase does not affect the relative growth rates of these muscles and total muscle when considered on either an age or a weight basis.

(4) It follows, therefore, that the rate of growth of an animal will be found not to have influenced the relative growth rate of any muscles when animals containing the same muscle weight are being compared. It must be noted, however, that this statement relates only to actively growing animals and cannot be extended to cover weight loss from the musculature. However, this is probably an important consideration only during or immediately following weight loss, for, as shown by Yeates (1964) and Butterfield (1966), recovery of muscle weight following weight loss restores the original characteristics of the musculature to a high degree.

Because of the less distinct patterns of some of the anatomical groups, precise, direct demonstration of the effects of rate of growth on all groups must await the dissection of very large numbers of calves.

However, having developed the principles set out above, it is now possible to apply them to the growth patterns of the anatomical groups (Table 2), to indicate how these groups may be affected by different rates of growth when compared at the same age.

In these discussions the immediate postnatal phase is regarded as being the period during which the muscle-weight is about doubled. The length of time over which this extends varies of course with the rate of growth but is about 12 weeks in well-fed calves reared by the bucket feeding of milk.

The results in Table 3 may be expected when calves are grown slowly compared with calves grown more rapidly.

219

Table 2

Growth pattern of anatomically defined Standard Muscle
Groups (From Butterfield and Berg, 1966b, by courtesy
of Blackwell Scientific Publications)

Muscle group	Growth pattern
Proximal pelvic limb	High—average or low
Distal pelvic limb	Low
Surrounding spinal column	Average
Abdominal wall	High—average or high
Proximal thoracic limb	Low—average
Distal thoracic limb	Low—average or low
Thorax to thoracic limb	High
Neck to thoracic limb	Average—high
Neck and thorax	Low—average

Table 3

Effect of a slow rate of growth on the relative growth of muscle groups compared
with the relative growth of these same groups of muscles in calves grown rapidly.
All comparisons are made at the same age

Number	Muscle group — Location	Immediate postnatal period*	Remainder of growth period
1	Proximal pelvic limb	Retarded	Nil or slightly increased
2	Distal pelvic limb	Increased	Increased
3	Surrounding spinal column	Nil	Nil
4	Abdominal wall†	Retarded	Nil or slightly retarded
5	Proximal thoracic limb	Increased	Nil
6	Distal thoracic limb	Increased	Nil or slightly increased
7	Thorax and thoracic limb	Retarded	Retarded
8	Neck to thoracic limb	Nil	Retarded
9	Neck and thorax	Increased	Nil

* From birth until birth weight doubled
† This group is likely to be more affected by the physical form of the diet than by the rate of growth
(Butterfield and Johnson, unpublished)

During the immediate postnatal phase there is a large amount of
differential growth with some groups increasing, some decreasing,
and only two groups growing at the same rate as the total muscle.

Once this phase is over, most muscles and groups of muscles grow at the same rate as the total muscle.

It is apparent, therefore that differences in muscle-weight distribution caused by differences in rate of growth, will be greatest soon after birth and will disappear as the musculature gains weight.

DISCUSSION

All available evidence indicates that the growth rate has no effect on the relative weights of muscles when comparisons are made at equal total muscle weights. Differences observed in the conformation of cattle of similar weight cannot be ascribed to the effects of the growth rate of the total muscle on relative development of muscle.

It seems that differences in the distribution of muscle weight caused by any factor are small except those associated with weight and with starvation. The changes which occur with age (Butterfield, 1963, 1965) and with starvation (Butterfield, 1966) have been discussed elsewhere, but it is apparent from recent investigations (Elsley, McDonald and Fowler, 1964; Butterfield and Berg, 1966c), that changes related to age in these experiments should more truly have been related to weight. Closer attention to Wallace's (1948) findings would have suggested the use of weight instead of age as a basis for the study of relative growth changes.

It is apparent that the tendency within the cattle industry to ascribe observed differences in conformation to the differences in distribution of muscle-weight, is based more on the possible financial advantages to the owner of an individual animal or to the protagonists of a breed, or on the idealistic philosophies of nutritionists, than it is on anatomical facts. Many statements have been recorded which suggest superior muscle-weight distribution of muscular hypertrophied cattle based on the external assessment of conformation (Black, 1936; Schafer, 1937; Shrode and Lush, 1947; Kidwell, Vernon, Crown and Singletary, 1952). The only available information from total dissection of these anomalous animals indicates that although total muscle and muscle:bone ratio are increased, their muscle-weight distribution closely resembles that of normal cattle (Pomeroy and Williams, 1962; Butterfield, 1966; Butterfield and Johnson, unpublished). *Figure 5* shows an exaggerated case of muscular hypertrophy in a Santa Gertrudis cross bull. Dissections of similar animals have revealed only small differences in the distribution of muscle-weight compared with normal animals. Such evidence suggests that there are more important factors than distribution of muscle-weight that alter the external appearance of animals.

Thus, at standard age in very young calves we may expect to find differences in the distribution of muscle-weight which may be related to the growth rate and which may be responsible in part for

Figure 5. Muscular-hypertrophied bull. The muscle-weight distribution of these animals varies little from normal animals
By courtesy of The Editor, *Queensland Country Life*

the observed differences in the external appearance. However, because these differences in muscle-weight distribution mostly disappear by the time the calf has doubled its birth weight and because calves that are grown slowly retain their characteristic appearance well beyond this time, it follows that differences in muscle-weight distribution contribute little to the differences noted in conformation.

REFERENCES

Black, W. H. (1936). Cited by Kidwell, J. F., Vernon, E. H., Crown, R. M. and Singletary, C. B. (1952). *J. Hered.* **43**, 63

Butterfield, R. M. (1963). *Symp, Carcase Composition and Appraisal of Meat Animals,* Paper 7-1. (Ed. by D. E. Tribe.) Melbourne, C.S.I.R.O.
— (1965). *Proc. N.Z. Soc. Anim. Prod.* **25**, 152
— (1966). *Res. vet. Sci.* **7**, 168
— and Berg, R. T. (1966a). *Res. vet. Sci.* **7**, 326
— — (1966b). *Res. vet. Sci.* **7**, 389
— — (1966c). *Proc. Aust. Soc. Anim. Prod.* **VI**, 298
— Pryor, W. J. and Berg, R. T. (1966). *Res. vet. Sci.* **7**, 417
Elsley, F. W. H., McDonald, I. and Fowler, V. R. (1964). *Anim. Prod.* **6**, 141
Hammond, J. (1935). *Emp. J. exp. Agric.* **3**, 1
— (1960). *J. Inst. Corn agric. Merch.*, Autumn, 1960
Huxley, J. S. (1932). *Problems of Relative Growth,* 1st edn. London; Methuen
Kidwell, J. F., Vernon, E. H., Crown, R. M. and Singletary, C. B. (1952). *J. Hered.* **43**, 63
McMeekan, C. P. (1940). *J. agric. Sci., Camb.* **30**, 387
Pálsson, H. (1963). *Proc. 38th Meet. Br. Soc. Anim. Prod.* Mimeo
Pomeroy, R. W. and Williams, D. R. (1962). *Anim. Prod.* **4**, 302
Schafer, W. (1937). *Züchtungskunde* **12**, 423
Shrode, R. R. and Lush, J. L. (1947). *Advances in Genetics.* Vol. I. New York; Academic Press
Wallace, L. R. (1948). *J. agric. Sci., Camb.* **38**, 367
Yeates, N. T. M. (1964). *J. agric. Sci., Camb.* **62**, 267

THE EFFECT OF GROWTH RETARDATION ON POSTNATAL DEVELOPMENT

E. M. WIDDOWSON

*Medical Research Council Infant Nutrition Research Division,
Dunn Nutritional Laboratory, University of Cambridge*

To GROW is to increase in size, according to the Oxford English dictionary, and development is 'a gradual unfolding'. I interpret my assignment, therefore, as a consideration of what happens to the 'gradual unfolding' of the morphology, chemistry and physiology of the body that takes place after birth if the normal increase in size of the young organism is for some reason inhibited. There are many different reasons for such a retardation of growth, of which nutritional and hormonal causes are among the most important. Within the category of nutritional causes there are many different dietary essentials, a shortage of almost any one of which would hinder growth. Considerations here will be limited to retardations due to a shortage of food, no matter how good the food, and hence to one in which calories are the limiting factor; it will be assumed that there is enough food to enable some growth to take place, so that the question of complete or nearly complete starvation will not arise.

It can be safely said that the earlier in development the animal suffers from undernutrition the more likely is the undernutrition to have a permanent effect. The food supply of the rat, which is born in a very immature state, can be altered experimentally after birth much earlier in its developmental career than it can in the guinea-pig, which is much more mature at birth. Differences in maturity at birth explain why species differ in their resistance to a period of undernutrition soon after birth and the guinea-pig can be affected at the same stage of development as the rat only by undernutrition *in utero*.

Species differ greatly in their rate of growth after birth. The human baby takes 5 or 6 months to double its birth weight, while the piglet weighs one third as much as the baby at birth but twice as much as a man by the time it is a year old. The rat comes somewhere in between. This is illustrated in *Figure 1*, which shows the growth of the baby, the pig and the rat during the first year after birth. To make the three species comparable all subsequent weights have been related to a birth weight of unity.

The rat is normally weaned at 3 weeks, when it is six times its birth weight, and it reaches puberty at 6 or 7 weeks. The pig is weaned at 8 weeks, when it is 15 times its birth weight, and it reaches puberty in about 7 months. The baby is weaned at say 6 months, when it is twice its birth weight, and the child does not reach puberty for 12 or 13 years. It is obvious that we can retard the physical

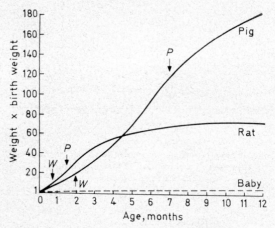

Figure 1. *Gain in weight of the pig and rat during the first year after birth compared with growth of the human baby.* Birth *weight= 1.* W= Weaning. P= Puberty

growth of a pig or a rat far more than that of the human infant could ever be retarded in terms of its final size. Nevertheless, there are certain general principles which apply to every retardation of growth, whatever its magnitude. It has been known for a long time that all parts of the body do not grow equally fast or at the same time. This is illustrated in Table 1, which shows how the weight of a human foetus is made up at 20 weeks gestation, how it has changed at birth, and by adult life. Particularly striking is the increase in skeletal muscle with age and the decrease in brain as a percentage of the body weight. The skin contributes more to the baby's weight because the baby has a bigger surface area in relation to its weight.

Not only do the relative weights of the organs change with age; their chemical compositions change also and at different rates. Some are much nearer their adult composition by the time of birth than others. Table 2 shows that the nitrogen in human heart

Table 1

Percentage contribution of organs and tissues to the body
weight of man at different ages

Tissue	Foetus (20–24 weeks)	Newborn baby	Adult
Skeletal muscle	25	25	43
Skin	13	15	7
Skeleton	22	18	18
Heart	0·6	0·5	0·4
Liver	4	5	2
Kidneys	0·7	1	0·5
Brain	13	13	2

Table 2

Concentration of nitrogen in organs and tissues at birth
expressed as a percentage of the adult value

Heart	86
Liver	80
Kidney	78
Muscle	67
Skin	50
Whole body	67

muscle is already 86 per cent of its adult value at the time of birth, whereas that in skin is only 50 per cent. It is tempting to relate this to functional activity, for the foetal heart has been working actively for nearly 9 months and can be heard doing so from the twentieth week of gestation, whereas skin is relatively quiescent in a metabolic sense till after the baby is born.

Some consideration of these general principles of normal development is necessary as a prelude to the next point, which is that undernutrition does not delay the development of all parts of the body to an equal extent. Organs that develop early, like the heart, the kidneys, the brain, are less affected than the parts that develop late, such as the skin and the skeletal muscles.   The skeleton is in rather a special position and will be referred to again later.

Some experiments which we have made on pigs and rats, may serve to illustrate these and other points.   First is the method of doing the experiments.   To undernourish rats from birth two litters of, say, 9 born on the same day are mixed and 15 returned to one mother and 3 to the other.   The mother suckling 15 provides more milk than the mother suckling 3, so the total weight of the litter increases more, but she does not provide five times as much milk and each rat in the large group grows more slowly than each rat in the group of 3 and weighs only one third or one half as much at weaning (Widdowson and McCance, 1960).   To undernourish rats after weaning at 3 weeks, they are given just enough of the stock diet to keep them alive and to gain weight as slowly as possible (Widdowson and McCance, 1963).   Since male rats normally grow much faster than females, this is a greater deprivation for the males.

Undernutrition of pigs is begun at 10 days of age by weaning them on to a commercial food and then including equal parts of a cereal—fish meal mixture so that the food contains 19 per cent protein.   The animals are given roughly 90 g of this food each day for a year (McCance, 1960; Čabak, Gresham and McCance, 1962).   On this amount of food they gain weight very slowly so that at the end of the year weigh only between 5 and 6 kg.   Their well-nourished litter-mates weight 150–180 kg by this time, so these undernourished pigs are only 3 per cent of their expected weight for age.

This great reduction in growth rate has a profound effect on the development of the bodies of the pigs.   The bodily proportions are abnormal, for some parts of the body continue to grow, albeit slowly, during the year of undernutrition, for example the brain, the skeleton and the teeth, while others, notably the skeletal muscle, have scarcely grown at all and become a smaller percentage of the body weight (Dickerson and McCance, 1964).   These animals have no subcutaneous fat and the subcutaneous tissue is filled with a curious watery gel.   The skin contains a quite abnormal percentage of water.   The muscle also contains far too much water, which is mostly extracellular (Table 3).   Consequently it contains too much of the extracellular ions sodium and chloride, sometimes more than the muscle of a new-born pig, and much less potassium because the cell mass is so small.   The cellular protein nitrogen is no higher than it is at birth (Widdowson, Dickerson and McCance, 1960); the concentration of total nitrogen, however, is somewhat higher than at

227

birth because of the large amount of extracellular protein in the connective tissue of the undernourished muscle.

The bones of the pigs continue to grow a little during the year of undernutrition. The development of bone is different from the development of other tissues in that active accretion on the outside and resorption on the inside go on at the same time. In the undernourished animals resorption outstrips accretion, which results in a

Table 3

The effect of undernourishing pigs from an early age on the composition of their skeletal muscle
(Values expressed per kg of fat-free muscle)

	Well-nourished pigs			Undernourished pigs
	New-born	4 weeks	1 year	1 year
Body weight (kg)	1·3	5	150	5
Composition of muscle				
Water (g)	823	784	735	838
Total N (g)	15	28	31	22
Cellular protein N (g)	8·7	21·2	26·3	8·5
Extracellular protein N (g)	2·3	2·5	1·0	14·4
K (m-equiv.)	73	106	93	53
Na (m-equiv.)	54	30	24	85
Cl (m-equiv.)	37	21	21	64

very thin cortex and a large marrow cavity (Dickerson and Mc-Cance, 1961). Instead of being filled with the normal fatty marrow it is filled with an extracellular gel, which is denser and heavier per unit volume than normal marrow. The total weight of the skeleton, therefore, gives no real indication of the amount of true bone tissue.

Bones are always less highly calcified at birth than they are in adult life and one of the characteristics of normal development is an increase in the calcification of the skeleton. Bone consists of an organic matrix consisting largely of the protein, collagen, on which the calcium salts are deposited. The ratio of calcium to collagen in the bone, therefore, is a good index of the degree of calcification of the protein framework. Table 4 shows the calcium:collagen ratio in the cortex of the humerus of an undernourished pig compared with that in a normal pig of the same weight but much younger and in a pig of the same age but much heavier (Dickerson and McCance, 1961). There is a rise in the ratio with normal development and the bone of the undernourished pig has a very high calcium:collagen ratio; in other words, it has the chemical characteristics of old bone although it is so small.

These observations on bone refer to that of pigs after one year of undernutrition. We have also compared skeletal development in the rats suckled in small and large litters, growing rapidly and slowly respectively (Dickerson and Widdowson, 1960). It was found that calcification of the centres of ossification in the paws was

Table 4

The effect of undernutrition on the calcium:collagen ratio
in the cortex of the pig's humerus

Undernourished animals	1·63
Animals of the same size (but much younger)	1·16
Animals of the same age (but much larger)	1·39

farther advanced in a rapidly growing animal suckled in a small group than in a smaller slow growing animal suckled in a large group, when animals of the same age were compared, but less advanced when the comparison was made between animals of the same body weight. Undernutrition retards skeletal maturation, but not so much as it retards growth in weight or length, and an animal that has reached a given weight slowly has a more mature skeleton than one which has reached the same weight more rapidly.

The teeth have a very high priority of growth. In the rats retarded by being suckled in large groups eruption of the teeth was scarcely delayed at all, and the same was true for the opening of the eyes (Widdowson and McCance, 1960); these aspects of development go on almost unchecked by the milder degrees of undernutrition. In the severely undernourished pigs, however, both the development and the eruption of the teeth were considerably delayed, but far less so than the growth of the jaw. Consequently, there was not enough room for the teeth in the jaw; the teeth were crowded together and many of them impacted (Tonge and McCance, 1965).

It is a common observation in all species that retardation of growth due to a low plane of nutrition delays sexual development. The opening of the vagina is a useful objective sign of puberty in the female rat. In the experiment on rats suckled in large and small groups, the period of undernutrition was finished by the time of normal puberty, as all the rats were given unlimited food from the age of 3 weeks. Even so, the rats that were small at weaning continued to grow more slowly than the others. The majority of the fast growing rats were about 36 days old and weighed about 90 g when their vaginae opened; most of the slow-growing rats were around 46 days old and they weighed almost the same (Widdowson and McCance, 1960). In these circumstances, therefore, functional sexual development was more related to the size and development of the body than with chronological age.

Although undernutrition and retardation of growth delay sexual development, they do not prevent some development taking place in the primary and secondary sexual organs. It has been known for a long time that if a young male rat is undernourished the testes increase in weight relatively more than the body. When male rats were undernourished for 8 weeks from weaning their testes were smaller in absolute terms than those of well-nourished control animals but they formed a higher percentage of the body weight, and as the undernutrition went on the difference increased (Widdowson, Mavor and McCance, 1964). When rehabilitation began the animals gained weight rapidly and although the testes gained weight also, their rate of gain was not as fast as that of the body as a whole, so their weight per 100 g of body weight fell.

Some sexual development goes on in the pigs during the year of severe undernutrition (Dickerson, Gresham and McCance, 1964). In the males, the testes become abnormally large for the size of the body, just as they do in the undernourished rats. The tubules in the testes develop but the interstitial cells regress, and there are no signs of spermatogenesis. In the females, the uterus and vulva enlarge, the primordial follicles in the ovaries develop, and some become large and cystic, but there is no sign of ovulation.

In conclusion some mention should be made of the effect of undernutrition on the growth and development of the brain. *Figure 2* shows the normal growth of the rat's brain compared with the growth of the body.

In order to compare the values directly, both brain and body weights have been expressed as a percentage of the value at 18 weeks. The brain grows extremely rapidly during the first 2 weeks after birth and by the end of this time it has reached 66 per cent of its mature size, whereas the body is then only 8 per cent of its mature weight. The growth of the brain then slows down and remains slow throughout the time when the body is growing most rapidly, that is, between 6 and 10 weeks, which corresponds to the puberty spurt in man. These curves illustrate that the extent to which the size of the brain is affected by nutritional deprivation depends upon the age when the nutritional deprivation begins. When rats are undernourished from birth, by being suckled in large groups, the undernutrition coincides with the time when the brain is growing very rapidly relative to the body as a whole. Under-nutrition at this age hinders the growth of the body much more than that of the brain, in fact it may not hinder the growth of the

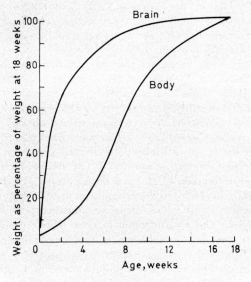

Figure 2.   Weight of brain and body of rat as percentage
of value at 18 weeks

brain at all, so that it comes to form an abnormally large percentage of the body weight; this is true both when the undernourished animal is compared with a well nourished one of the same age but much heavier, and also with a well nourished animal of the same weight but much younger (Dobbing, 1964). If undernutrition begins at 3 weeks of age, when the rate of growth of the brain has already slowed down but that of the body is accelerating, the growth of the body is still hindered more than that of the brain but the difference is quantitatively less (Dobbing and Widdowson, 1965). The brains are, of course, small in comparison with the brains of

231

well nourished animals of the same age, but they are very large in proportion to the body both for age and size; the situation is similar to that in rats undernourished from birth.

At birth the concentration of cholesterol in the brain, which is an index of the structural development and myelination, is about 30 per cent of the mature value. While the brain is growing rapidly in size during the first 2 weeks after birth the concentration of cholesterol is also rising. In rats retarded in growth from birth the rate of deposition of cholesterol in the brain does not quite keep pace with growth in size, and the brain ends up with a slightly subnormal concentration of cholesterol (Dobbing, 1964). If undernutrition begins after 3 weeks of age the concentration of cholesterol tends to be high, not low, and more appropriate to the age than the size of the animal (Dobbing and Widdowson, 1965).

In pigs growing normally, the concentration of cholesterol in the brain rises, as in rats, as the brain grows (Dickerson and Dobbing, 1967). The concentration of cholesterol in the brains of severely undernourished pigs rises during the year of undernutrition to the same extent as in a well-nourished animal (Dickerson, Dobbing and McCance, 1967). Chemical development, therefore, goes on in the brain of the undernourished pigs even though the growth of their bodies is retarded. The development, however, does not go on equally in all parts of the brain. In the cerebrum the deposition of cholesterol does not keep pace with growth in size, as in the whole brain of the rats undernourished from birth, but in the cerebellum and basal ganglia it more than keeps pace, so the concentration tends to be high rather than low, as in the rats undernourished from 3 weeks of age. The explanation of this is probably that the cerebellum and basal ganglia develop earlier than the cerebrum and they are nearer their mature proportions and chemical anatomy when the undernutrition is imposed.

It will now be appreciated that an apparently simple retardation of growth due to insufficient food sets off a highly complicated chain of responses. Some aspects of development continue relatively unaffected by the retardation of somatic growth; the growth of the brain and teeth are good examples. Other aspects of development keep more or less in step with the size of the animal; sexual development does this. Yet other aspects of development seem to be reversed and the skeletal muscles may finish with as much extracellular fluid as is in those of a new-born pig.

Although rehabilitation is not strictly within the scope of this paper, it might be mentioned in conclusion that rehabilitation after even such severe undernutrition as was produced in the pigs is not only possible but easy and the animals grow very nearly, if not quite, to the size they would have been had they never been undernourished (Lister and McCance, 1967). Their bodily proportions also become normal, as does the composition of their tissues. Careful measurement of the bones caused us to think that the bones of the rehabilitated animals do not grow to quite the same length as those of the well-nourished ones, but the astonishing thing is that they very nearly do. Functionally the animals become normal too; they mate and produce good litters and the females lactate normally. There is no evidence of any effect of the year of undernutrition on the first generation, and the young, if fed normally, grow normally and become full sized adults.

REFERENCES

Čabak, V., Gresham, G. A. and McCance, R. A. (1962). *Br. J. Nutr.* **16**, 635

Dickerson, J. W. T. and Dobbing, J. (1967). *Proc. R. Soc.* **B 166**, 384

— — and McCance, R. A. (1967). *Proc. R. Soc.* **B 166**, 396

— Gresham, G. A. and McCance, R. A. (1964). *J. Endocr.* **29**, 111

— and McCance, R. A. (1961). *Br. J. Nutr.* **15**, 567

— — (1964). *Clin. Sci.* **27**, 123

— and Widdowson, E. M. (1960). *Proc. R. Soc.* **B 152**, 207

Dobbing, J. (1964). *Proc. R. Soc.* **B 159**, 503

— and Widdowson, E. M. (1965). *Brain* **88**, 357

Lister, D. and McCance, R. A. (1967). *Br. J. Nutr.* **21**, 187

McCance, R. A. (1960). *Br. J. Nutr.* **14**, 59

Tonge, C. H. and McCance, R. A. (1965). *Br. J. Nutr.* **19**, 361

Widdowson, E. M., Dickerson, J. W. T. and McCance, R. A. (1960). *Br. J. Nutr.* **14**, 457

— Mavor, W. O. and McCance, R. A. (1964). *J. Endocr.* **29**, 119

— and McCance, R. A. (1960). *Proc. R. Soc.* **B 152**, 188

— — (1963). *Proc. R. Soc.* **B 158**, 329

INFLUENCE OF GROWTH RATE ON REPRODUCTION AND LACTATION IN DAIRY CATTLE*

S. AMIR, J. KALI and R. VOLCANI

Volcani Institute of Agricultural Research, Rehovot, Israel

As THE dairy industry changes to more intensive production systems involving higher feeding levels, information on optimal growth rates and feeding systems for dairy heifers, as well as on optimal age for first calving, becomes increasingly more necessary.

Biological and agricultural research has amassed a great deal of information concerned with the basic processes of growth in farm animals; Waters (1908), Hammond (1932) and many others (e.g., Brody, 1945; Callow, 1948; Pálsson and Vergés, 1952) provided information and theory on the development of tissues and organs for different classes of livestock on different planes of nutrition. Reproductive processes of farm animals, especially sexual maturity and development of the mammary gland, were investigated by Hammond (1927) and others (Swett, Book, Matthews and Fohrmann, 1955; Turner, 1952; Ziegler and Mosiman, 1960). Numerous experiments with dairy heifers were also carried out; Ragsdale (1934) and Steensberg (1947) provided feeding standards and optimal growth rates for dairy heifers. Experiments with different levels of feeding for dairy heifers and their influence on growth rate and subsequent production were carried out by Eckles (1953) and many others (Crichton, Aitken and Boyne, 1959a, 1959b and 1960; Hansson, 1956; Reid, Loosli, Trimberger, Turk, Asdell and Smith, 1964; Sorensen, Hansel, Hough, Armstrong, McEntee and Bratton, 1959).

From the various experiments reported it is clear that increasing the plane of nutrition accelerates the growth rate of farm animals, including dairy heifers, but most authors agree that adult size is influenced not by the plane of nutrition but by genetic determination (Hansson, 1956; Hansson, Brannang and Claesson, 1953). Sexual maturity occurs earlier in heifers with higher levels of feeding (Reid, 1960).

* This investigation was supported by a grant from the U.S. Dept. Agric.

There is conflicting evidence on the influence of feeding on subsequent milk production. Most authors (Crichton *et al.*, 1960; Eskedal and Clausen, 1958; Hansen and Steensberg, 1950; Reid *et al.*, 1964) did not find any relationship between level of feeding during growth and subsequent milk yield. Other workers found a negative relationship, milk yields being lower for amply-fed heifers than for controls. There is also some evidence for a positive relationship.

Experiments with different calving ages are too few to allow any conclusions to be drawn and most experiments cited are relevant to more extensive dairy farming. Feeding levels varied between 30 per cent below and 20 per cent above prevailing standards (Crichton *et al.*, 1959a; Hansson, 1956; Breirem, Ekern and Homb, 1960); age at calving varied between 25 and 35 months. Reports of higher feeding levels are few. Reid *et al.* (1964) did not find any appreciable difference in milk yield between heifers of different ages reared according to the Morrison standard. Herman and Ragsdale (1946) found that heifers reared on concentrates *ad lib*. did not have a higher milk yield than heifers reared on standard feeding but the former grew heavy and coarse and developed fleshy udders. Swanson (1960) reported that high level concentrate feeding impaired udder development and caused low milk yields. Wickersham and Schulz (1963) experimented with early calving at 20·4 and 24·2 months and did not find any significant differences for heifers bred at different ages, but concluded that calving difficulties were the main obstacle to breeding early.

The present studies were concerned with investigating rearing methods for heifers adapted to intensive farming practices by experimenting with planes of nutrition 70–80 per cent above recommended standards. In the light of present knowledge, as cited above, it was concluded that this would accelerate growth considerably and, especially, sexual maturity. However, it might cause extreme fattening with possible harm to physiological development. In the experiments cited above, high feeding levels for dairy heifers were coupled with insemination at the usual recommended age, which under these conditions was a late physiological age. It was reasoned that probable harm to udder development and fertility could be overcome by prompt utilization of early attained puberty through very early insemination, if possible at first oestrus. This would mean utilizing high feed consumption for higher growth rate, early pregnancy and milk production and in this way excessive fattening and probably harmful physiological manifestations might be avoided. The experiment here described was carried out to test this hypothesis (Amir, Volcani and Kali, 1967).

EXPERIMENTAL PROCEDURE

The animals used were Israeli-Friesian heifers and these were subjected to one or other of three feeding regimes. The Standard Level was according to Steensberg's allowances (1947), the High Level was *ad lib*. feeding with concentrates, while Milk Feeding

Table 1

Plan of the experiment and feed consumptions

Plane of nutrition	Insemination	Feed intake to 72 weeks (Percentage of Scandinavian Standard)
'Standard'	(1) 18 months	95
	(2) 1st oestrus	
'High'	1st oestrus	170
'Milk'	1st oestrus	183

consisted of large amounts of whole milk or milk replacers up to 5 months of age and then the same as High Plane. Average consumption of milk during these 5 months was equivalent to 2,000 kg milk. Comparative feed intakes up to 72 weeks are presented in Table 1.

Average daily weight increases during this period were 619, 790 and 862 g for 'standard', 'high' and 'milk' treatments respectively. In all treatments a basic ration of roughage was given (green fodder and hay) to allow 1–2 Scandinavian Feed Units per day, the remaining feed being a commercial mixture containing 15 per cent DCP. After calving all animals were fed alike a basal ration of roughage and concentrates given *ad lib*. for 2–3 months. Later the feed consumption was rationed according to milk yield but an additional allowance for further growth of 1–2 S.F.U. was fed during the first lactation.

A number of animals from each treatment were slaughtered at various ages for examination of organs and carcass composition.

SEXUAL MATURITY

There are two concepts as to how to determine the onset of sexual maturity. One defines it as the age at which an animal has attained the full potential of reproduction; the other considers two different stages, (1) the time when reproduction becomes possible, and (2) the time at which the full potential of reproduction is reached. The first stage is commonly regarded as 'sexual maturity' (Joubert, 1963), the second stage is reached sometime later (Dukes, 1955). The plane of nutrition exerts a strong influence on the onset of sexual maturity. Howe, Folley and Black (1963) found Graffian follicles of various sizes and up to a diameter of 20 mm in dairy calves between 1–6 months of age. They did not detect ovulation at these ages but Marden (1953) reported that ova from very young dairy calves were capable of fertilization. Many experiments were carried out with cattle to test the influence of the plane of nutrition on age at onset of puberty, but as different breeds and different planes of nutrition relevant to different feeding standards were employed, comparisons are not easy. Within each experiment, however, a higher plane of nutrition was related to an earlier onset of oestrus. There are conflicting results as to whether earlier puberty is attained at higher or lower body weights and heights at withers. For 'normal' level of feeding, ages at puberty ranged between 317 and 372 days and body weights between 250 and 270 kg. For levels of feeding 20 per cent higher than 'standard', age at puberty was between 261 and 276 days, an average reduction of 77 days as compared to 'normal' feeding, while body weights were 262–272 kg; similar to that at sexual maturity on 'standard' feeding.

In our studies the onset of puberty was examined in two ways, (1) by examination of ovaries of slaughtered animals, and (2) by external observation, i.e. animals standing when mounted.

OVULATION

Table 2 shows the number of corpora lutea and corpora albicantia found with different feeding levels and at different ages. Up to 6 months of age no corpora were found and diameters of the largest follicles did not exceed 12–14 mm; the diameter of most was not greater than 2 mm. Corpora lutea were found at the age of 9 months on the 'high' and 'milk' levels but there was no sign of ovulation at this age in the 'standard' fed heifers. When age at first ovulation was calculated according to number of corpora present in the

Table 2

Corpora lutea in relation to plane of nutrition and age

Plane of nutrition	Age (months)								
	9			12			16		
	Animal	No. of C.L.*	Calculated age at 1st oestrus (days)	Animal	No. of C.L.*	Calculated age at 1st oestrus (days)	Animal	No. of C.L.*	Calculated age at 1st oestrus (days)
'Standard'	641 642 607 621	— — —	— — —	59 60	4 —	282 365	56 57	8 8	340 335
'High'	643 637 606 624	1 3 1	— 212 211 229	25 35 38	6 7 7	188 202 199	41 96	9 14	280 229
'Milk'	8 10	3 6	219 182	3	9	180			

* Plus corpora albicantia

238

ovaries, it was found that the average age on the 'standard' level was 331 days, on the 'high' level 219 days and on the 'milk' level 193 days.

AGE, LIVE WEIGHT AND HEIGHT AT WITHERS AT FIRST OESTRUS

The data from external observations are presented in Table 3. Plane of nutrition exerts a strong influence on the age at which sexual maturity is attained; the higher the level of nutrition, the earlier the onset of puberty. On the 'high' level first oestrus

Table 3

Age, live weight and height at withers at first oestrus and total food intake

Plane of nutrition	No. of animals	Age (days)	Live weight (kg)	Height at withers (cm)	Feed intake (SFU)
'Standard'	26	357*	260*	113*	944
'High'	38	235*	227	107	986
'Milk'	11	177*	226	106	719*

* Significant difference ($P = 0.01$)

appeared 122 days earlier and on the 'milk plane' 180 days earlier, or just half the time, than on the 'standard' plane. The correlation coefficients between feed intake and age at first oestrus for 'standard', 'high' and 'milk' levels were 0·21, 0·63 and 0·60 respectively, indicating that higher feed intakes are associated with earlier puberty. The earliest oestrus was detected in a heifer on the 'milk' treatment at the age of 138 days, at a weight of 220 kg and following a total feed intake of about 800 SFU.

Differences in live weight at first oestrus were smaller than differences in age. This supports the contention that size rather than chronological age determines onset of puberty. There were, however, significant differences in body weight at onset of maturity between the treatments, and on a higher level of nutrition maturity was attained at a lighter body weight. These differences would be greater if body weights were compared on a fat-free basis. This seems to suggest, as we shall also see later, that sexual organs develop under the influence of high level feeding at a faster rate than other organs and tissues.

DEVELOPMENT OF THE UTERUS

The effect of level of nutrition on the growth of the uterus is indirect through the initiation of puberty. Development of the uterus is under the influence of the ovarian hormones and the most rapid phase of the growth of this organ takes place at the time of puberty. If growth of the uterus at the time of puberty is compared with growth in other parts of the body, it is evident that a high level of

Table 4

Influence of plane of nutrition on growth of various organs and tissues
(From birth to 9 months)

Plane of nutrition	'Standard'		'High'	
	(g)	(%)	(g)	(%)
Brain	179	100	187	104
Cannon bone	176	100	185	105
Heart	750	100	930	124
Liver	2,825	100	3,650	129
Body weight (kg)	195	100	258	133
Fat in carcass (kg)	11	100	235	214
Uterus	55	100	139	253

nutrition has the most striking effect on this organ (Table 4). This is clearly indicated in *Figure 1*, which shows two uteri of the same age but the bigger one is from a heifer on the 'high' plane of nutrition and the smaller one from a heifer on 'standard' feeding. Yet more important to note is the difference in maturity, one uterus being from a heifer before puberty, the other from a heifer after first ovulation. It is also noteworthy that whereas there is considerable difference in size of the uterus, there is no big difference in the size of the vagina.

Uterine differences on different planes of nutrition are not only in size but also in the development of the endometrial tissue. *Figure 2* shows cross-sections from the middle of the uterine horn; the higher the level of nutrition the more developed are the endometrium and endometrial glands.

240

Weight of uterus

The influence of plane of nutrition on the uterus at different ages is indicated in Table 5. Again it is evident that the highest growth rate on each level of nutrition was associated with age of puberty at each level; between 9–12 months on the 'standard' plane and between 6–9 months on the 'high' plane. During this period the

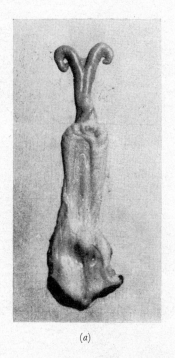

(a)

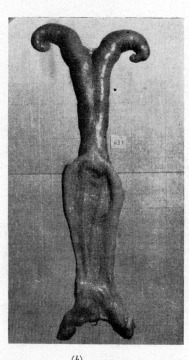

(b)

Figure 1. *Growth of uterus following two planes of nutrition;* (a) '*standard*' (at 234 days *of age before ovulation*) and (b) '*high*' (at 242 days *of age after first ovulation*)

uterus enlarged by two or three times. After first ovulation, growth rate declined and each additional oestrous cycle was associated with a declining rate of growth, so that the differences between varying planes of nutrition tended to disappear and were quite small by 16 months of age. These results confirm the findings of Sorenson, Hansel, Hough, Armstrong, McEntee and Bratton (1959), who give very similar data on uterine weights for Holstein Friesian heifers at comparable planes of nutrition. They stated that 'the most striking changes that occurred in the reproductive organs were those that took place in the uterus at puberty'. Allen and Lamming (1961) noted a similar effect in sheep but, conversely, Pálsson and Vergés

Figure 2. (*a*)

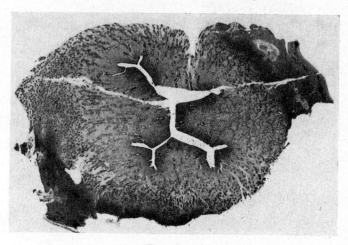

Figure 2. (*b*)

(1952), who slaughtered sheep at 9 and 41 weeks of age on 'normal' and 'high' level feeding, did not find a greater effect of plane of nutrition on the uterus than on body weight in general. In fact, Hammond (1957) states on the basis of these observations that body

weight is affected more by plane of nutrition than are the reproductive organs. But it is possible that 9 and 41 weeks of age in
sheep are not suitable for finding this effect. It may be noted also
that Pálsson and Vergés used the combined weight of uterus and
vagina as the basis of their comparison, whereas growth rates of
these two organs may be different.

Our studies show that the uterus is more responsive than any
other organ to an increased level of feeding at the time of puberty.

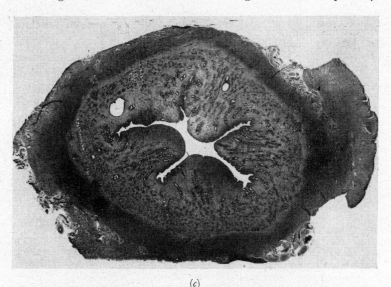

(c)

*Figure 2. Cross-sections of uterine horn at 9 months of age following one or other of three
levels of nutrition; (a) 'standard', (b) 'high' or (c) 'milk'*

At puberty the weight of the uterus is approximately 100 g, whereas
data from 7 adult cows slaughtered in this experiment showed the
adult weight of the uterus to be in the range of 460–630 g. As in
this experiment heifers became pregnant at first oestrus at young
ages, this implies that functional maturity of the uterus is reached at
approximately 20 per cent of its mature weight.

PLANE OF NUTRITION AND CONCEPTION

It is generally believed that a high plane of nutrition impairs fertility
of cattle. Asdell (1955) sums up the situation as follows: 'It is
common belief that ample feeding that causes extreme fattening

Table 5

Weight of uterus in relation to plane of nutrition and age

Plane of nutrition	Age (months)									
	3		6		9		12		16	
	No. of animals	Weight of uterus (g)	No. of animals	Weight of uterus (g)	No. of animals	Weight of uterus (g)	No. of animals	Weight of uterus (g)	No. of animals	Weight of uterus (g)
'Standard'					4	55	2	131	2	215
'High'			2	40	4	139	3	232	3	250
'Milk'	8	47	4	66	2	114	1	267		
			Milk > High*		High > Standard* Milk > Standard*		High > Standard*		N.S.	

* Significant difference ($P = 0.05$)

impairs fertility. It is believed that it causes fatty degeneration of reproductive organs or clogging of the oviduct by fat'. Newton (1939) claims that rapid fattening causes deposition of fat around the ovaries and prevents the ova from sliding into the oviduct, so causing infertility. Actually, none of these effects has been demonstrated. Sorensen et al. (1959) in their investigations with heifers on three levels of feeding, namely, 60, 100 and 130 per cent of the Ragsdale standard, could find no differences between them in a thorough histological examination of the reproductive tracts. Nevertheless, in many experiments carried out with dairy heifers reared on different planes of nutrition, conception rates were found to be somewhat lower on 'high' planes than on 'standard' planes of nutrition. Reid et al. (1964) in an experiment involving 30 animals on each of three planes of nutrition (60, 100 and 140 per cent of Ragsdale) found insemination per conception to be 1·55 on the low level, 1·41 on the medium and 1·48 on the high level, but these differences were slight and non-significant and it was concluded that there was no difference in conception rate between the different treatments. Asdell (1952) found the number of services per conception for 'low', 'medium' and 'high' planes of feeding to be 1·38, 1·40 and 1·64 respectively; Joubert (1954b), comparing different breeds, reached similar conclusions.

On the relationship between age of animal and conception rate information is scarce, but there is some on the effect of postponement of insemination. Thus, Joubert (1954a and b) found that postponing insemination, as compared with insemination at the time of puberty, led to a lower conception rate, while Wohanka (1960) demonstrated an extremely good conception rate of 1·13 inseminations per conception in heifers inseminated for the first time 18 months after the onset of puberty.

On breeding at very young ages only one experiment has been reported. Wickersham and Schultz (1963) reared heifers at approximately 15 per cent above the Ragsdale Standard and inseminated at either 10, 14 or 18 months of age. Conception occurred at the average ages of 11·4, 15·4 and 19·1 months respectively and conception rates were 3·25, 2·50 and 2·17. These differences were not statistically significant. The Cornell group (Sorensen et al., 1959) reported that in heifers on a 'high' plane of nutrition which reached puberty at 8·5 months of age there was no sign of any abnormality of the ovaries; in fact the Cornell workers were the first to propose breeding heifers at that age. In our studies also, we could discover no abnormalities of the ovaries or any of the extreme effects of fattening of the reproductive tract or clogging of the oviduct.

Conception data for the different treatments are presented in Table 6. All heifers conceived except one on the 'milk' treatment. Fewer inseminations were needed on 'standard' levels than on the 'high' level, but the differences were not significant. Initially, a significantly lower rate of conception was found on the 'milk' treatment but when another group was placed on the same treatment the following year, conception rate was then the highest of all.

Table 6

Conception rate in relation to level of nutrition

Plane of nutrition	Age at breeding months	No. of animals	No. of animals conceiving	Inseminations per conception	Age at conception (days)	Abortions
'Standard'	18	13	13	1·9	569	1
'Standard'	12	10	10	1·85	372	
'High'	1st oestrus	27	27	2·5	302	5
'Milk'	1st oestrus	8	7	4·1*	284	1
'Milk'	1st oestrus	7	7	1·7	216	

* Significant difference ($P = 0.05$)

It is difficult to arrive at conclusions with regard to the influence of plane of nutrition and age at insemination on performance on the basis of these results, because young age is associated with a high plane of nutrition, but some general conclusions may be drawn. It seems to be established that conception and pregnancy can be attained at very young ages; 27 heifers conceived at the age of 10 months and the heifers on 'milk' treatments at 9 and 7 months, so it seems apparent that early attained puberty can be utilized. The conception ages given in Table 6 are the averages for each treatment, however, and there was actually a distinct grouping within the 'high' and 'milk' treatments. On the 'high' plane treatment 6 heifers conceived at the average age of 13·7 months and 21 heifers at the average age of 8·6 months. In the 'milk' group also, 10 animals conceived at the average age of 6·6 months and 8 animals at 11·3. Earliest conception followed by normal pregnancy was found in a heifer on the 'milk' treatment which conceived at the age of 138 days and at a body weight of 220 kg. Altogether 4 animals from this treatment conceived at between 4·5 and 5·5 months with body weights ranging from 175 to 220 kg.

This variation within groups meant that some animals conceived at first or second insemination whereas others were served up to 8 times. It is difficult to account for this phenomenon, especially as there were no signs of differences in development. In fact, body weights at the beginning of puberty and insemination were the same for those which conceived early as for those which did not.

Furthermore, the difference cannot be explained by influence of excessive fattening on fertility, since all except one animal conceived at a later age and at a higher stage of fat deposition.

Most abortions occurred on the 'high' plane and these were in the 6–9 months of pregnancy. As during this period frequent abortions occurred in the herd, this could probably be ascribed to infection rather than some functional defect.

Between first and second calvings there were no differences in conception rate between animals on the 'high' and 'standard' planes; a small but insignificant difference occurred between the 'milk' plane and other treatments.

MAMMARY GLAND

Various workers have reported lower milk yields from cows raised on a 'high' plane of nutrition. Herman and Ragsdale (1946) compared heifers reared on *ad lib.* concentrate feeding with others on a conventional system and found no marked differences in milk yield, although the udders of cows reared on high plane feeding tended to be larger and more fleshy than commonly thought desirable in the dairy cow. Swanson (1960) found lower milk yields in Jersey cows reared on *ad lib.* feeding than in the limited fed group. Udders of these cows were dissected after the second lactation and found to lack development of the secretory tissue.

In the present studies of the development of the mammary gland interest was centred on finding whether any harmful effects could be detected of fattening on udder development. Table 7 shows weight of udder on different treatments and at different ages. The most rapid development took place during puberty on each feeding level, but as puberty was earlier on the higher planes of nutrition, this stage was reached earlier on these treatments. This is indicated more clearly by the amount of secretory tissue found upon dissection (Table 8) and significantly more secretory tissue was found in the udders of heifers from the higher feeding levels. Increase in amount of secretory tissue was closely related to the onset of puberty on each feeding level; between 12 and 16 months on the 'standard' level, and between 6 and 9 months on the higher levels.

Table 7

Weight of udder in relation to plane of nutrition and age

Plane of nutrition	Age (months)							
	6		9		12		16	
	No. of animals	Weight (g)	No. of animals	Weight (g)	No. of animals	Weight (g)	No. of animals	Weight (g)
'Standard'			4	729†	2	1,228*	2	1,420*
'High'	2	863*	4	1,874†	3	3,017*	3	4,900*
'Milk'	2	1,592*	2	2,720†	1	3,550*		

* Significant difference $(P = 0.05)$
† Significant difference $(P = 0.01)$

Table 8

Secretive tissue in mammary gland in relation to plane of nutrition and age

Plane of nutrition	Age (months)							
	6		9		12		16	
	No. of animals	Weight (g)	No. of animals	Weight (g)	No. of animals	Weight (g)	No. of animals	Weight (g)
'Standard'			4	204†	2	342*	2	630†
'High'	2	128	4	359†	3	461*	3	833†
'Milk'	3	125	2	585†	1	440*		

* Significant difference $(P = 0.05)$
† Significant difference $(P = 0.01)$

There were significant differences in deposition of fatty tissue in the udder between treatments at all ages (Table 9) and fat deposition was clearly dependent on plane of nutrition. This large amount of fat deposited in the udders of heifers on the higher levels did not prevent the simultaneous development of the secretory tissue, according to the stage of sexual maturity. On the higher planes of feeding, fat deposition in the udder occurred at a rate somewhere between that in the carcass as a whole and the areas of highest deposition, namely kidney and mesenteric fat (Table 10).

Table 9

Fat in udder in relation to plane of nutrition and age

Plane of nutrition	Age (months)							
	6		9		12		16	
	No. of animals	Weight (g)	No. of animals	Weight (g)	No. of animals	Weight (g)	No. of animals	Weight (g)
'Standard'			4	524†	2	872*	2	790†
'High'	2	635†	4	1,515†	3	2,555*	3	4,067†
'Milk'	2	1,467†	2	2,135†	1	3,110*		

* Significant difference ($P = 0.05$)
† Significant difference ($P = 0.01$)

Table 10

Fat deposition in the dairy heifer in relation to plane of nutrition as percentage of 'standard' (at age 12 months)

Plane of nutrition	Udder	Kidney fat	Mesenteric fat	Carcass
'Standard'	100	100	100	100
'High'	293	425	302	222
'Milk'	356	421	336	253

Comparison of the chemical composition of secretory tissue from 'standard' and 'high' feeding levels, indicated no excessive infiltration of fat on the 'high' level of nutrition (Table 11). Histological examination further revealed a more developed tissue at the same age on the 'high' level of feeding.

On the basis of these results it would appear that a high plane of nutrition does not impair the development of the mammary gland and does not cause excessive or harmful fattening of the gland. The suggestions are that fat deposition in the udder and the development of the secretory gland are two independent processes, one related directly to level of energy intake and the other to sexual maturity.

Table 11

Chemical composition of secretory tissue in relation to plane of nutrition and age

Plane of nutrition	Age (months)			
	12		16	
	Ether extract (%)	N × 6·25 (%)	Ether extract (%)	N × 6·25 (%)
'Standard'	42·7	9·3	46·6	9·3
'High'	46·2	8·7	37·9	10·6
	(N.S.)	(N.S.)	(N.S.)	(N.S.)

It is difficult to reconcile these results with those of Swanson (1960), but in the present work development was followed only to the age of 16 months. As the main development of the mammary gland takes place during pregnancy, possible harm to the gland from high level feeding could have occurred during this later period.

MILK YIELDS

Many experiments have been carried out in order to investigate the relationship between plane of nutrition and subsequent milk yield. In addition to the experiments already mentioned of Swanson (1960) and Herman and Ragsdale (1946), reports by Hansson (1956) and Starcev and Fillipson (1957) also show a tendency for a decline in milk yield with a plane of nutrition higher than 'standard'. Levels of feeding lower than 'standard' resulted in lower milk yields according to Bonnier, Hannson and Skejrvold (1948) and Steensberg (1947), and in higher milk yields according to Hansen and Steensberg (1950), Eskedal and Clausen (1958), and

Stahl and Mudra (1961). However, in the experiments of Crichton *et al.* (1960) and in an extensive experiment by Reid *et al.* (1964) employing 35 animals on each feeding level, namely at 40, 100 and 140 per cent of the Ragsdale standard, no differences in milk yield from the first to the fourth lactation were found. The evidence is conflicting, therefore, but even in those experiments where effects of different planes of nutrition could be demonstrated, those differences were small; generally not more than 10 per cent of the annual milk yield.

Table 12

Milk yields (mean daily yield for 2 weeks at peak of lactation) in relation to plane of nutrition during rearing

Treatment (and age in months at breeding)	1st lactation		2nd lactation	
	No. of animals	Yield (kg)	No. of animals	Yield (kg)
'Standard' (18·3)	13	23·8	9	30·9
'Standard' (12·2)	10	20·5	7	24·5
'High' (13·7)	5	21·6	3	28·0
'High' (8·6)	13	19·0	6	26·9
'Milk' (11·3)	8	17·8	6	23·5
'Milk' (6·6)	10	14·2	5	18·8

With regard to the influence of age at calving on subsequent milk yield, analysis of population data by Clark and Touchberry (1962), Deaton and McGilliard (1964) and Bar-Anan (1961) revealed an increase in milk yield for each additional month of age of 13, 36 and 40 kg respectively.

Few experiments have been conducted in this field. Stahl and Mudra (1961) compared calving ages of 2 and 2½ years and at 'standard' feeding level found a difference of 80 kg in annual milk yield. Wickersham and Schultz (1963) compared calving ages of 20·4, 24·2 and 27·9 months; respective annual milk yields were 9,500, 9,590 and 11,067 lb but the differences were not significant.

In the present experiment, within each treatment later insemination was related to higher milk yield (Table 12). There seemed to be no depression of milk yield with high plane feeding when comparison was made between 'standard' at 12·2 months and the two 'high' treatments; there seemed to be some depressing effect on 'milk' treatment at 11·3 months.

The most promising treatment seemed to be 'high' at 8·6 months, especially on the basis of milk yield in the second lactation compared with 'standard' at 18·3 months in the first lactation, when the difference in age between the two groups was roughly 3 months.

DYSTOCIA

Dystocia is probably the main obstacle to a system of early breeding. Van Dieten (1964), for example, found that mortality of calves born to adult cows was 3·9 per cent but of calves born to heifers it was 12·5 per cent.

Table 13

Calving difficulties (dystocia) in relation to plane of nutrition during rearing

Plane of nutrition	Age at breeding (months)	No. of animals	Normal calving	Dystocia (No. and %)
'Standard'	18·3	13	13	—
'Standard'	12·2	10	9	1 (10)
'High'	13·7	6	6	—
'High'	8·6	14	10	4 (28)
'Milk'	11·3	8	6	2 (25)
'Milk'	6·6	10	7	3 (33)

Many authors confirm these findings and stress particularly the calving difficulties in the Friesian breed. Most investigators found that dystocia and resulting mortality was caused by the large size of the new-born calves. Length of gestation, weight and age of mother, and plane of nutrition are factors affecting the size of the new-born at birth, but correlation between the size and age of dam and size of the new-born is low.

In the present experiment, dystocia, which refers to death of the new-born during calving, was related to age of the dam (Table 13). In fact, dystocia was found only in heifers less than 700 days old or of body weight less than 500 kg after calving.

Comparison between normal and difficult calving for this age group reveals no significant difference in the measurements of dams for age, live weight after calving or external skeletal measures, nor in length of gestation (Table 14).

A striking and significant difference was found, however, in the weight of the new-born calves and in their relative weight to that of their dams. It would seem that for heifers calving at a young age, the crucial factor is body weight of the new-born; there was not a single case of dystocia when the weight of the new-born did not exceed 35 kg. The correlation coefficient between weight of new-

Table 14

Comparison between normal and difficult calving in animals less than 700 days old

| Calving | Measurements of dams | | | | | | |
	No. of animals	Age (days)	Live weight (kg)	Width at hooks (cm)	Width at pinbones (cm)	Weight of calf (kg and % of dam's weight)	Length of gestation (days)
Normal	35	585	432	47·4	31·7	34·6 (8·1)	278·5
Difficult	10	543	420	47·0	30·0	39·5* (9·7*)	278·0

* Significant difference ($P = <0·05$)

born and weight of dam was 0·12 and not significant, but the correlation coefficient between weight of new-born and age of dam of 0·31 was significant. This would suggest a high priority of the embryo for nutrients and that the weight of the new-born of even very young heifers is only slightly reduced. In fact, the earliest calver in this experiment, calving at the age of 13·5 months, bore a normal male calf weighing 40 kg with resulting severe dystocia.

It may be concluded that dystocia in young heifers can be overcome by selecting sires known for small offspring or using sires from other smaller breeds, such as Aberdeen-Angus or Jersey. At second calving no dystocia occurred, even though new-born calves were 10 per cent heavier.

GROWTH

At breeding there were large and significant differences in live weight between treatments but by third calving these differences had disappeared (Table 15). Animals bred over a wide range of ages (from 6·6 to 18 months) reached the same live weight. As the estimated adult live weight of an Israeli-Friesian cow is 612 kg, there is reason to believe that heifers from all the treatments will attain this weight. Apart from body weight, skeletal measurements taken monthly during these experiments did not reveal any significant differences between the treatments at $2\frac{1}{2}$ years of age. These results suggest that pregnancy and lactation imposed on even very young animals do not stunt ultimate body development. It should be

Table 15

Live weights at breeding, first calving and third calving
in relation to plane of nutrition during rearing

Treatment (and age in months at breeding)	Live weights (kg)		
	At breeding	1st calving	3rd calving
'Standard' (18·3)	374	498	598
'Standard' (12·2)	274	401	602
'High' (13·7)	351	501	601
'High' (8·6)	262	428	586
'Milk' (11·3)	346	484	644
'Milk' (6·6)	218	397	603

remembered, however, that these animals were kept on a high level of feeding for 2–3 months after calving. Stunting of ultimate body size might occur under conditions of restricted feeding, as indeed has been demonstrated by Eckles (1953).

CONCLUSION

It would seem that the results of this study show the possibility of great acceleration of the reproductive cycle in the dairy heifer. As the ultimate aim of agricultural research is eventual applicability, it seems worth while to reflect on the possible application of these results in farming practice.

Obvious economic advantages would result from shortening the time between birth and first calving, as overhead expenses per animal would be much reduced.

Surplus female calves of the dairy herd could probably compete successfully with male calves on a 'barley beef' system. Heifers raised on the 'high' level and inseminated at 8·6 months could be slaughtered 2 months after calving; milk yield in the first 2 months and the additional calf would compensate for the poorer growth rate and feed efficiency than in the male animal.

A system of heavy feeding and early breeding would also allow for more rapid sire evaluation and genetic progress.

On the other hand many problems present themselves before such a system could be adopted, including longevity and survival in the herd, dystocia and abortion. In Israel the system of high level feeding and early breeding under field conditions is currently being tested.

REFERENCES

Allen, D. M. and Lamming, G. E. (1961). *J. agric. Sci. Camb.* **57**, 1, 87

Amir, S., Volcani, R. and Kali, J. (1967). *Publ. Volcani Institute of agric. Res.*, Rehovot, Israel (in Hebrew)

Asdell, C. B. (1952). *Proc. 2nd Congr. int. Physiol. Path. Reprod. anim.* Copenhagen **2**, 7

— (1955). *Cattle Fertility and Sterility.* Boston and Toronto; Little, Brown & Co.

Bar-Anan, R. (1961). *Hassadeh* **41**, 11 (in Hebrew)

Bonnier, G., Hannson, A. and Skejrvold, H. (1948). *Acta agric. suec.* **3**, 1

Breirem, K., Ekern, A. and Homb, T. (1960). *Proc. 5th Int. Congr. Nutr.*, Washington

Brody (1945). *Bioenergetics and Growth.* New York; Reinhold

Callow, E. H. (1948). *J. agric. Sci. Camb.* **38**, 174

Clark, R. D. and Touchberry, R. W. (1962). *J. Dairy Sci.* **45**, 1500

Crichton, J. A., Aitken, J. N. and Boyne, A. W. (1959a). *Anim. Prod.* **1**, 145

— — — (1959b). *Anim. Prod.* **2**, 45

— — — (1960). *Anim. Prod.* **2**, 159

Deaton, O. W. and McGilliard, L. D. (1964). *J. Dairy Sci.* **47**, 1004

Dukes, H. H. (1955). *The Physiology of Domestic Animals.* New York; Constock, Ithaca

Eckles, C. H. (1953). *Dairy Cattle and Milk Production.* New York; Macmillan

Eskedal, H. W. and Clausen, S. (1958). *Beretn. Forsøgslab.* 305–40

Hammond, J. (1927). *The Physiology of Reproduction in the Cow.* Cambridge; Univ. Press

— (1932). *Growth and Development of Mutton Qualities in the Sheep.* Edinburgh; Oliver and Boyd

— (1957). *Züchtungskunde.* **29**, 483

Hansen, K. and Steensberg, V. (1950). *Beretn. Forsøgslab* 246.

Hansson, A. (1956). *Proc. Br. Soc. Anim. Prod.* 1956, 51

— Brannang, E. and Claesson, O. (1953). *Acta agric. scand.* **3**, 61

Herman, H. A. and Ragsdale, A. C. (1946). *J. Anim. Sci.* **5**, 398

Howe, G. R., Folley, R. C. and Black, D. L. (1963). *J. Anim. Sci.* **23**, 3

Joubert, D. M. (1954a). *J. agric. Sci.* **44**, 5
— (1954b). *J. agric. Sci.* **45**, 164
— (1963). *Anim. Breed. Abstr.* **31**, 295
Marden, W. G. H. (1953). *J. agric. Sci. Camb.* **43**, 381
Newton, O. R. (1939). Cited by Mixner, J. P. (1959). In: *Reproduction in Domestic Animals.* Vol. II, Ch. 7. New York and London; Academic Press
Pálsson, H. and Vergés, J. B. (1952). *J. agric. Sci. Camb.* **42**, 1 and 93
Ragsdale, A. C. (1934). *Bull. Mo. agric. Exp. Stn.* 336
Reid, J. T. (1960). *J. Dairy Sci.* **43**, 103
— Loosli, J. K., Trimberger, G. W., Turk, K. L., Asdell, S. A. and Smith, S. E. (1964). *Bull. Cornell Univ. agric. Exp. Stn.* 987
Sorensen, A. H., Hansel, W., Hough, W. H., Armstrong, D. T., McEntee, K. and Bratton, R. W. (1959). *Bull. Cornell Univ. agric. Exp. Stn.* 936
Stahl, W. and Mudra, K. (1961). *Arch. Tierz.* **4**, 71
Starcev and Fillipson (1957). Cited by Coleov, J. (1964). *Report Eur. Ass. Anim. Prod.*, Lisbon
Steensberg, V. (1947). *J. Nutr.* **1**, 295
Swanson, E. W. (1960). *J. Dairy Sci.* **43**, 377
Swett, W. W., Book, I. H., Matthews, C. A. and Fohrmann, M. H. (1955). *Dep. Agric. Tech. Bull. U.S.* 1111
Turner, C. W. (1952). *The Mammary Gland.* Columbia, Mo.; Lucas Bros.
Van Dieten, S. W. C. (1964). *Veeteelt-en Zuivelberichten* **7**, 20 (*Anim. Breed. Abstr.*, 1965, **33**, 221)
Waters, H. J. (1908). *Proc. Soc. Promot. agric. Sci.* **29**, 71
Wickersham, E. W. and Schultz, L. H. (1963). *J. Dairy Sci.* **46**, 544
Wohanka, K. (1960). *Mh. VetMed.* **15**, 113
Ziegler, H. and Mosiman, W. (1960). *Anatomie und Physiologie der Kindermilch drüse.* Berlin; Paul Parey

DISCUSSION

DR. RERAT (Jouy-en-Josas) asked what were the weights and ages of the animals from which the weight coefficients were derived, as work with pigs in which growth of two muscles was measured, indicated that they had a positive allometry until 4 months of age, isometry for 1 month, followed by negative allometry. DR. FOWLER replied that the growth coefficients were measured soon after birth and thought it quite probable that the allometric relationship in certain muscles would change relative to total weight of the animal.

To a further question by Dr. Rerat on whether there were any differences between sexes in the effects of undernutrition, DR. FOWLER replied that there were sex differences but the numbers of animals were too small to allow firm conclusions.

DR. EVERITT (Ruakura) remarked that presenting the relationships on a logarithmic basis gave a precise-looking line but there was still some variation which might be of biological importance, to which DR. FOWLER replied that variation in the magnitude of differences along a logarithmic scale were accounted for in analysis of variance; the variation was not due to the nutritional differences and was presumed to be largely of genetic origin.

DR. WILSON (Unilever, Ltd.) commented that correction of the data by subtraction of the fat content of both muscle mass and skeleton would give a better approximation to the regression line than would the fresh weight data; this had been done with goats down to fat-free dry matters and had produced very good regressions.

DR. PÁLSSON (Reykjavik) commented that while it was acceptable to use old data to test new theories if they could be of help, there were certain criticisms in this instance;

(a) there was no biological justification for excluding adipose tissue from growth studies as it was an integral part of the body; although adipose tissue was primarily a storage depot for fat it had also a structural function and all body tissues, except nervous, could be regarded as stores to some extent, particularly the skeleton.

(b) when considering the plane of nutrition, it was not acceptable to change the original data and omit fat, because to justify that the experiment would have had to have been designed in that way, i.e. the low-plane lambs at 41 weeks would have to have had the same fat-free weight as the high-plane lambs at 9 weeks; similarly the low–high and high–low lambs should have had the same fat-free weight at time of slaughter but in fact did not. Therefore, these data were not valid for this kind of study and new experiments would have to be done.

(c) logarithmic graphs should not be used in these studies because they reduced the magnitude of differences from normal plotting.

(d) he had never found great differences between treatments in muscle but always had in skeleton which showed, even using few lambs, that differences in plane of nutrition affected certain bones differently from the total skeleton. This needed either disproving or confirming with larger numbers and could not be done by using these original data.

DR. EVERITT, in relation to the controversial question of whether fat tissue should or should not be included in assessing growth, commented that in the foetal stages, where fatty tissue formed a very small proportion of total weight, one could account for nearly all differences by relating weight of a particular component to foetal body weight *per se*; so if body weight were disregarded large significant differences due to nutrition could be obtained in almost any characteristic, but if these were related to total body weight then, almost without exception, these differences would disappear. He went on to ask Professor Butterfield whether the problem resolved itself as the influence of nutrition, or other treatment, on the total muscle mass in relation to body weight at some particular stage.

PROFESSOR BUTTERFIELD (Sydney) replied that it was only necessary to think about total muscle; distribution was relatively constant, as was to be expected in view of the fact that the muscle mass had a specific function to perform as an anatomical unit, so it was necessary only to decide where to assess muscle on the live animal. Rather than look at loin or rump, where there was considerable fat cover, it was more logical to look at, for example, shin where true muscle development could be seen.

To a question from DR. TAYLOR (Edinburgh) on which external measurement was the best to take as an indicator of total muscle, PROFESSOR BUTTERFIELD was unable to say, as so little had been done to relate external measurements to internal components; CALLOW had stated that one-third of live weight was muscle, which was perhaps as close as anything.

MR. HOUSTON (Leeds) asked Professor Butterfield whether in the classification of the musculature into different groups, in which the weights of individual muscles were expressed in relation to the whole musculature, account had been taken of the relative rates of fat deposition within different muscles; if fat within muscles had been excluded would the differential character have persisted, because fat deposition in muscle changed with time and location in the animal? PROFESSOR BUTTERFIELD replied that they had not assessed fat deposition in individual muscles but, on the other hand, most of the changes which they were measuring took place in the 'non-fattening' phase in the young calf; also muscle-weight distribution did not change very greatly during the fattening phase. There were differences, however, in that an Aberdeen-Angus steer with 20 per cent dissectible fat had about 33 per cent of its total muscle in the hind limbs, whereas if the fat increased to 30 per cent then the muscle in that group dropped to 30 per cent of the total.

MR. HOUSTON observed that the carcass was only some 50 per cent of the animal's total weight and that not only should carcass fat be included in assessments of growth in the whole animal but also the other parts, such as skin, which contained appreciable fat and protein. The ratio of water to dry fat-free residue took into account the fat content and could be used in growth rate studies in animals. DR. PÁLSSON agreed that it was best to study the whole animal, including internal organs, blood, hide, head and feet. Empty body weight was the best starting point but standardization of gut contents was a problem in arriving at the true live weight.

DR. TIMON (Dublin) asked Professor Butterfield, in fitting allometric equations to muscle distribution data what was the residual variation as a proportion of the total? This would measure, apart from the experimental errors, the inherent variability of the animals. PROFESSOR BUTTERFIELD couldn't remember exact values but knew that they were very small and were presented in published results.

DR. BLAXTER (Aberdeen) asked whether Professor Butterfield could comment on the fact that the muscles with a positive growth impetus tended to be the superficial ones. PROFESSOR BUTTERFIELD agreed with this observation and said that the low impetus muscles were the small, deeply situated ones and the more distal ones on the limbs, whereas the larger, superficial muscles had a higher growth impetus. Some amongst the larger, however, had different peaks; under the influence of androgens, for example, there was differential growth of the neck muscles, especially the *spinus* in bulls.

MR. VIAL (Dublin) stated that MASON had examined *m. longissimus dorsi* cross-sections in Friesian cattle and classified them on a scale from 0 to 7 on the basis of marbling fat. When subjected to ether extract analysis, however, the range was only 4 per cent of total wet weight of muscle. It appeared, therefore, that the incorporation of intermuscular fat had, in fact, little effect on overall composition.

DR. BRAUDE (Reading) asked whether there were any differences between muscles in water content, but PROFESSOR BUTTERFIELD was unable to say, other than to presume that this was inversely related to their fat content; he thought that the amount of connective tissue would be even more important in influencing total chemical composition.

DR. WILSON commented on the high correlations obtained in some of the Australian work between *semitendinosus* and *semimembrinosus* muscles and total musculature. There could be danger in extrapolating such coefficients to other groups of animals under quite different conditions, so what was the variation among animals used, and would the results be applicable to other circumstances? PROFESSOR BUTTERFIELD replied that the earlier work, on the first hundred dissections, was based on very variable animals which were a cross-section of the Australian cattle population; the very first two dissected revealed the striking similarity in muscle distribution despite extreme difference in type. It would seem, therefore, that there was little risk involved in relating the results on muscle distribution to other breeds. The *semitendinosus* muscle, which had been mentioned, was one of the least reliable because it did vary quite widely; the shin was used for estimating total muscle in, for example, progeny test work because it was cheap, readily accessible and contained 10 individual muscles plus a bone, which could be used for total bone estimation.

DR. TAYLER (Hurley) asked Professor Butterfield if he could enlarge upon his earlier statement that nutrition had a greater effect upon abdominal muscle than on the general rate of growth, and, further, did he think that the grazing animal, because of its greater activity, would show any difference in muscle development to one confined indoors. PROFESSOR BUTTERFIELD replied that the *nature* of the diet could have a greater effect upon development of abdominal muscles than on rate of growth; thus when some calves were given whole milk and others dry diets, the former had a lower proportion of abdominal muscles irrespective of rate of growth. On the question of function on muscle development, calves given 6 weeks of exercise over jumps were in no way different at the end from calves standing in stalls the whole time; it would seem that extreme differences in function were necessary to reveal differences in muscle growth.

DR. MILLS (Aberdeen) asked for further information on the fivefold increase in extra-cellular nitrogen in under-nourished animals, compared with the normal; was this a general increase in all characterized proteins or was it a specific change? DR. WIDDOWSON replied that they had not studied this in detail but the increase was true of all extra-cellular fractions studied and was primarily collagen. DR. RERAT asked whether the limiting factor to growth had been energy or protein, to which DR. WIDDOWSON replied that it was energy; the animals were given very small amounts of a first class diet on which they would have grown well if given enough of it. The protein intake, of course, was low but calories were the limiting factor in these particular studies.

DR. SHAW (Merck Sharpe and Dohme, Ltd.) referred to malnutrition in children and said that it had been found in Peru that children stunted from birth could remain permanently stunted but if malnutrition was only from 6 weeks of age then the effect was not permanent. DR. WIDDOWSON commented that this illustrated the point that the earlier in life the growth retardation, the more likely was it to be permanent. It was very difficult, however, to follow the progress of malnourished children through into adult life to confirm this.

DR. WARNER (Cornell) asked Dr. Widdowson whether they had done any studies in which early-retarded rats had been allowed to complete the ultimate life span. DR. WIDDOWSON replied that they had and, unlike McCay, they had found that animals under-nourished for the first 3 weeks and then given unlimited food always remained smaller than those not so restricted; also they suffered greater mortality, mainly from respiratory diseases, in the first year. From the first year onwards there was no difference between treatments but during the first year the expectancy of life was much less for the under-nourished.

DR. FREDERICK (Ottawa) asked for further data on the reproductive performance of the rehabilitated sows; DR. WIDDOWSON replied that they reached puberty at a very small size but all reproduced normally when mated to rehabilitated boars and produced young of normal size.

DR. MITCHELL (Reading) enquired whether this severe growth retardation and rehabilitation had any effect on subsequent body composition, to which DR. WIDDOWSON replied that they would get very fat if allowed to, but this was prevented by limiting feed in order to avoid leg troubles through bone lesions; body composition did eventually become quite normal.

DR. TAYLOR asked whether it was possible to assess the age of the animal at the end of the period of retardation from the width of the epiphyses, but DR. WIDDOWSON thought it was not.

DR. TOPPS (Aberdeen) enquired whether, if a large animal was subjected to several periods of under-nutrition, this would be more likely to cause permanent stunting than merely one period of restriction, as with store cattle. DR. WIDDOWSON thought not, as older animals compensated very rapidly for periods of undernutrition but it would depend upon whether they were able to catch up in the time available before the next period of restriction.

DR. EVERITT found it difficult to reconcile the permanent retardation in mature size of rats following 3 weeks restriction immediately after birth, with the situation in the pig where there was no restriction in mature size even though they were undernourished from 10 days of age for long periods. Did this imply that the first 10 days were critical in the pig? DR. WIDDOWSON replied that this species difference was due to the difference in stage of maturity at birth; the rat was very immature at birth and so to replicate the rat effect in the pig it would be necessary to impose the undernutrition *in utero*.

To a question from DR. BURT (Unilever Ltd.) on which system of rearing heifers showed the most promise in the field work being carried out, DR. AMIR replied that it was high-plane feeding followed by breeding at 9 months; it was probable that the highest possible plane of feeding, with *ad lib*. milk, might have some depressing effect on subsequent milk yield.

DR. BURT then asked whether there was any evidence that efficiency of food conversion during, say, the first two lactations, was better on this system than on standard feeding and serving at 12 months. DR. AMIR replied that there was no difference in total feed intake at time of calving between 'high-plane' served at 9 months and 'standard' served at 18 months; the only saving was in time and not in feed.

DR. TAYLER asked whether the rates of gain in the early period were linear and would variations in the gain, through perhaps lower planes of nutrition in the first 3 months, even though the total over the whole period was the same, have any effect on the results. DR. AMIR replied that the variations were larger on the higher planes of feeding but did not know whether variations in themselves would affect the results.

MR. LITTLE (B.O.C.M. Ltd.) asked whether the 'high-plane' calves had been early-weaned, to which DR. AMIR replied that they had not, but had been given 300–500 kg milk to provide optimal conditions.

MR. THURLEY (Weybridge) asked, first, in view of the fact that undernutrition in early life affected the ratio of brain and other organs to body weight, whether it could be assumed that a high brain to body weight ratio at birth indicated previous malnutrition, and second, what was the limiting factor to even faster growth in a healthy animal well fed and housed? DR. WIDDOWSON (Cambridge) replied that after rehabilitation the brain became an absolutely normal proportion of body weight, so if this was not found it would indicate that rehabilitation was not yet complete. On the question of what limited growth, presumably infection was one factor but there must also be a genetic limitation. DR. LODGE (Nottingham) asked whether during rehabilitation pigs grew faster than was normal for animals of that weight, to which DR. WIDDOWSON replied that their growth was no faster than normal for their weight but very fast for a year-old animal.

DR. ELSLEY (Aberdeen) commented, in relation to leg weakness, that this had been encountered in early work with early-weaned pigs in which high weight gains had followed a period of low feed intake; in more recent work on leg weakness there had been no clear correlation between high growth rates in the early stages and later defects but the leg lesions obtained in these pigs at about 120 lb live weight were very similar to those found in the rehabilitated pigs of Professor McCance.

Dr. Armstrong (Newcastle) asked Dr. Widdowson for further comment on the possibility of differences in the body composition and biochemistry of the pig born as a runt compared with one subjected to calorie restriction post-natally. Dr. Widdowson replied that runt pigs (500 g) had been compared chemically and anatomically with larger pigs (1,800 g) from the same litter; the runt did have less chemically mature muscle and less calcified bone but none of these had been carried through for examination at a later stage. To a question from Dr. Armstrong on whether exercise in cattle had resulted in any increase in total muscle weight, even though there had been no differential muscle growth, Professor Butterfield replied that there was no increase in total muscle either.

Dr. Blaxter (Aberdeen) commented on Dr. Widdowson's finding that the restricted rats had shown a higher incidence of respiratory infection and wondered how this compared with McCay's finding of eventual death from degenerative diseases and with the general finding in human medicine of an association between excess weight and respiratory infection. Dr. Widdowson replied that respiratory infections killed the young whereas degenerative diseases killed the old. In the Cambridge experiments rats had been undernourished only in the first 3 weeks of life; McCay had used a much longer period of restriction and it was the few survivors of this which died at a much greater age from degenerative diseases.

Dr. Message (Cambridge) asked how early in gestation in the sow it would be possible to remove a pig by Caesarian section in order to effect a more accurate comparison with the effect of early feed restriction in the rat. Secondly, was the nitrogen content of muscle a valid criterion of its maturity? Dr. Widdowson, in reply, did not know at what stage the foetal pig's eyes opened but restriction would have to be before this to be relative to the rat. On the question of muscle maturity, total nitrogen was a deceptive criterion and the relationship between intracellular and extracellular components of the muscle was much more accurate; with maturity the proportion of extracellular fluid declined and this was indicated by the relationship between K and Na.

Dr. McCarthy (Dublin) asked whether 'runting' in pigs was caused by shortage of blood supply, as had been demonstrated in the rat, to which Dr. Widdowson replied that it probably was, as in mice those parts of the uterus with a poor blood supply produced the smallest animals at birth; the experiment did not seem to have been done in the pig.

Mr. Vial asked Professor Butterfield what advice he would give to the beef breeders in Australia, as a result of his dissection work and knowledge of the range which occurs in muscle/bone ratio in cattle. Professor Butterfield thought that although there was a wide range in muscle/bone ratio over the cattle population as a whole, there was probably little variation within breeds but information on this was sparse. He felt that far too much emphasis was placed by breeders on bone instead of on muscle.

Dr. McCarthy asked whether complete dissection would be practised in progeny testing or, alternatively, some indirect assessment of carcass composition, to which Professor Butterfield replied that assessments based on shin dissection in progeny groups of ten would be adequate.

Dr. Mills commented that Platt and co-workers, working with puppies, had found indications of degenerative changes in the central nervous system with calorie and protein deficient diets, and asked whether DR. WIDDOWSON had encountered a similar effect in rats or pigs and whether there might be species differences here, related to stage of development at birth. DR. WIDDOWSON replied that in Platt's work high carbohydrate/low protein diets had been given to the bitches during pregnancy also and this was mainly an effect apparently of protein deficiency; no nervous defects had been found in pigs.

DR. MITCHELL asked DR. FOWLER if he could relate his own observations on the effects of feed restriction on growth to the extreme situation as described by Dr. Widdowson, to which DR. FOWLER replied that Dr. Widdowson's data did support the point he had attempted to make, namely that form was related to function and that pigs even as deprived as those still retained the ability to develop into normal functional animals. DR. WIDDOWSON then commented on the fact that these under-fed animals were healthy, they were extremely hungry and adapted in a variety of ways to undernutrition, such as low metabolic rate, low body temperature and the development of vital organs at the expense of muscle; only when these adaptations broke down did they succumb to infection or cold.

DR. WARNER referring again to longevity, thought that most experiments had indicated that limited underfeeding for fairly extended periods in early life had increased longevity, and wondered whether in Dr. Widdowson's experiments with rats either the underfeeding had been too severe or for too short a period to demonstrate this effect. DR. WIDDOWSON thought that exposure to infection may be relevant here; in their own experiments no attempt had been made to reduce this and, therefore, the results were more typical of what may be expected to occur in practice. She thought that the general indications were that children or young animals should be fed as well as possible but as they became older then food was better limited.

DR. HARTE (Dublin) asked whether total feed intake of restricted pigs, later rehabilitated, had been compared with that of unrestricted animals, to which DR. WIDDOWSON replied that the food conversion efficiency of the rehabilitated pigs was no different to what would be expected in a normal pig of that weight but a year younger, so as they had to be maintained for a whole year longer then they were obviously less efficient overall. DR. FOWLER (Aberdeen) asked DR. WIDDOWSON if she would comment on the idea that the 'degenerative clock' was set at zero at puberty and that, therefore, delayed puberty increased longevity at least to that extent. DR. WIDDOWSON agreed that this appeared to be true. DR. TOPPS (Aberdeen) asked whether any observations had been made on the activity of under-fed pigs, as lethargy was a striking feature of under-fed cattle, to which DR. WIDDOWSON replied that although the pigs appeared to be extremely active when observed, in anticipation of being fed, no continual observations were made.

MR. LITTLE enquired whether any deficiency of vitamins or minerals during the period of feed restriction could have accounted for any later increased susceptibility to infection, but DR. WIDDOWSON replied that the diets had been made adequate in all respects other than calories.

DR. BICHARD (Newcastle upon Tyne) asked DR. FOWLER whether he could comment further on the wide variation which was found in muscle/bone ratio in relation to the concept of a function/form relationship. DR. FOWLER replied that the degree of variation which could occur in bone/muscle ratio before function was impaired must be quite small and, therefore, there was probably little to exploit in breeding selection. DR. WILSON pursued this aspect further by enquiring how it had been possible for dog breeders to produce such extremes of bone/muscle ratio, to which DR. FOWLER replied that this was probably the same effect as occurred between species, in that, as animals increased in size, weight increased as the cube whereas tensile strength of bone increased as the square and so relatively more bone was required in the larger species.

V. GENETIC INFLUENCES

GENETIC VARIATION IN GROWTH
AND DEVELOPMENT OF CATTLE

ST. C. S. TAYLOR

A.R.C. Animal Breeding Research Organization, Edinburgh

TWENTY YEARS ago, in a comprehensive review of the genetics of cattle, Shrode and Lush (1947) stated that 'attempts to explain cattle inheritance in Mendelian terms began almost with the rediscovery of Mendel's laws—the postulation of a dominant gene for polledness by Bateson and Saunders in 1902 being perhaps the first specific attempt to apply these laws to inheritance in farm animals'. Mendelian genetics in cattle now embraces genes associated with blood groups (Stormont, 1967), specific blood and milk proteins (Kiddy, 1964), coat colour (Berge, 1965), and hereditary defects (Johansson, 1965). Some significant associations between growth and blood group genes have been found (for example Salerno, 1964) but so far these have not been widely confirmed. Hereditary defects, on the other hand, almost invariably involve gross differences in growth and development. Although the genes causing these defects are normally regarded by breeders as highly undesirable, there are, nevertheless, one or two situations in which such major genes have been favoured.

MAJOR GENES WITH SOME ECONOMIC INTEREST

The most investigated gene affecting growth in cattle is probably that associated with 'snorter dwarfism'. It is almost, but not completely, recessive and breeders appear to have shown a preference for heterozygotes (Bovard and Hazel, 1963; Marlowe, 1964). In homozygotes, it produces all the symptoms of achondroplasia and markedly shortens the length of the legs and head and also the body without affecting the width of the body, head, or cannon bone (Bovard and Hazel, 1963). Heterozygotes for another dwarfing gene, in this case an incomplete dominant, are variously referred to as 'comprest' Herefords or 'compact' Shorthorns and for a time were highly favoured by some breeders (Washburn, Matsushima, Pearson and Tom, 1948). These and other forms of hereditary dwarfism have been reviewed by Bovard (1960) and Gregory, Tyler and Julian (1966), while Young (1953) has described the extremely interesting situation in the Dexter breed which, in order to survive, has to remain heterozygous for a dwarfing gene.

The only other widely reported gene in cattle with a potentially desirable major effect on growth is that associated with 'double-muscling'. Its mode of inheritance is not yet entirely clear, since its degree of dominance appears to vary widely in different breeds (Lauvergne, Wissac and Perramon, 1963). A double-muscled animal has a higher total muscle content than a normal, a somewhat finer skeleton, a lower percentage of total fat (Raimondi, 1962) and a greatly increased muscle–bone ratio (Butterfield, 1966). Extreme forms of double-muscling, however, can cause serious difficulties at parturition (Raimondi, 1962).

BREED DIFFERENCES

Apart from major gene effects, the dominant role of genetics in the growth of cattle is most obvious in the wide range of breed sizes. In very large breeds such as the Charolais, Chiana or South Devon, the mean birth weight of calves is around 45 kg and mature weights can be from 700–1,000 kg for cows and 1,100–1,400 kg for bulls. In the giant Chiana breed, mature cows and bulls attain a height at withers of about 155 and 170 cm respectively and 8 outstanding bulls are reported with a mean mature height of 180 cm, a mean weight of 608 kg at 12 months of age and a mean adult weight of 1,586 kg (Bonadonna, 1959). In very small breeds such as the West African Dwarf Shorthorn (Montsma, 1959) or any of the dwarf breeds of the Iberian type, for example the Degestan Mountain (Kolesnik, 1940), calves weigh about 12–14 kg at birth and mature animals have a body weight of less than 200 kg and a withers height of less than 100 cm.

There is an enormous variety of breeds lying between these extremes of size. Mason (1951 and 1957), in his world dictionary of breeds, has listed about 250 'recognized' breeds of cattle together with almost twice as many minor breeds. Innumerable books have been written describing breed characteristics, qualities and standards; that by Bonadonna (1959) being one of the most comprehensive.

Relative to the masses of growth data collected on many different breeds, less than might be imagined is known about their comparative growth. Environmental conditions are very variable from country to country and locality to locality. For example, a comparison of the growth curve given by Matthews and Fohrman (1954) for American Jersey females at Beltsville with that given by Hodges, O'Connor and Clark (1960) for British Friesians on commercial farms would suggest that these two breeds were alike in body weight between 9 months and 2 years of age. Furthermore, no breed is genetically uniform and many different strains of the same breed usually exist. In addition, selection over the years is continuously altering mean growth curves. Mason (1961), for example, described the growth in body weight of several British beef breeds as they were presented at the Smithfield Shows in the 1950s and, in relation to results from a similar survey made at the beginning of the century, mean weight had either increased or decreased considerably in the majority of breeds.

Although standardized feeding has been advocated frequently over the last 30 years, it has not yet become an integral part of experimental practice. At present, breed comparisons in cattle are still mainly geared to detect gross size differences associated with gross differences in food intake. In the absence of any standardization either of genetic material or of food intake and composition, only a very rough ranking of breeds for size is possible. Such a ranking might be worked out in terms of any one of a variety of size characteristics, for example birth weight, mature height or mature body weight or total weight of muscle. It can be anticipated that most absolute measures of breed size will be highly correlated and that the ranking will therefore be roughly the same whatever size characteristic is used. In other words, body proportions and their development, like body composition whether determined chemically or in terms of the distribution of bone, muscle and fat in the carcass (Barton, 1967; Cole, Ramsey, Hobbs and Temple, 1964), are roughly the same in large and small breeds. A single scaling of the whole of growth and development of each breed would thus reduce the three to fourfold range of breed differences to a very much smaller order of magnitude. There would probably be several specific exceptions. Apart from these, residual variation is likely to be greatest for height:width ratios and for percentages of body fat.

The difficult problem of just *when* to make breed comparisons has also to be considered. When immature animals are involved, comparisons of carcass characteristics and efficiency of food utilization are often made at a fixed age or fixed weight. Such procedures have some economic significance but no biological justification.

269

For comparisons to be valid, they must be made at an equal degree of fatness (Guilbert and Gregory, 1944) at the same physiological age (Brody, 1945), over equivalent growth curve segments (Guilbert and Gregory, 1952; Kidwell and McCormick, 1956) or at the same percentage of mature weight (Roy, 1966).

Increased growth rate in larger breeds is, in general, offset by an increased time required to attain a given stage of maturity. Although this relationship has never been investigated thoroughly for different breeds of cattle, a comparison of the percentages of mature size attained at various ages by Hereford and Aberdeen-Angus females (Brown, Ray, Gifford and Honea, 1956a and b) clearly shows that the relatively smaller Angus breed matured earlier than the Hereford. Similarly in Brody's (1945) studies, Jerseys matured earlier in live weight than Ayrshires, and Ayrshires earlier than Holsteins. In Brody's (1945) growth curve analyses of 21 different body measurements, the Jersey:Holstein ratio for the growth rate parameter, k, is greater than unity in 17 of the body measurements. The average value of the ratio is $1 \cdot 115$, indicating that the age scale for Jerseys would have to be magnified by $11 \cdot 5$ per cent in order to make it comparable, on average, with that for Holsteins.

An alternative procedure for growth comparisons is to use 'metabolic' age scales based on the $0 \cdot 27$th power of the ratios of mature body weights (Taylor, 1965). Since the Holstein:Jersey ratio for mature weight is about $1 \cdot 4$–$1 \cdot 5$, this and Brody's method agree reasonably well in this particular case. When the live weight growth curves of Holsteins and Jerseys given by Brody (1945) in his Table $16 \cdot 2$ are expressed as a percentage of mature weight (taken as the mean weight between 6 and 8 years of age) and are then compared on a metabolic age scale, it is found that Holsteins are born relatively earlier than Jerseys (the gestation length in days is virtually the same for both breeds) and are relatively more mature from birth until first calving (*Figure 1*). This result, however, may be specific to the nutritional régime in operation at the time.

Breed comparisons made at the same degree of maturity are not quite equivalent to those made at the same metabolic age. More generally, different results are liable to be obtained (particularly in the case of immature carcass characteristics and efficiency of food utilization) depending on whether breeds are compared at the same body weight, carcass weight, age, degree of fatness, degree of maturity or metabolic age. Complete comparisons require entire growth curves. Likewise in carcass studies, allometric relationships of fat, muscle and bone, obtained from serial slaughter experiments,

270

allow breeds to be compared from several different standpoints (Berg and Butterfield, 1966; also pages 212 and 429 in the present volume).

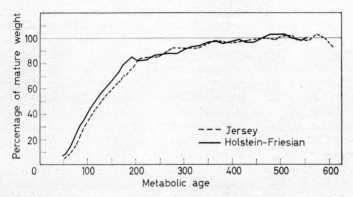

Figure 1. Comparative growth of Holstein-Friesian and Jersey females on a metabolic age scale (Data from Brody, 1945)

GENOTYPE-ENVIRONMENT INTERACTIONS

Even in carefully controlled experimental conditions, the ranking of breeds for growth and development is not unique. Genotype-environment interactions preclude any definitive ranking. Such interactions are most marked in comparisons of cattle and zebu. Thus Ragsdale, Cheng and Johnson (1957) found that at a controlled environmental temperature of 50°F, Shorthorns and Brahman did not differ greatly in their live weight growth curves, but that at 80°F the curve for Shorthorns was considerably below that for Brahman.

CROSS-BREEDING

The majority of experimentally controlled breed comparisons in cattle have been made in connection with cross-breeding. The results of cross-breeding on growth and carcass characteristics have been reviewed recently by Mason (1967). In general, the amount of hybrid vigour is small, rarely exceeding 5 per cent, except for crosses of British beef breeds with Brahman where heterosis for body weight may be around 15 per cent.

271

INBREEDING

Relatively few experiments on the effects of inbreeding on growth have been carried out since the early 1950s, when the method of improving livestock by crossing inbred lines gave way to other methods (Robertson, 1949). In general, inbreeding depression on mature skeletal size appeared to be slight, becoming somewhat greater for body weight and growth at early ages. The effects of mild inbreeding in association with some form of selection however, have continued to be investigated, to some extent. In beef cattle, Clark, Brinks, Bogart, Holland, Roubicek, Pahnish, Bennett and Christian (1963) reported that inbreeding of calf and dam had a marked detrimental effect on birth and weaning weight and on post weaning weights and gains, while Swiger, Gregory, Koch and Arthaud (1961a) found that inbreeding depressed both food consumption and growth rate. In dairy cattle, Sutherland and Lush (1962), from their results on body weights and measurements at several ages from birth to maturity, concluded that the effect of inbreeding was to depress body size generally without affecting body proportions to any noticeable extent. Their results for body weight were in general agreement with those obtained later by Hillers and Freeman (1964).

SEX DIFFERENCES

Interest in the effect of different techniques of full or partial castration and in the effect of sex hormones as growth stimulants has led to many studies involving sex differences in growth and carcass characteristics in cattle. The main sex difference appears to be that males grow faster than females, have a greater food intake, a better efficiency of food conversion over a fixed weight or age interval, and, at slaughter, their carcasses tend to contain more muscle and less fat. The relative merits of bulls and castrates for meat production have been reviewed by Turton (1962), and the general subject of sex differences has been dealt with by Bradfield in Chapter 6 of the present volume.

Many of the conclusions drawn about sex differences are subject, however, to the criticism that sexes are usually compared at different stages of maturity (Guilbert and Gregory, 1952). Thus, for example, at birth females are about 7 per cent but males only 4–5 per cent of their mature weight (Meyer, 1964). It is therefore of interest to evaluate a relative age scaling factor that might be applied to the growth of males and females. The ratio of mature weights for bulls and cows is difficult to ascertain accurately since the sexes tend to be treated differently. A ratio of 1·5 would be representative of much of the published data, but this may be an overestimate. Marlowe (1962), in a study of growth from birth to maturity in the Aberdeen-Angus and Hereford breeds, gives mean weights for bulls and cows that have been adjusted to a constant flesh condition, and here the sex ratio for mature weight is nearer 1·4. The 0·27th power of the mature weight ratio (cf. the Friesian–Jersey breed comparison considered above) gives a relative age scaling factor for growth in the two sexes of about 10 per cent.

WITHIN-BREED VARIATION

In the Galtonian or Fisherian field of correlations between relatives, genetic coefficients of variation for growth and carcass characteristics are rarely greater than 5–10 per cent. Although genetic differences between individuals of the same breed and sex are thus usually smaller in average magnitude than those associated with major genes, breeds or sexes, their distributions often show considerable overlap.

A wide variety of genetic categories are used in studying quantitative genetic differences in cattle, by far the most widely employed being paternal half sibs, daughter-dam pairs and also, with a large literature of its own, monozygotic and dizygotic twins. In addition, several workers have used the combined information from a number of different genetic categories in studying growth characteristics, for example Donald (1958) in a genetic analysis of body weights and measurements, or Everett and Magee (1965) in a study of genetically determined maternal effects on birth weight.

Within-breed genetic parameters are usually limited to heritabilities and genetic correlations. Very few cattle experiments have been both specifically designed and large enough to measure non-additive genetic variation or within-breed genotype-environment interactions with the possible exception of some twin experiments (for example Donald, 1958 and Meadows, 1960). Heritabilities and genetic correlations for growth and carcass characteristics in cattle are summarized and discussed in general terms by Clark *et al.* (1963), Gregory (1961) and Warwick (1958), and with emphasis on growth and conformation scores in young beef cattle by Petty and Cartwright (1966) and with emphasis on scores and carcass traits by Barton (1967), while Kunkel (1961) included many cattle results in his more general framework of an analysis of genetically variable growth of animals.

Body weight

Throughout the whole of the pre-natal period, when the major part of growth and differentiation takes place, our ignorance about genetic differences is almost total. Pre-natal studies in cattle with some genetic import were carried out by Lyne (1960), Swett, Mathews and Fohrman (1948) and Winters, Green and Comstock (1942). Their results on foetal body weights and linear measurements, however, are confined to breed differences which could be detected as early as 90 days after conception.

As if to compensate for the lack of pre-natal information, a large body of quantitative genetic results have accumulated on birth weight. Body size at birth may be put into perspective by recalling that female calves are very roughly about 50 per cent of their final height, about 40 per cent of their final length, and about 30 per cent of their final width. Birth weight (agreeing with the product of these three dimensions) is around 6 per cent of mature weight. Estimates of the heritability of birth weight range from 0·1 to 0·7 with an average of about 0·4 (Meyer, 1964). Genetically controlled maternal effects are involved and render a proper evaluation of the genetic variation in birth weight somewhat complex (Koch and Clark, 1955b; Everett and Magee, 1965). Indeed, to elucidate the problem thoroughly, transfer of fertilized ova between specifically chosen relatives would be required. Good reviews of birth weight are given by Meyer (1964) and Anderson and Plum (1965), who also deal with the genetic relationship of birth weight to weight of dam.

A very thorough summary and review of the heritabilities of body weight and growth rate in young beef cattle of the Aberdeen-Angus, Hereford and Shorthorn breeds has been given by Petty and Cartwright (1966). They give an overall average estimate of 0·28 for the heritability of weaning weight, of 0·41 for yearling pasture weight, and of 0·58 for final feedlot weight.

The inheritance of post-natal body weights in cattle from birth to maturity, including changes in heritability with age, inter-age genetic correlations, and correlated responses to selection for body weight at different ages, will be examined further in a later section.

Linear body measurements

Heritabilities for linear body measurements in cattle are normally high (for example Brown and Franks, 1964 or Johansson, 1964). With the exception of a study by Blackmore, McGilliard and Lush (1958) and another by Tyler, Hyatt, Chapman and Dickerson (1948), all results on changes in the heritability of linear body measurements with age have come from studies on twin cattle. These were examined and discussed in a recent paper by Taylor and Craig (1967). There appeared to be fairly general agreement that these heritabilities increased with age, at least from weaning up to about 2 years of age. Whether or not there was a depression in heritability at weaning was uncertain since the data point at birth was missing in both twin and non-twin experiments. Changes in heritability during growth were also examined in relation to the corresponding average degree of maturity of the linear measurement. On this scale, most body measurements showed a similar trend. Heritabilities appeared to approach zero at low degrees of maturity. On average they increased slowly until a body part was about two-thirds mature, when the heritability was about 0·4. Thereafter there was a steady and strong increase right up till full maturity.

The genetic correlation between body size at some given age and body size at some later age has been investigated and discussed by Taylor and Craig (1965). Inter-age genetic correlations for linear body measurements were in general very high but they tended to decrease as the age interval increased. Results on about 300 different estimates of genetic correlation were summarized as an exponential decline in genetic correlation with increasing difference in degree of maturity between the ages correlated.

Genetic correlations between different body measurements at the same age are generally found to be high (for example Johansson, 1964). In addition to evaluating these, several interesting techniques have been used to investigate the genetic inter-relationship between different linear measurements, namely, allometry, principal component analysis and discriminant function analysis.

An allometric and genetic analysis on body measurements in cattle was carried out by Kidwell, Gregory and Guilbert (1952), who found that for growth of individual Hereford cattle between 4 and 16 months of age the allometric relationship was satisfactory, but that their data at birth had to be omitted. Between 4 and 16 months, several significant sex differences were found in the allometric constants. Average values were always greater for males, indicating a continued divergence in the allometric lines of males and females. The main point of interest, however, was that the allometric constants, as measured on individual animals, were highly heritable in several instances, particularly so for the growth of withers height, body length and hook width relative to that in heart girth. This appears to be the only genetic study of the allometry of body measurements within a breed, although some allometric relationships for different breeds and sexes are included in a review by Johansson and Hildeman (1954) of the relationship between heart girth and live and slaughter weights.

Several principal component and factor analyses have been carried out on genetic material in cattle. Tanner and Burt (1954), using 15 body measurements, found that two or three factors accounted for most of the differences in body size between the Ayrshire and Shorthorn breeds. Touchberry (1951), working within the Holstein breed, and Weber (1957) within the Simmental, also found two or three predominant genetic factors. Touchberry delineated a general size factor, a skeletal factor and a flesh factor, and these three together accounted for between 67 and 100 per cent of the genetic variance in each of the separate body measurements. More recently, Lefebvre (1966) has presented an analysis with a review and some discussion of such multivariate analysis of linear body measurements in cattle. Finally in this connection, De Groot (1965) investigated the possibility of using a discriminant function analysis of linear body measurements to distinguish between one-egg and two-egg twins and concluded that this method provided a very useful diagnostic check on zygosity. The difficulty at present with all types of genetic multivariate analyses in cattle, however, is simply that not enough have been done to assess whether the same types of factor are consistently turning up.

Efficiency of food utilization

Several results concerning genetic variation in food intake and efficiency of food utilization in cattle appear to be reasonably well established. The heritability of food intake is quite high, with estimates of 0·43 by Brown and Gacula (1964), 0·76 by Brown and Gifford (1962), 0·64 by Koch, Swiger, Chambers and Gregory (1963), 0·46 by Swiger, Gregory, Sumption, Breidenstein and Arthaud (1965), and from 0·07 to 0·77, depending on the length of feeding period, by Swiger, Koch, Gregory and Arthaud (1961b). The heritability of efficiency of food utilization is also moderately high, estimates from eleven different experiments yielding an average value close to 0·50, which may be compared with the average value of 0·46 given by Kunkel (1961), of 0·40 given by Gregory (1961) and of 0·39 given by Warwick (1958). The genetic correlation between food intake and efficiency of food utilization, however, does not appear to be very high (Koch *et al.*, 1963). Moderate to high genetic correlations ranging from 0·3 to 1·0 with an average of about 0·6 have been found between efficiency of food utilization and rate of gain in live weight (Brown and Gifford, 1962; Carter and Kincaid, 1959; Koch *et al.*, 1963; Lickley, Stonaker, Sutherland and Riddle, 1960; Swiger, Koch, Gregory and Arthaud, 1962).

These results, however, are not all based on the same method of calculating efficiency of food utilization, and in addition they come from experiments carried out over various age or weight intervals. All apply to only a part, and in many cases a short part, of the total period of growth, and very little is known about the genetic correlation between efficiency of food utilization in one period and efficiency in a subsequent period. None of the experiments deal with the efficiency with which adult animals maintain themselves.

Gregory (1965) has emphasized the difficulties that stem from lack of information on the genetic relationship between feed efficiency and appetite, maintenance requirements and body composition. In addition, problems of standardized feeding and age scaling that were more apparent at the level of breed comparisons are equally and more insidiously present when individuals within a breed are being compared. Nevertheless, some of the difficulties associated with measuring and interpreting genetic differences in efficiency of food utilization are being tackled. Different measures of efficiency of food utilization have been investigated by Koch *et al.* (1963) and their relative merits compared on an empirical genetic basis. These and other measures were later examined in mathematical terms by Sutherland (1965), while the complexities that arise due to genetic variation being dependent on the feeding system used have been examined by Taylor and Young (1966), who concluded that some general functional relationship between an animal's sequence of ages, weights and food intakes was a prerequisite of genetic analysis of efficiency of food utilization.

Carcass traits

Genetic variation in carcass traits has been summarized by Clark *et al.* (1963), Gravert (1962–1963), Gregory (1961), Skjervold (1962), and Warwick (1958), and methods of progeny testing for beef production were examined by the E.A.A.P. Commission on Cattle Production (1962). Several references are also cited by Cundiff, Chambers, Stephens and Willham (1964), who found heritabilities for rib-eye area and carcass grade to be around 0·6–0·7 and those for backfat thickness, carcass yield grade, percentage retail cuts and carcass weight per day of age to be around 0·4. They also give genetic correlations among these traits. Scheper (1965) estimated the heritability of daily gain of lean to be 0·35–0·56. Further studies of the inheritance of carcass traits have been carried out *inter alia* by Bogner (1966), Heidler (1966), Liljedahl and Lindhé (1964), Mason (1964), and Swiger *et al.* (1965).

SELECTION

Selection for desirable growth characteristics in cattle has been going on for centuries but has become intensified in the last few decades in step with the increase in genetic knowledge. Thus tests have been introduced for assessing genetic merit that can be based on an individual's own records or on those of its progeny, collateral relatives or ancestors, or on a weighted combination of these (for example, Searle, 1963). This trend is likely to be greatly accelerated when full use is made of the powerful selection index which can either select for a diverse set of characteristics, the total change adding up to maximum economic improvement; or which can combine the genetic information from a wide variety of different traits and bring it all to bear in a concentrated assault on a single trait; or which can change a specific set of traits while leaving another set unaltered (Kempthorne and Nordskog, 1959; Tallis, 1962). These selection techniques, however, are of limited practical value without reliable estimates of the genetic parameters required in their evaluation and application. Such estimates are not always available. Often they are simply not known or are known only at one particular age, and information on correlated responses at other ages and in other traits is usually lacking.

The most difficult problem, however, is not how to select but what to select for. Analysis of this problem would require a comprehensive understanding not only of the genetic and physiological basis of growth and development together with how it is affected by different feeding systems and environments, but also of the economic aspects of herd structure, size and management as well as market requirements. It is, therefore, hardly surprising that there is no generally acceptable solution.

Selection based on visual appraisal (type and conformation score) for desirable growth and carcass characteristics has been widely practised in the beef breeds. Ability to fatten at lighter weights and younger ages appears to have increased over the last 50 years, but there has been no positive evidence of consistent improvement in either gaining ability or efficiency of gain (Warwick, 1958). At one stage a preference for stocky (that is, wide relative to their height) early maturing animals probably led in the first place to a reduction in size, thence towards 'comprest' Herefords and 'compact' Short-horns and eventually to the problems of dwarfism. At that point, small size began to lose some favour and selection was directed towards the elimination of dwarfing genes.

In the same way, selection has generally been directed towards eliminating the gene for double-muscling. In the Piedmont breed, however, breeders are selecting not against double-muscling but against the parturition difficulties and neonatal weakness associated with it, and double-muscled animals now predominate in the breed (Raimondi, 1962).

Selection for increased efficiency of food utilization has also been strongly advocated but, because of the practical difficulty of measuring food intake and because of the high genetic correlation found between growth rate and efficiency of food conversion, it has tended to take the form of selection for growth rate over some fixed interval. Thus Koch *et al.* (1963), for example, in line with earlier suggestions by Knapp and Baker (1944) or Winters and McMahon (1933), have calculated that selection for live weight gain over a fixed interval would lead to 81 per cent as much genetic improvement in efficiency as would direct selection for efficiency.

At present, however, inaccuracies associated with gain in body weight (Lewis, 1966) and recommendations by Clark *et al.* (1963), Gregory (1965) and others have resulted in the breeding programmes at some centres now being based on selection for body weight at the end of a test period usually around 1 year of age, although other traits are unlikely to be ignored entirely. Further justification for this procedure comes from findings such as those of Rollins, Carroll, Pollock and Kudoda (1962) or Swiger *et al.* (1965), who have shown that body weight at the end of a test period is a more efficient selection criterion than live weight gain for improving productive merit. The attractiveness of such a simple indirect approach is undeniable. Its ultimate effectiveness, however, is difficult to assess, since the results of Swiger *et al.* (1965) for example, might be interpreted as implying that 80–90 per cent of the genetic variation in net productive merit is directly dependent on body weight and therefore, according to Kleiber (1947), irrelevant to efficiency of production, while the remaining 10–20 per cent, which is the important part representing improvement independent of size, is being entirely disregarded in any selection programme based on body weight alone.

Berg and Butterfield (1966) have proposed muscle:bone ratio and fat percentage as the two most important components in carcass assessment. Selection based on progeny tests for these traits would ideally involve the slaughter of progeny at a series of ages but this, the authors suggest, might be replaced by suitable statistical adjustment of the muscle:bone ratio to a constant carcass weight. An alternative procedure, that of selecting for deviations from the muscle:bone regression line, was suggested by Vial (1966) in a discussion of selection objectives in beef breeding.

Selection would become much more effective, of course, if body composition could be measured in the live animal and serial slaughter of progeny avoided. No technique, however, is yet fully operational in cattle. The genetic relationship of live-animal measurements to carcass composition is being widely investigated and has been reviewed recently by Barton (1967).

GENETIC VARIATION IN BODY WEIGHT

The main fields of genetic investigation in cattle, namely, major genes, breed differences, breed-environment interactions, cross-breeding, inbreeding, sex differences, within-breed variation, and selection, have now been discussed briefly in general terms for growth and carcass characteristics. In this final section, one or two aspects of genetic variation in the body weight of individuals of the same breed and sex will be examined in greater detail.

Changes in heritability with age

A representative but far from exhaustive collection of heritability estimates reported for body weight are plotted against age in *Figure 2*. No distinction is made as to breed or method of calculation. Birth, 6 months and 12 months of age are all reasonably well represented, but thereafter estimates became less numerous.

The outstanding feature is the enormous range of values at any one age. At birth, the range is from 0·11 to 0·72, at 6 months from 0 to 0·81, and at 1 year from 0 to 0·84. Moreover, if the standard errors reported can be accepted as valid, many of the estimates at the same age differ significantly from one another. Nevertheless, an unweighted average for each age is shown by the heavy black line. The average at birth is 0·38, the same as the average value given by Meyer (1964). The average heritability at 6 months of age is 0·30, in good agreement with Petty and Cartwright's (1966) average value of 0·28 for weaning weight in beef cattle. The heritability of body weight thus appears to be lower at 6 months of age than at birth. This decline was also present in the majority of individual experiments where both birth and 6 months were represented, 10 experiments agreeing with the average trend and 5 showing an increase. There is somewhat more consistent agreement that heritability at 1 year of age is higher than at earlier ages, although

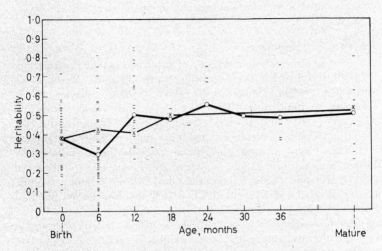

Figure 2. Changes with age in the heritability of body weight in cattle. ─⊖─ *mean of reported estimates;* × *results of Brinks* et al. (*1964*)

the average value of 0·50 is somewhat higher than the corresponding average (0·41) given by Petty and Cartwright (1966) and much higher than the average (0·28) of estimates summarized by Clark *et al.* (1963). Beyond 1 year of age, there is apparently no further average increase. The impression from a few individual experiments, however, was that heritability continued to increase slowly until maturity, although both Blackmore *et al.* (1958) and Tyler *et al.* (1948) found maximum heritability at around 2 years of age.

The experimental results of Brinks, Clark, Kieffer and Urick (1964), also shown in *Figure 2*, will be used again shortly in another connection. This set of results ranges from birth to maturity and is based on a large body of data (over 3,000 Hereford females). It also agrees quite closely with the average trend. It may be considered therefore as a representative set of results even though the mean growth curve is not smooth, exhibiting relatively slow growth in the winter period.

Much higher heritabilities than the averages shown in *Figure 2* are, of course, attainable by increased statistical or experimental control. The effect of averaging weights at different ages or stages, for example, can be quite marked. Thus an estimate of 0·57 was reported by Brinks, Clark, Kieffer and Quesenberry (1962) for mature weight taken on a single occasion, but when mature weight was averaged over several consecutive years, the heritability increased to 0·73.

Inter-age genetic correlations

Experiments yielding estimates of inter-age genetic correlations for body weight have been reported by Brinks *et al.* (1964), Koch and Clark (1955a), Swiger *et al.* (1965), Swiger, Gregory, Koch, Rowden, Arthaud and Ingalls (1963), Wilson, Dinkel, Ray and Minyard (1963) and several others cited by Taylor and Craig (1965). Genetic correlations involving gain in body weight are included in most of these studies and have also been examined by Brown and Gifford (1962), Carter (1957), Carter and Kincaid (1959), Kidwell (1954), Knapp and Clark (1947), Lehman, Gaines, Carter, Bovard and Kincaid (1961) and Urick, Flower, Willson and Shelby (1957).

Very few estimates have been made of the genetic correlation between mature weight and early weights or gains. Lickley *et al.* (1960), however, reported a genetic correlation of 0·64 between mature weight and daily gain from weaning to 1 year of age; Marlowe (1962) found sire-offspring correlations of 0·22 and 0·30 between mature weight and preweaning growth rate in Aberdeen-Angus and Herefords respectively; while Brinks *et al.* (1964) gave estimates ranging from 0·34 to 0·84 for the genetic correlation of mature weight with various weights and gains up to 18 months of age.

Overall, estimates of inter-age genetic correlations tend to be very variable, but there is an undoubted preponderance of very high values. Several authors (for example Kumazaki, Mori and Kihara, 1962) conclude from their high genetic correlations that body weight is controlled by the same set of genes throughout growth. Whatever the underlying situation is, the empirical implications are of considerable importance in connection with selection for size.

Since weight at every age appears to be fairly highly genetically correlated with mature weight, a large part of the genetic variation in body weight at any age will be removed when mature weight is held constant. Nevertheless, the partial genetic correlations between body weight at two ages with mature weight held constant still appear to be positive and moderately high. For example, this partial genetic correlation can be evaluated from the data of Brinks *et al.* (1964) as 0·84 for body weight at 12 and 18 months of age, although for birth weight and weight at 12 months it is only 0·24. Thus it appears that the animal that is most mature at birth tends to be somewhat more mature than average at all later ages, and that beyond one year of age this genetic association is strong.

Selection and changes in the mean growth curve

Brinks *et al.* (1964) evaluated the changes that might be expected as a result of direct or indirect selection for body weight, growth rates and scores at several different ages. It is of interest to separate the predicted changes in the mean growth curve into those due to changes in mature weight and those due to changes in the shape of the growth curve. This can be effected in a simple way by expressing each immature weight as a proportion of mature weight, such a proportion being referred to as the average degree of maturity of body weight at the stage being considered. To make the predicted responses to selection more obvious, a selection differential of 10 standard deviation units was applied in all cases. In practical terms, this would be equivalent to at least 7 or 8 generations or 20–30 years of selection.

First, consider the predicted change in mature weight brought about by applying selection to the various body weight and growth rate traits listed in Table 1. The largest change in mature autumn weight is achieved by direct selection, which increases mean mature weight by about 44 per cent. Selection on birth weight, however, is also expected to produce a very marked increase of about 25 per cent in mature weight. The main point to be made is that whatever trait is selected, there is a large predicted increase in mature body weight.

The predicted effect of this relatively large amount of selection on the shape of the mean growth curve is shown in *Figure 3* where all results are expressed in terms of the average degree of maturity of body weight at each age.

The heavy line represents the unselected mean maturity curve. None of the selected maturity curves are very different from the unselected curve; they all cluster around it fairly closely. Some are above the unselected line and some below while others cut across it,

Table 1

Direct and correlated responses to selection for various growth traits expressed as a percentage of mature weight
(derived from Tables 1 and 3 of Brinks et al., 1964)

Direct or correlated response	Trait selected										
	Birth wt.	Gain: birth–wean.	Wean. wt.	Wean. score	Gain: wean.–12 month	12 month wt.	Gain: 12–18 month	18 month wt.	18 month score	Mature spring wt.	Mature autumn wt.
Birth wt.	0·8	0·2	0·3	−0·1	−0·7	−0·0	−0·0	−0·0	0·1	−0·6	−0·6
Gain: birth–weaning	−0·9	7·5	6·5	2·4	−6·2	1·8	1·1	1·2	−1·4	−2·5	−4·4
Weaning wt.	−0·1	7·9	6·8	2·3	−7·0	1·9	1·2	1·2	−0·9	−2·8	−4·7
Weaning score	(0·9)*	(5·7)	(5·4)	(16·0)	(−8·3)	(−3·7)	(1·8)	(−1·7)	(1·1)	(2·2)	(2·0)
Gain: weaning–12 month	−0·9	−3·4	−3·4	−4·3	10·4	4·0	−1·7	2·0	1·0	1·2	1·1
12 month wt.	−1·1	4·1	3·3	−3·7	3·4	5·7	−0·6	2·9	0·2	−3·2	−4·7
Gain: 12–18 month	−0·5	1·9	1·5	0·3	−3·3	−0·4	6·1	2·2	3·8	−0·7	−2·4
18 month wt.	−1·6	6·1	4·9	−3·3	0·1	5·4	5·4	5·0	5·4	−1·9	−5·5
18 month score	(2·7)	(−0·5)	(0·9)	(−1·0)	(2·4)	(2·5)	(8·4)	(8·2)	(6·8)	(3·8)	(2·8)
Mature spring wt.	−3·8	1·9	0·7	0·0	−1·5	−0·8	1·8	0·6	1·1	−0·8	−5·3
% Increase in mature autumn wt.	24·4	14·9	19·4	2·6	13·6	22·9	16·0	30·4	4·2	38·7	43·9
Unselected means as % of mature autumn wt.	6·9	28·6	35·5	(74·6)	7·8	43·3	22·3	65·6	(74·7)	100·3	1,116 lb

284

the net effect of all types of selection on mean degree of maturity being a very small increase. Age scales, however, were not transformed (as they were for breed comparisons) to allow for differences

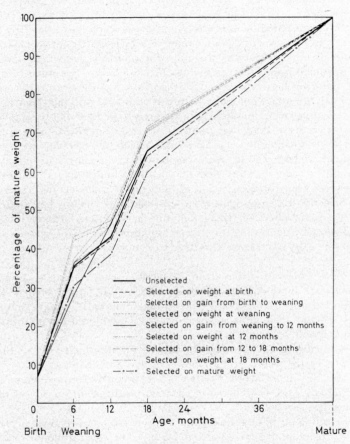

Figure 3. Predicted changes in the mean maturity curve for body weight resulting from selection for various growth traits (Derived from Tables 1 and 3 of Brinks et al., 1964)

in mature weight, since ages were too widely separated to give smooth curves. Such an adjustment would slightly increase the degree of maturity of selected curves relative to the unselected.

Predicted changes in the mean maturity curves may be examined in much more detail by reading down the columns in Table 1.

Selecting on birth weight is expected to change birth weight from its unselected value of 6·9 per cent to 7·7 per cent of mature weight. At all other stages the selected animals become slightly less mature on average. Mature weight, however, is increased by almost 25 per cent.

Selection based on gain from birth to weaning has little effect in terms of the direct response, most of which is a result of making animals more mature at weaning (by almost 8 percentage units). Subsequent rate of maturing, however, suffers by −3·4 units. Rate of maturing between 12 and 18 months is slightly more rapid and animals are somewhat more mature at both 12 and 18 months of age.

Selecting on weaning weight apparently gives a slightly smaller direct increase than does indirect selection on growth rate from birth to weaning. In general, the effects of selection are very similar in both cases. Selection based on weaning score greatly improves weaning score and puts a slight bump on the maturity curve at preweaning ages and a slight dip at post-weaning ages.

Selection based on gain from weaning to 12 months has a fairly large direct effect on rate of maturing, but two-thirds of this is achieved by depressing the degree of maturity at weaning and only one-third by increasing the degree of maturity at 12 months of age.

When selection is based on 12 month weight, animals become somewhat more mature at all early ages other than birth. An expected improvement of 5·7 percentage units at 12 months and of 5·4 units at 18 months is close to the most achieved by any of the traits listed.

Selecting on gain between 12 and 18 months of age very slightly depresses the curve at 12 months of age and increases it by 5·4 percentage units at 18 months. The predicted improvement in 18 month score is greater for selection on this than on any of the other traits.

Selecting on 18 month weight is expected to increase mean degree of maturity at all ages other than birth. The effect is similar to but less than that expected from selection on 12 month weight, although 18 month score shows a greater increase.

Perhaps surprisingly, selecting on 18 month score is almost as effective as selecting on any of the other traits in increasing degree of maturity at 18 months of age. (This expectation is interesting when considered in conjunction with Warwick's (1958) findings on changes due to selection actually practised in beef cattle.) The maturity curve prior to 18 months of age is scarcely affected and mature weight is increased by only 4·2 per cent. The direct response in 18 month score is an increase of 6·8 units.

Selection based on mature weight, either spring or autumn, is expected to produce animals that are on average less mature at all immature ages.

The main conclusion to be drawn from the predicted effect of the various types of selection on the mean curve for degree of maturity in body weight is that, while various changes in the shape of the growth curve can be brought about, in no case is the expected change of very great magnitude in comparison to the amount of selection pressure exerted. Almost the whole of the selection effort on these particular growth traits (with the exception of scores) has gone towards changing mature body weight. All the traits listed must therefore be regarded as highly inefficient for altering degree of maturity, and it is obvious that in any efficient selection procedure for producing animals that are more mature at some required age, the correlated increase in mature body weight must be minimized.

Waiting until mature weight can be measured is, however, time consuming, and an indirect approach for augmenting the type of selection achieved by 18 month score might be desirable. Firstly, it should be possible to make use of the fact that some immature skeletal measurements are much more highly correlated genetically with final size than is immature body weight. This suggests a return to a measure something like the muscle-skeletal index put forward by Gregory (1933) or some modern equivalent such as the muscle:bone ratio, but so constructed as to be genetically uncorrelated with final size. Secondly, it should be possible to make use of the mature body size of an animal's sire and dam on the grounds that they have some predictive value for mature size of offspring. Nowadays the evaluation of a complicated selection index incorporating these features as well as many others relating to efficiency of food utilization and carcass merit would be among the least of our practical problems. The determination of just what the index should be, however, is a research product not yet on the cattle market.

REFERENCES

Anderson, H. and Plum, M. (1965). *J. Dairy Sci.* **48**, 1224
Barton, R. A. (1967). *Anim. Breed. Abstr.* **35**, 1
Berg, R. T. and Butterfield, R. M. (1966). *Anim. Prod.* **8**, 1
Berge, S. (1965). *Meld. Norg. LandbrHøigsk* **44** (16)

Blackmore, D. W., McGilliard, L. D. and Lush, J. L. (1958). *J. Dairy Sci.*
41, 1045
Bogner, H. (1966). *Bayer. landw. Jb.* **43** (Spec. No. 2), p. 87
Bovard, K. P. (1960). *Anim. Breed. Abstr.* **28**, 223
— and Hazel, L. N. (1963). *J. Anim. Sci.* **22**, 188
Bonadonna, T. (1959). *La Razze Bovine.* Milan; Progresso Zootechnico
Brinks, J. S., Clark, R. T., Kieffer, N. M. and Quesenberry, J. R. (1962).
J. Anim. Sci. **21**, 501
— — — and Urick, J. J. (1964). *J. Anim. Sci.* **23**, 711
Brody, S. (1945). *Bioenergetics and Growth.* New York; Reinhold
Brown, C. J. and Franks, L. (1964). *J. Anim. Sci.* **23**, 665
— and Gacula, M. (1964). *J. Anim. Sci.* **23**, 321
— and Gifford, W. (1962). *J. Anim. Sci.* **21**, 388 (abstr.)
— Ray, M. L., Gifford, W. and Honea, R. S. (1956a). *Bull. Ark. agric.
Exp. Stn.*, No. 570
— — — — (1956b). *Bull. Ark. agric. Exp. Stn.*, No. 571
Butterfield, R. M. (1966). *Aust. vet. J.* **42**, 37
Carter, R. C. (1957). *Abstr. Doct. Thesis No. 1858* (1956). *Iowa St. Coll.
J. Sci.* **32**, 151
— and Kincaid, C. M. (1959). *J. Anim. Sci.* **18**, 331
Clark, R. T., Brinks, J. S., Bogart, R., Holland, L. A., Roubicek, C. B.,
Pahnish, O. F., Bennett, J. S. and Christian, R. E. (1963). *Stn. tech.
Bull. Ore. agric. Exp. Stn.*, No. 73: p. 72
Cole, J. W., Ramsey, C. B., Hobbs, C. S. and Temple, R. S. (1964). *J.
Anim. Sci.* **23**, 71
Cundiff, L. V., Chambers, D., Stephens, D. F. and Willham, R. L. (1964).
J. Anim. Sci. **23**, 1133
De Groot, B. (1965). *Genetica* **36**, 277
Donald, H. P. (1958). *Proc. 10th int. Congr. Genet.* **1**, 225
E.A.A.P. Commission on Cattle Production (1962). *Eur. Ass. Anim. Prod.*
Rome [Mimeograph]
Everett, R. W. and Magee, W. T. (1965). *J. Dairy Sci.* **48**, 957
Gravert, H. O. (1962–1963). *Z. Tierzücht. ZüchtBiol.* **78**, 43
Gregory, K. E. (1961). *Res. Bull. Neb. agric. Exp. Stn.* No. 196, p. 32
— (1965). *J. Anim. Sci.* **24**, 248
Gregory, P. W. (1933). *Genetics* **18**, 221
— Tyler, W. S. and Julian, L. M. (1966). *Growth* **30**, 393
Guilbert, H. R. and Gregory, P. W. (1944). *J. Anim. Sci.* **3**, 143
— — (1952). *J. Anim. Sci.* **11**, 3
Heidler, W. (1966). *Arch. Tierz.* **9**, 179
Hillers, J. and Freeman, A. E. (1964). *J. Dairy Sci.* **47**, 894
Hodges, J., O'Connor, L. K. and Clark, W. M. (1960). *Anim. Prod.* **2**, 187
Johansson, I. (1964). *Anim. Breed. Abstr.* **32**, 421
— (1965). *Wld Rev. Anim. Prod.* **1**, 19
— and Hildeman, S. E. (1954). *Anim. Breed. Abstr.* **22**, 1
Kempthorne, O. and Nordskog, A. W. (1959). *Biometrics* **15**, 10

Kiddy, C. A. (1964). *J. Dairy Sci.* **47**, 510

Kidwell, J. F. (1954). *J. Anim. Sci.* **13**, 54

— Gregory, P. W. and Guilbert, H. R. (1952). *Genetics* **37**, 158

— and McCormick, J. S. (1956). *J. Anim. Sci.* **15**, 109

Kleiber, M. (1947). *Physiol. Rev.* **27**, 511

Knapp, B. and Baker, A. L. (1944). *J. Anim. Sci.* **3**, 219

— and Clark, R. T. (1947). *J. Anim. Sci.* **6**, 174

Koch, R. M. and Clark, R. T. (1955a). *J. Anim. Sci.* **14**, 775

—— (1955b). *J. Anim. Sci.* **14**, 979

— Swiger, L. A., Chambers, D. and Gregory, K. E. (1963). *J. Anim. Sci.* **22**, 486

Kolesnik, N. N. (1940). Cited in *Anim. Breed. Abstr.* **15**, 14

Kumazaki, K., Mori, T. and Kihara, Y. (1962). Cited in *Anim. Breed. Abstr.* **33**, 191

Kunkel, H. O. (1961). *Misc. Publs. Tex. agric. Exp. Stn.*, No. MP-499

Lauvergne, J. J., Vissac, B. and Perramon, A. (1963). *Annls. Zootech.* **12**, 133

Lefebvre, J. (1966). *Thèses Doct. Ing.*, Fac. Sci., Univ. Caen

Lehmann, R. P., Gaines, J. A., Carter, R. C., Bovard, K. P. and Kincaid, C. M. (1961). *J. Anim. Sci.* **20**, 53

Lewis, W. H. E. (1966). *Tech. Rep. Beef Rec. Ass.*, No. 3

Lickley, C. R., Stonaker, H. H., Sutherland, T. M. and Riddle, K. H. (1960). *J. Anim. Sci.* **19**, 957 (abstr.)

Liljedahl, L. E. and Linhé, B. (1964). *LantbrHögsk. Meddn*, A, No. 19

Lyne, A. G. (1960). *Proc. Aust. Soc. Anim. Prod.* **3**, 153

Marlowe, T. J. (1962). *Bull. Va. agric. Exp. Stn.*, No. 537

— (1964). *J. Anim. Sci.* **23**, 454

Mason, I. L. (1951) (1957). *Tech. Commun. Commonw. Bur. Anim. Breed. Genet.*, No. 8. *Ibid.*, No. 8 (Suppl.)

— (1961). *Agriculture, Lond.* **68**, 71

— (1964). *Anim. Prod.* **6**, 31

— (1967). *Anim. Breed. Abstr.* **34**, 28

Matthews, C. A. and Fohrman, M. H. (1954). *Tech. Bull. U.S. Dep. Agric.*, No. 1098

Meadows, C. E. (1960). *Diss. Abstr.* **20**, No. 9, 3479

Meyer, H. (1964). *Züchtungskunde* **36**, 299

Montsma, G. (1959). *Trop. Agric., Trin.* **36**, 306

Petty, R. R. and Cartwright, T. C. (1966). *Dept. Tech. Rep. Tex. agric. Exp. Stn.*, No. 5 [Mimeograph]

Ragsdale, A. C., Cheng, C. S. and Johnson, H. D. (1957). *Res. Bull. Mo. agric. Exp. Stn.*, No. 642

Raimondi, R. (1962). Abridged translation cited in *Anim. Breed. Abstr.* **32**, 2614

Robertson, A. (1949). *Anim. Breed. Abstr.* **17**, 1

Rollins, W. C., Carroll, F. D., Pollock, J. W. T. and Kudoda, M. N. (1962). *J. Anim. Sci.* **21**, 200

Roy, J. H. B. (1966). *9th int. Congr. Anim. Prod., Edinb.* scient. Progm. Abstr. Eng. edn.

Salerno, A. (1964). Cited in *Anim. Breed. Abstr.* **33**, 2257

Scheper, J. (1965). *Züchtungskunde* **37**, 251

Searle, S. R. (1963). *Heredity* **18**, 351

Shrode, R. R. and Lush, J. L. (1947). *Adv. Genet.* **1**, 209

Skjervold, H. (1962). Article in E.A.A.P. Commission on Cattle Production (1962) (*q.v.*)

Stormont, C. (1967). *J. Dairy Sci.* **50**, 253

Sutherland, T. M. (1965). *Biometrics* **21**, 739

— and Lush, J. L. (1962). *J. Dairy Sci.* **45**, 390

Swett, W. W., Mathews, C. A. and Fohrman, M. H. (1948). *Tech. Bull. U.S. Dep. Agric.*, No. 964

Swiger, L. A., Gregory, K. E., Koch, R. M. and Arthaud, V. H. (1961a). *J. Anim. Sci.* **20**, 626

— Koch, R. M., Gregory, K. E. and Arthaud, V. H. (1961b). *J. Anim. Sci.* **20**, 802

— — — — (1962). *J. Anim. Sci.* **21**, 588

— Gregory, K. E., Koch, R. M., Rowden, W. W., Arthaud, V. H. and Ingalls, J. E. (1963). *J. Anim. Sci.* **22**, 514

— — Sumption, L. J., Breidenstein, B. C. and Arthaud, V. H. (1965). *J. Anim. Sci.* **24**, 418

Tallis, G. M. (1962). *Biometrics* **18**, 120

Tanner, J. M. and Burt, A. W. A. (1954). *J. Genet.* **52**, 36

Taylor, St. C. S. (1965). *Anim. Prod.* **7**, 203

— and Craig, J. (1965). *Anim. Prod.* **7**, 83

— — (1967). *Anim. Prod.* **9**, 35

— and Young, G. B. (1966). *J. agric. Sci., Camb.* **66**, 67

Touchberry, R. W. (1951). *J. Dairy Sci.* **34**, 242

Turton, J. D. (1962). *Anim. Breed. Abstr.* **30**, 447

Tyler, W. J., Hyatt, G., Chapman, A. B. and Dickerson, G. E. (1948). *J. Anim. Sci.* **7**, 516

Urick, J., Flower, A. E., Willson, F. W. and Shelby, C. E. (1957). *J. Anim. Sci.* **16**, 217

Vial, V. E. (1966). *Occ. Symp. Br. Grassld, Soc.*, No. 2, 62

Warwick, E. J. (1958). *J. Anim. Sci.* **17**, 922

Washburn, L. E., Matsushima, J., Pearson, H. E. and Tom, R. C. (1948). *J. Anim. Sci.* **7**, 127

Weber, F. (1957). *Z. Tierzucht. ZüchtBiol.* **69**, 225

Wilson, L. L., Dinkel, C. A., Ray, D. E. and Minyard, J. A. (1963). *J. Anim. Sci.* **22**, 1086

Winters, L. M., Green, W. W. and Comstock, R. E. (1942). *Tech. Bull. Minn. agric. Exp. Stn.*, No. 151

— and McMahon, H. (1933). *Tech. Bull. Minn. agric. Exp. Stn.*, No. 94

Young, G. B. (1953). *J. agric. Sci., Camb.* **43**, 369

GENETIC VARIATION OF BODY WEIGHT IN SHEEP

JOHN C. BOWMAN

Department of Agriculture, The University, Reading

COMPARED WITH some other species the sheep has a relatively small mature body weight range. The total range for all domesticated breeds under comparable environments is probably only of a four-fold order, exemplified by adult females of the Welsh Mountain breed weighing 90–110 lb and adult males of the Lincoln and some European breeds weighing 300–350 lb. Some other species, such as the chicken with an eightfold range and the dog with an eighty-fold range, obviously have very much greater genetic variation.

Considering the very large number of sheep breeds, it is surprising, on first assessment, that the variation is so small, particularly in view of the fact that sheep have been selected separately for milk, meat and wool production and have been selected to produce in a very wide range of climatic and nutritional environments. However, it must be remembered that the majority of present-day breeds can be traced back in relatively recent times to but four or five independent breeding origins (Ryder, 1964), which may not have been very different in body weight.

The objective here is to review the information available on the genetic differences in body weight at different ages both between and within breeds. There is a considerable quantity of non-comparative information on the body weights of single breeds but comparative data on weight for age for several breeds, maintained contemporaneously in the same environment, have been published only recently and, it must be stressed, are totally inadequate for current applied breeding requirements. Comparative data on pure-breeds and cross-breeds from the same pure-breeds are even more limited, but are valuable in indicating the degree of dominance involved in body weight variation.

Intra-breed genetic variation of body weight in sheep has been estimated as heritability by several methods, for several ages and for many breeds. It is now generally accepted that heritability estimates are probably fairly specific to the animals from which the data

291

were collected, and generalizations such as 'body weight is a character with medium heritability' are not only not helpful but can be positively misleading. A discussion of single heritability estimates, even in relation to several previous estimates, is not usually conclusive or productive. However, a survey of many such estimates, particularly calculated by different methods, can be quite revealing about the relative magnitude of additive and non-additive genetic variation and maternal effects. Further evidence on such variation and effects can be obtained from observations on body weight changes caused by inbreeding. The main purpose in obtaining estimates of genetic variation is obviously to optimize design of selection programmes and to be able to predict the outcome of such programmes. It is intended here to review the few selection experiments involving body weight in sheep and to discuss the implications of the closeness of observed and predicted responses.

The animal breeder is not solely interested in body weight *per se* but is concerned also with body composition. Little research has been carried out on the genetic variation of the constituents of body weight but one or two valuable conclusions can be drawn from the work already done.

At this point it is pertinent to reiterate the purposes for which sheep have been and are now bred, since these have had and will have some effect on the genetic variation to be found in different breeds. The early pioneer breeders were principally interested in either wool or milk production with meat largely derived from mature animals. More recently, emphasis has switched from wool to meat with large quantities of the latter being derived from animals which are not mature and which in many cases are slaughtered at or shortly after weaning. Wool from several breeds is now a minor or non-existent by-product and selection is based on the production of suckled meat animals, which to some extent must place some indirect selection pressure on milk production in the dam. A new trend is already evident, in that as early weaning becomes more widespread, selection emphasis on ewe milk production will decline and even greater selection will be placed on growth rate of the young lamb fed on concentrate or semi-concentrate diets.

INTER-BREED GENETIC VARIATION

Throughout this paper, apart from the section on body composition, body weight refers to weight at specific ages of the animals concerned.

Since there are few data on the relative mature sizes of breeds it is not possible to discuss body weight in terms of maturity. This is a disadvantage in that comparisons between or within breeds of genetic variation of body weight at a specific age, or perhaps at a specific managemental stage such as weaning, may represent a comparison of genetic variation between breeds which are at different stages of maturity. So far as possible the discussion has been pursued with the aim of avoiding this criticism. Some idea of the problem can be gauged from the data presented by Boyd, Doney, Gunn and Jewell (1964) for the weights up to 6 years of age of four breeds of sheep (Rambouillet, Scottish Blackface, Welsh Mountain and Soay). Though the data are not strictly comparable, in that they were gathered from the four breeds kept in different places at different times, they indicate that at 1 year of age the breeds reached 0·65, 0·72, 0·67 and 0·58 respectively of their mature adult weight. These fractions do not correlate with the mature sizes of the breeds involved.

Comparative data on body weight by age for several pure-breeds are summarized in Table 1. It cannot be stressed too highly that breed comparisons must be made strictly contemporaneously under the same environmental conditions, for it is obvious from the Southdown data in Table 1 that differences between breeds in different environments can be almost as large as between-breed differences in the same environment. The data of Wiener (1967) are particularly comprehensive but do not give a comparison of rates of maturity for the breeds because the birth and weaning data were collected from the progeny of the ewes on which the adult data were collected. Further, the ewes were brought in from several locations so that differences in adult weights between breeds may be partly attributable to prenatal and postnatal environmental effects specific to the locations on which the ewes were bred and partly reared. The paucity of data in Table 1, which represent the bulk of published information on body weight comparisons, fully emphasizes the need for much unexciting but worth-while research. Not only is this necessary to obtain relative rates of maturity for breeds but also to obtain body weight variation estimates for breeds at different ages and stages of maturity.

The comparisons by Sidwell, Everson and Terrill (1964), Vesely and Peters (1964) and Lambe, Bowman and Rennie (1965) indicate that variation between breeds is relatively less at birth than at weaning, but as the latter weight reflects not only the breeds' genetic differences for this character but also the differences in milk pro-

293

Table 1

Comparative mean body weights of sheep at different ages

Reference and country	Breed	Weight (lb)		
		Birth	Weaning	Later
Phillips et al. (1940) U.S.A.	Columbia Corriedale Rambouillet			yearling ewes 87·8 82·5 87·4
King and Young (1955) U.K. (High plane of nutrition)	Blackface Cheviot Wiltshire			ewes 45 (6 mths) 95 (14 mths) 64 (7 mths) 113 (15 mths) 93 (8 mths) 132 (16 mths)
Sidwell et al. (1964) U.S.A.	Hampshire Shropshire Southdown Merino Columbia- Southdale	mixed sexes 8·9 7·5 6·1 7·0 7·8	64·4 50·7 42·0 42·5 56·5	
Vesely and Peters (1964) Canada	Rambouillet Romnelet Can. Corriedale Romeldale	mixed sexes 10·1 9·6 9·8 9·4	70·6 63·3 60·4 61·6	
Botkin and Paules (1965) U.S.A.	Corriedale Suffolk	mixed sexes	74·1 84·7	
Lambe et al. (1965) Canada	Suffolk Hampshire Southdown	mixed sexes 7·3 6·8 5·6	88·5 69·2 54·7	
Wiener (1967) U.K.	Cheviot Lincoln Southdown Blackface Welsh Mountain	single lambs born and reared from first parity ewes 9·6 15·1 7·8 8·1 8·1	67·0 94·9 57·4 68·0 58·3	ewes 3 years 145 195 130 150 105

duction and other postnatal maternal effects between breeds, this finding is hardly surprising.

Sidwell *et al.* (1964) carried their comparison of breeds to include many but not all two-, three- and four-way crosses of the breeds. The data on birth weight and weaning weight (not defined as an age) for the two-way cross lambs from pure-breed ewes are presented graphically in *Figure 1* as deviations from midparent values. It can be seen that, without exception, the cross-breed weights are above

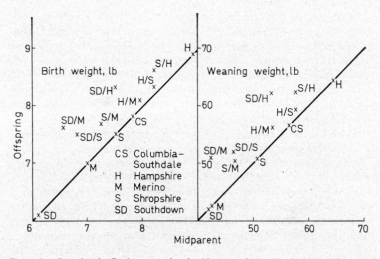

Figure 1. Cross-breed offspring—pure-breed midparent relationship for birth and weaning weight (From Sidwell et al. *(1964))*

By courtesy of The Editor, *J. Anim. Sci.*

the midparent values, which indicates that some dominance variation is responsible for between breed body weight differences. The same authors reported that the birth and weaning weights for three- and four-breed cross lambs, which involved matings of pure-breed and two-breed cross rams to two-breed cross ewes, were also always higher than the average parental value for the breeds involved. The average deviation of two-, three- and four-breed crosses over pure-breeds was 0·52, 0·78 and 0·82 lb respectively for birth weights and 5·2, 9·5 and 10·4 lb respectively for weaning weight. These results all indicate some dominance variation for between-breed differences, though comparisons involving cross-breed females are complicated by changes in body weight resulting from changes in maternal environment. Sidwell *et al.* (1964), however, found that though three-breed crosses were always heavier than the average of the pure-breeds and the two-breed crosses, three-breed crosses derived from two-breed cross rams mated to pure-breed ewes were lighter than the corresponding three-breed crosses derived from a pure-breed ram mated to a two-breed cross ewe. These results confirm that body weight differences between pure-breeds and cross-breeds resulted from a combination of genetic variation for body weight *per se* and from genetic variation for maternal environment including milk production.

In *Figure 1* it is interesting to note that the birth weight of lambs from crosses of Southdown × Merino and Shropshire × Merino exceeded that of either parent and the weaning weight of lambs from crosses of Southdown × Merino and Southdown × Shropshire exceeded that of either parent. Weights of all other crosses are either equal to or less than the value for the larger parent. The greater weights of these three crosses at birth and weaning might be explained in terms of overdominance but other explanations would be equally valid and there is not sufficient evidence for a critical decision as to the correct explanation.

Further comparison of *Figure 1* and pure-breed parental weaning weights in sheep have been reported by Botkin and Paules (1965). They compared Corriedale × Suffolk lambs with pure-breed lambs and found that the weaning weights of *Figure 1* did not exceed but were closer to the weaning weight of the heavier (Suffolk) parent than the lighter (Corriedale) parent. This is more supporting evidence of dominance variation for body weight differences between breeds.

INTRA-BREED GENETIC VARIATION

There are many published analyses of genetic variation of body weights within breeds. All these analyses have been made to estimate heritability in the narrow sense and they vary considerably in magnitude depending on method of estimation, breed, source of data, and the age and sex of animals involved. The early heritability estimates were calculated on data which had not been corrected for differences in body weight due to number of lambs born per parturition, sex, age of dam, year of birth, season of birth, type of rearing and farm of origin. There is now much evidence, summarized by Bowman (1966), that some or all of these factors have significant effects on body weight at different ages. The correction factors applicable (to reduce bias in heritability estimates) are specific to the data from which they are calculated, so that they must be estimated separately for any new set of data. Examples of methods of correcting data for the environmental factors mentioned are given by Hazel and Terrill (1945), Yalcin and Bichard (1964) and Bowman, Marshall and Broadbent (1968).

Heritability estimates have been calculated by three methods. These are the half sib and full sib correlations and the offspring–parent regression technique. The latter can take one of three forms, which are the offspring–dam, offspring–sire and offspring–mid parent regressions. Whilst all these methods largely estimate additive genetic variation as a proportion of total variation, it is necessary to realize that there are sources of bias in all cases. These biases are caused by common maternal effects and non-additive genetic variation. The composition of the genetic variation estimates by the three heritability methods is as follows:

Full sib correlation: $\quad\quad\quad\sigma_A^2 + \frac{1}{2}\sigma_D^2 + \frac{1}{2}\sigma_{AA}^2 + \frac{1}{4}\sigma_{AD}^2 + \frac{1}{8}\sigma_{DD}^2$

Half sib correlation: $\quad\quad\quad\sigma_A^2 \quad\quad\quad + \frac{1}{4}\sigma_{AA}^2$

Parent-offspring regression: $\quad\sigma_A^2 \quad\quad\quad + \frac{1}{2}\sigma_{AA}^2$

where σ_A^2 = genetic variation attributable to additive effects.

$\quad\quad\sigma_D^2$ = genetic variation attributable to dominance effects.

$\quad\quad\sigma_{AA}^2$ = genetic variation attributable to additive × additive effects.

$\quad\quad\sigma_{AD}^2$ = genetic variation attributable to additive × dominance effects.

$\quad\quad\sigma_{DD}^2$ = genetic variation attributable to dominance × dominance effects.

The estimate from the full sib correlation is biased upwards by any common maternal effects within full sib groups and the estimate from the parent-offspring regression, particularly the dam-daughter regression, is also biased upward by common maternal and environmental effects between paired individuals of the two generations. The half sib correlation estimate is expected to be the least biased because common environmental effects within half sib groups are rather less likely than for the other groupings. The offspring-sire regression is not likely to be biased by common environmental effects between generations but the heritability estimate derived from this regression contains twice the additive × additive genetic variation contained in the half sib correlation estimate.

There are several reports where heritability has been calculated by different methods from the same data source, and these give a guide to the relative importance of additive and non-additive genetic variation and environmental-maternal effects on body weight. The relevant material has been summarized in Table 2. The comparison of full and half sib correlation estimates by Botkin (1964), showing the former to be double or more than double the

Table 2

Comparisons of heritability calculated by different methods from the same source data

Reference	Character	Method of calculating heritability*				
		FS	HS	OD	OS	OMP
Botkin (1964)	Weaning wt.	0·59	0·21			
	Initial wt. on test	0·68	0·24			
	Final wt. on test	0·56	0·20			
	Gain from weaning to end of test	0·32	0·08			
	Gain on test	0·24	0·12			
Carter and McClure (1962)	Gain from birth to weaning		−0·02		0·03	
	120 day wt.		0·12		0·08	
Ensminger et al. (1943)	Shropshire birth wt.		0·10	0·10		
	Shropshire 140 day wt.		0·12	0·12		
	Southdown birth wt.		0·08	0·40		
	Southdown 140 day wt.		0·12	0·06		
Hazel and Terrill (1946)	Corriedale weaning wt.		0·32	0·45		
	Targhee weaning wt.		0·08	−0·01		
	Columbia weaning wt.		0·16	0·21		
Hazel and Terrill (1945)	Rambouillet weaning wt.		0·27 ± 0·05	0·34 ± 0·08		
Nelson and Venka-tachalam (1949)	Birth wt.		0·29 ± 0·14	0·72 ± 0·10		
	Weaning wt.		0·42 ± 0·21	0·15 ± 0·17		
Warwick and Cart-wright (1957)	120 day wt.		0·56	0·77	0·27	0·41
Yao et al. (1953)	Birth wt.		0·18	0·35		0·25

*Key: FS—full sib, HS—half sib, OD—offspring-dam, OS—offspring-sire, OMP—offspring-midparent

latter, suggests considerable dominance and non-additive genetic variation and large maternal effects on body weight even in the postweaning stages of growth. These conclusions are supported by the comparisons of half sib correlation estimates with offspring-dam regression estimates (Ensminger, Phillips, Schott and Parsons, 1943; Hazel and Terrill, 1945, 1946; Nelson and Venkatachalam, 1949; Warwick and Cartwright, 1957; Yao, Simmons and Schott, 1953). The fact that offspring-dam regression estimates are larger whilst offspring-sire estimates are smaller, generally, than half sib correlation estimates can be explained if there is little additive × additive genetic variation but large common environmental factors between daughter and dam. The intermediacy of the offspring-midparent estimate between the offspring-dam and offspring-sire estimates is in agreement with this hypothesis. It can be concluded from these results that for body weight at birth, weaning and postweaning, additive × additive genetic variation is relatively unimportant, maternal and common environmental factors are important and the importance of dominance variation cannot be estimated, though bearing in mind the interbreed cross results and the comparison of the full and half sib regression estimates, it is to be expected.

It has been mentioned already that sex has significant effects on body weight and Young, Turner and Dolling (1960) investigated the possibility of differences in heritability between sexes for 15–16 month body weight in Australian Merinos. These authors found no significant difference. This subject is worth further investigation along with the genetic correlation of weight in the two sexes. The sex differential for body weight has important applications to the meat production industry. There is evidence in poultry (Horton and McBride, 1964) that heritability may not be the same in the two sexes and that the genetic correlation is less than unity. Selection techniques for weight change and for sex differential might be different from those currently applied if the true situation was known.

As maternal effects, including ewe milk production, are obviously major influences on body weight and as the influences may be differential between age periods, the genetic variation may change for different phases of growth. Several reports deal with this aspect of body weight and the results are summarized in Table 3. These data are consistent in that, for appropriate comparisons, heritability at ages of a year and more are without exception higher than those at weaning and with only one exception higher than those at birth. It is to be expected that heritability would be highest at later ages because milk production and other maternal effects of the ewe would have least influence at that age. The comparative values of heritability at weaning and at birth do not show a consistent trend, with some values higher at weaning and some higher at birth. An explanation for this inconsistency may be that the age of weaning and the dependence of the lamb on the ewe's milk supply at weaning may be different for the several breeds and environments from which the data were collected. Thus, for those lambs which were still heavily dependent on their ewe's milk at weaning the heritability may be lower than at birth, whereas in those cases where the lambs have effectively weaned themselves before physical separation from their ewes, the heritability may be equal or higher at weaning than at birth. The data are not informative on this hypothesis, but it could be tested easily by comparing heritability from split half-sib groups, in which one group was reared artificially to weaning and the other group reared on the ewe.

Table 3

Comparisons of heritability of body weight and growth calculated
at different ages of the same animals

Reference	Breed	Heritability for weight at		
		Birth	Weaning	Later
Blackwell and Henderson (1955)	Down breeds	0.33 ± 0.11	0.07 ± 0.19	
Cassard and Weir (1956)	Suffolk	0.09	0.41	
Doney (1958b)	Welsh	0.30	0.68	0.80 (18 months)
Karam (1959)	Rahmani	0.16 ± 0.12	0.18 ± 0.12	0.19 ± 0.16 (1 year)
MacNaughton (1957)	Rambouillet	0.27	0.33	0.52 ⎱ Shearling
	Corriedale	0.36	0.45	0.46 ⎰
Osman and Bradford (1965)	Targhee (high nutrition)	0.45 ± 0.17	0.40 ± 0.17	1.06 ± 0.35 (450 days)
	Targhee (low nutrition)	0.19 ± 0.07	0.19 ± 0.08	0.40 ± 0.16
Ragab et al. (1953)	Ossimi	0.34 ± 0.13	0.10 ± 0.14	0.29 ± 0.20 (market wt)
Tallis (1960)	Merino	0.33	0.09	
Vesely and Slen (1961)	Romnelet		0.28 ± 0.09	0.37 ± 0.11 (1 year)

Some further evidence on this early decline of heritability after birth followed by increasing heritability estimates close to weaning is presented in Table 4. Apart from the estimates of Cassard and Weir (1956), which were higher for growth rate from birth to 70 days of age than from 70 to 120 days of age, the other reports show a decline in heritability postnatally. Presumably, the decline in heritability which is contemporaneous with the period of maximum dependence for nutrition of the lamb on the ewe, is caused by the variation in ewes milk supply. The conclusion is that additive genetic variation is quite large for birth weight and for postweaning weights but that weights between birth and weaning, or possibly only for 6–8 weeks postnatally, are reduced by variation in maternal environment. To digress briefly, it would appear that though heritability is of a comparable magnitude for birth and postweaning weights, the genetic correlation between birth weight and later weights declines the greater the time between the two weights.

Table 4

Comparisons of heritability of preweaning body weight and growth
calculated at different ages of the same animals

Reference	Breed	Heritability of weight at				
Broadbent and Watson (1967)	Suffolk × Welsh	2–16 weeks 0.32 ± 0.21 8–16 weeks 0.52 ± 0.28				
Bichard and Yalcin (1964)	Suffolk × Half bred	Birth 0.08 ± 0.05	6 weeks 0.02 ± 0.04	9 weeks 0.08 ± 0.05	12 weeks 0.09 ± 0.06	15 weeks $0.10 \pm .06$
Cassard and Weir (1956)	Suffolk	0–70 days 0.40			70–120 days 0.18	
Harrington et al. (1962)	Rambouillet	0–50 lb 0.10 ± 0.07			50–90 lb 0.38 ± 0.13	
				0–90 lb 0.35 ± 0.12		

Data published by MacNaughton (1957) and Bichard and Yalcin (1964) supporting this conclusion are presented in Table 5.

Finally, in reviewing heritability estimates derived from different sources it is productive to look at data from several breeds. The Down breeds have been primarily bred for lamb production and rapid early growth whereas the Merino, the hill breeds such as the Welsh Mountain and the range sheep of the United States of America have been bred for wool, hardiness and prolificacy. Under these circumstances the heritability of body weight, particularly in early postnatal life, might be expected to be lower in the Down breeds than in the other breeds mentioned if selection has been applied and been effective. A summary of half-sib correlation and within-sire dam-offspring regression heritability estimates for several breeds are given in Table 6, and these tend to be generally in accord with the expectations. Certainly heritability is low in the Down breeds and rather higher in range type sheep in the United States of America. Several of the American range breeds have been derived relatively recently by selection from crosses of several other breeds, and for this reason also, genetic variation would be expected to be higher than in the earlier derived Down breeds. However, as must be empha-

Table 5

Genetic and phenotypic correlations between body weights at different ages*

Reference						
MacNaughton (1957)		*Birth wt.*	*Weaning wt.*	*Shearling wt.*		
Corriedales:	Birth wt.	—	0·52	0·37		
	Weaning wt.	0·54	—	0·61		
	Shearling wt.	0·50	0·06	—		
Rambouillet:	Birth wt.	—	0·40	0·36		
	Weaning wt.	0·24	—	0·46		
	Shearling wt.	0·44	0·06	—		
Bichard and Yalcin (1964) Suffolk ×		*Birth wt.*	*6 wk. wt.*	*9 wk. wt.*	*12 wk. wt.*	*15 wk. wt.*
Half-bred:	Birth wt.	—	0·52	0·36	0·34	0·20
	6 wk. wt.	0·57	—	0·90	0·87	0·74
	9 wk. wt.	0·27	0·94	—	0·96	0·89
	12 wk. wt.	0·26	0·90	0·96	—	0·96
	15 wk. wt.	0·09	0·77	0·88	0·96	—

* Phenotypic correlations above and genetic correlations below the diagonals

sized again, the breed comparisons are confounded by differences in management affecting age at weaning and the dependence of the lamb on the maternal environment just prior to weaning. Comparisons of breeds based on standardized stages of maturity might lead to a different set of conclusions to those above.

INBREEDING

The available evidence on the effect of inbreeding on body weight in sheep has been reviewed by Rae (1956) and Bowman (1966). The experimental results all confirm that body weight declines with inbreeding, which is further evidence of dominance variation for this trait. Brown, Baugus and Sabin (1961) have reported a reduction due to inbreeding of all weights of lambs from birth to 120 days of age and found that inbreeding of both dam and lamb had an effect, though the effect was greater on the latter. After inbreeding without selection over several generations, Doney (1957, 1958a) found that Merinos in Australia were lower in body weight and had an increased variance for this character. Attempts have been made in the United States of America to produce many inbred lines of sheep

Table 6

A comparison of heritability of body weight at two ages in different breeds

| Reference | Breed | Heritability of weight at | | | |
| | | Birth | | Weaning | |
		HS*	OD	HS	OD
Blackwell and Henderson (1955)	Down breeds		0·33±0·11		0·07±0·19
Bichard and Yalcin (1964)	Down rams	0·08±0·05		0·10±0·06	
Bowman and Broadbent (1966)	Down rams			0·22	
	Suffolk rams			0·14	
Carter and McClure (1962)	Hampshire rams			0·12	
Cassard and Weir (1956)	Suffolk	0·09		0·41	
Ensminger et al. (1943)	Shropshire	0·08	0·10	0·12	0·12
	Southdown	0·10	0·40	0·12	0·06
Givens et al. (1960)	Hampshire rams			0·02	
Dalton (1962)	Welsh Mountain				0·51±0·07
Doney (1958b)	Welsh Mountain		0·30		0·48
Chang and Rae (1961)	Romney Marsh				0·35
Karam (1959)	Rahmani	0·16±0·12		0·18±0·12	
Ragab et al. (1953)	Ossimi		0·34±0·13		0·10±0·14
Tallis (1960)	Merino	0·34		0·09	
Hazel and Terrill (1945)	Range sheep			0·27±0·05	0·34±0·08
Ibid. (1946)	Columbia			0·16	0·21
	Corriedale			0·32	0·45
	Targhee			0·08	−0·01
MacNaughton (1957)	Rambouillet		0·27		0·33
	Corriedale		0·36		0·45
Osman and Bradford (1965)	Targhee	0·45±0·17		0·40±0·17	
Shelton and Campbell (1962)	Rambouillet			0·14	
Twombly et al. (1961)	Columbia × Rambouillet			0·35	
Vesely and Slen (1961)	Romnelet			0·28±0·09	

* HS—half sib   OD—offspring-dam

from but a few sources, but the data have not been published from either the inbreeding programme or from any crosses made from the lines which survived. Such data could provide useful information on the genetic control of body weight.

SELECTION PROGRAMMES

Reports of two short-term selection experiments for weaning weight in sheep have been published. An experiment covering 10 years and four selection generations (overlapping generations were involved in the mating scheme) for high and low weaning weight in the Australian Merino has been analysed by Pattie (1964, 1965a). For part of this experiment selection was based on weight at weaning which varied in age from 95 to 165 days for individual lambs. For the last 4 years of the experiment weaning weights were corrected for both age and type of birth (single or twin) by standard corrections. Apart from the two selection flocks a random bred control flock was also maintained. Heritability of corrected weaning weights was calculated for the base and random bred flocks by two methods (paternal half sib correlation and dam-offspring regression) and for ewes and rams separately. There was no significant difference between the sexes in heritability calculated by either method but though not significantly different heritability was nearly doubled when estimated by dam-offspring regression compared to the paternal half sib correlation. This is in accord with other reports summarized above.

Selection for high and low weaning weight was successful as judged by deviation from the control. Realized heritability was calculated from the selection differential and the selection response, both estimated as differences between the two selection flocks, for the two sexes separately. It is extremely interesting to note that realized heritability was significantly greater for ewes than for rams, being 0.31 ± 0.05 and 0.19 ± 0.05 respectively, and that for ewes the realized heritability almost equalled the dam-offspring regression estimate for ewes for the base and random bred flock. For rams the realized heritability equalled the half sib correlation estimate for the base and random bred flock. Pattie (1965a) offers no explanation for these results but states that they cannot be explained by differential inbreeding effects between sexes. Sex linked genes for body weight could be an explanation and such genes could be either on the sex chromosomes or on the autosomes but be sex dependent for their expression. Such an explanation indicates that the genetic correlation for body weight between sexes is less than one.

Other interesting findings in this experiment reported by Pattie (1965b), included the result that selection for weaning weight had caused both a change in the milk production of the ewe and a genetic change in the body weight of the lambs. The latter was considered to be the main factor causing the divergence of the high and low weaning weight flocks.

Selection led to correlated responses in weight at 17 months of age and the realized genetic correlation of 0·72, together with an estimated heritability of body weight at 17 months of between $0·93 \pm 0·03$ (dam-offspring regression estimate) and $0·50 \pm 0·18$ (paternal half sib correlation estimate) indicate that it would be as productive to select for changes in weaning weight by selection on 17 month weight as by direct selection. However, from applied viewpoints a larger adult sheep as well as a larger animal at weaning may not be always what is wanted.

Another three generation selection experiment for increased 120-day weight in two flocks at separate locations but with a cross comparison provided by a common control flock has been reported by Osman and Bradford (1965). The two locations represented a high and a low nutritional environment and throughout the experiment heritability for 120-day weight was higher in the high plane environment. Realized heritability was comparable to estimated heritability in the low plane environment but much lower in the high plane environment. However, Osman and Bradford found no evidence of large genotype-environment interactions for this trait. They did find a genetic correlation of 0·5–0·8, depending on the environment, between 120- and 450-day weight.

These two experiments both indicate that realized heritability over the short term for weaning weight is comparable to the estimated heritability from base populations. They also reveal a high genetic correlation between weaning weight and adult body weight.

BODY COMPOSITION

Studies of the genetic variation of the components, such as lean, fat and bone, of body weight have been hardly started, largely because of the difficulty of finding suitable quantitative measures of the components or even in deciding what components to measure. One study of the genetic variation of the components of a 'commercial' dissection of lamb carcasses has been reported by Bowman et al.

(1968). Lamb carcasses relating to 73 sire half sib groups of Down cross sheep were 'commercially' dissected into retail joints. The carcasses included had been selected for suitable fat cover both on the basis of a live animal assessment just before slaughter and on a visual appraisal of the whole hanging carcass. Carcasses weighing more than 45 lb were not included in the study.

It was found that heritability calculated from the paternal half sib correlation was 0·23, 0·49, 0·28 and 0·52 for percentage leg, percentage best end, age at slaughter and eye muscle area respectively. There was a significant negative correlation between percentage leg and percentage best end of neck and though no direct estimate was possible the results also suggested that there is still considerable genetic variation for fat percentage in sheep.

CONCLUSIONS

It is clear that the knowledge of genetic variation of body weight and composition of sheep is fragmentary and in many cases not sufficiently critical. In particular, data on the variation of rates of maturity both between and within breeds must be obtained. There is only limited information on the magnitude of genotype-environment interactions with respect to this trait. Prenatal effects on birth and postnatal body weight have not been mentioned mainly because there is little information and because the work by Hunter (1956) and Dickinson, Hancock, Hovell, Taylor and Wiener (1962) leads to different conclusions, which make generalizations invalid. Finally, body weight is very much a composite trait and knowledge of genetic variation of the components and how they change with age and under selection is urgently needed.

REFERENCES

Blackwell, R. L. and Henderson, C. R. (1955). *J. Anim. Sci.* **14**, 831

Bichard, M. and Yalcin, B. C. (1964). *Anim. Prod.* **6**, 179

Botkin, M. P. (1964). *J. Anim. Sci.* **23**, 132

— and Paules, L. (1965). *J. Anim. Sci.* **24**, 1111

Bowman, J. C. (1966). *Anim. Breed. Abstr.* **34**, 293

— and Broadbent, J. S. (1966). *Anim. Prod.* **8**, 129

— Marshall, J. E. and Broadbent, J. S. (1968). *Anim. Prod.* In press

Boyd, J. M., Doney, J. M., Gunn, R. G. and Jewell, P. A. (1964). *Proc. zool. Soc., Lond.* **142**, 129

Broadbent, J. S. and Watson, J. H. (1967). *Anim. Prod.* **9**, 99

Brown, C. J., Baugus, C. A. and Sabin, S. (1961). *Bull. Ark. agric. Exp. Stn.*, No. 646

Carter, R. C. and McClure, W. H. (1962). *J. Anim. Sci.* **21**, 970 (abstr.)

Cassard, D. W. and Weir, W. C. (1956). *J. Anim. Sci.* **15**, 1221 (abstr.)

Chang, T. S. and Rae, A. L. (1961). *N.Z. Jl agric. Res.* **4**, 578

Dalton, D. C. (1962). *Anim. Prod.* **4**, 269

Dickinson, A. G., Hancock, J. L., Hovell, G. J. R., Taylor, St. G. S. and Wiener, G. (1962). *Anim. Prod.* **4**, 64

Doney, J. M. (1957). *Aust. J. agric. Res.* **8**, 299

— (1958a). *Aust. J. agric. Res.* **9**, 252

— (1958b). *Aust. J. agric. Res.* **9**, 819

Ensminger, M. E., Phillips, R. W., Schott, R. G. and Parsons, C. H. (1943). *J. Anim. Sci.* **2**, 157

Givens, C. S. (Jr), Carter, R. C. and Gaines, J. A. (1960). *J. Anim. Sci.* **19**, 134

Harrington, R. B., Brothers, D. G. and Whiteman, J. V. (1962). *J. Anim. Sci.* **21**, 78

Hazel, L. N. and Terrill, C. E. (1945). *J. Anim. Sci.* **4**, 347

— — (1946). *J. Anim. Sci.* **5**, 371

Horton, I. E. and McBride, G. (1964). *Proc. Aust. Poult. Conv. 1964*, 79

Hunter, G. L. (1956). *J. agric. Sci., Camb.* **48**, 36

Karam, H. A. (1959). *Emp. J. exp. Agric.* **27**, 313

King, J. W. B. and Young, G. B. (1955). *J. agric. Sci., Camb.* **45**, 331

Lambe, J. W. (Jr), Bowman, G. H. and Rennie, J. C. (1965). *Can. J. Anim. Sci.* **45**, 1

MacNaughton, W. N. (1957). Abstr. *Doct. Thesis No. 1763* (1956). *Iowa St. Coll. J. Sci.* **31**, 465

Nelson, R. H. and Venkatachalam, G. (1949). *J. Anim. Sci.* **8**, 607

Osman, A. H. and Bradford, G. E. (1965). *J. Anim. Sci.* **24**, 766

Pattie, W. A. (1964). *Proc. Aust. Soc. Anim. Prod.* **5**, 152

— (1965a). *Aust. J. exp. Agric. Anim. Husb.* **5**, 353

— (1965b). *Aust. J. exp. Agric. Anim. Husb.* **5**, 361

Phillips, R. W., Schott, R. G., Lambert, W. V. and Brier, G. W. (1940). *Circ. U.S. Dep. Agric.* 580

Rae, A. L. (1956). *Adv. Genet.* **8**, 189

Ragab, M. T., Asker, A. A. and Kadi, M. R. (1953). *Emp. J. exp. Agric.* **21**, 304

Ryder, M. L. (1964). *Agric. Hist. Rev.* **12**, 1 and 65

Shelton, M. and Campbell, F. (1962). *J. Anim. Sci.* **21**, 91

Sidwell, G. M., Everson, D. O. and Terrill, C. F. (1964). *J. Anim. Sci.* **23**, 105

Tallis, G. M. (1960). *J. Anim. Sci.* **19**, 1208

Twombly, L. T., Sutherland, T. M., Esplin, A. L. and Stonaker, H. H. (1961). *J. Anim. Sci.* **20**, 678 (abstr.)

Vesely, J. A. and Peters, H. F. (1964). *Can. J. Anim. Sci.* **44**, 215

— and Slen, S. B. (1961). *Can. J. Anim. Sci.* **41**, 109

Warwick, B. L. and Cartwright, T. C. (1957). *J. Anim. Sci.* **16**, 1025
Wiener, G. (1967). *Anim. Prod.* **9**, 177
Yalcin, B. C. and Bichard, M. (1964). *Anim. Prod.* **6**, 73
Yao, T. S., Simmons, V. L. and Schott, R. G. (1953). *J. Anim. Sci.* **12**, 431
Young, S. S. Y., Turner, H. N. and Dolling, C. H. S. (1960). *Aust. J. agric. Res.* **11**, 604

GENETIC ASPECTS OF GROWTH AND DEVELOPMENT IN THE PIG

MAURICE BICHARD

School of Agriculture, The University, Newcastle upon Tyne

THIS SEEMS to be a particularly useful time for workers from many different disciplines to discuss the subject of mammalian growth and development with geneticists. There is an accepted view that research in quantitative genetics has finished with one era of development but has not yet decided which new direction will be most fruitful in the future. I do not think that any such dilemma faces those workers primarily interested in the application of existing genetic theory. The problem which they face is how to move our national animal populations forward, as rapidly as possible, in the direction of improved efficiency. During the past 25 years a body of theory dealing with different selection methods, and considerable knowledge of the parameters needed to apply this theory to farm animals have been developed. Other workers have continued to improve our understanding of growth, development and carcass quality. It is now important to co-ordinate all these aspects and to consider how we should define improved efficiency and which components to emphasize in our new and large-scale selection schemes.

This review deals first with an outline of current thought on growth and development in animals, based largely upon the work of Taylor (1965a, 1965b), Taylor and Craig (1965), Fowler (1966a) and Elsley, McDonald and Fowler (1964). This is necessary to provide the framework within which to discuss our knowledge of breed differences in economically important growth traits, for our established British pig breeds as well as for some recent arrivals in the U.K. and for their crosses, and also the question of selection procedures for improving muscle growth in pigs.

BASIC FRAMEWORK OF GROWTH AND DEVELOPMENT

An individual animal grows during postnatal life by taking in more nutrients than it requires to maintain its existing body mass. It grows skeletal, nervous and muscular tissue, with some associated essential fatty tissue. If it has more productive energy available than can be utilized at any one time for these types of growth (chiefly muscle) then it will store the surplus as additional fatty tissue. We may recognize an energy utilization potential of the lean tissues, or a lean tissue growth potential, which will vary in different periods of the individual's growth and development.

Since energy is required for maintenance almost irrespective of the rate of growth, the overall food cost per unit of growth (food conversion efficiency) will improve the higher the rate of growth. In addition, since lean tissue production is energetically much cheaper than fat production, food conversion efficiency is improved when lean is deposited. Thus to achieve the most efficient growth, and the most desirable carcass for today's requirements, we need to achieve the optimum intake of energy which will just satisfy the individual's lean tissue growth potential.

Under common pig keeping systems the growing animal is frequently fed *ad lib.*, or nearly so, on highly digestible diets, so that the supply of metabolizable energy is controlled by the amount of food the individual is able, or chooses, to consume. This we call appetite. Appetite may often be out of step with lean growth potential. In some cases appetite may be limiting so that the optimum nutrient intake cannot be achieved. We should then expect the animal to grow somewhat more slowly, and to be less efficient, but to remain lean. In other cases appetite may be excessive so that the animal's intake has to be limited artificially if we are to prevent the wasteful conversion of energy to fat. It might seem useful if our pigs were the sort which had high lean tissue growth potential and large enough appetites to satisfy this from easily available feeds, but not excessively large appetites, so that naturally lean carcasses were produced.

If we are to pursue the goal of increased efficiency of feeds into lean animal tissues then we must look in detail at the scope for change. It seems that among animals so far examined there is little evidence of variation in digestive or metabolic efficiency. If there is little overall variation then clearly there can be little genetic variation. Other studies show that appetite at a given weight does show a good measure of genetic variation, which may give a possible method of improving food conversion by selection for animals with high voluntary intakes. Some of this appetite variation, however, is positively associated with variation in the maintenance requirements of different animals at the same weight, and so some of the increased intake is absorbed by the higher 'running' costs. In addition, most of the remaining variation in intake between animals at one weight is positively associated with differences in the animals' genetically determined mature sizes.

Individuals with larger mature size, and correspondingly higher appetite, will in fact grow more quickly and more efficiently between fixed weights, which is desirable. Such animals, however, do have two other characteristics which may be less desirable. They will, on average, be later maturing, and this may be desirable for an animal being fattened for slaughter, with today's consumer requirements for lean meat, since the carcasses will tend to be leaner at any given weight. However, for the breeding population to become later maturing may be a disadvantage if it means that the unproductive rearing stage has to be extended. There is some evidence for pigs, as for other farm stock, that larger mature size implies puberty at a later age (Joubert, 1963) and an extension of this waiting period may be undesirable. The second change to be considered is the effect of an increase in genetic mature size on the maintenance requirement of parent stock. This may be particularly important where the reproductive rate is low and thus where the maternal charge is a significant fraction of the cost of producing weaned offspring. This applies to cattle more than pigs, and to pigs more than poultry, which point will receive further consideration later.

Thus, it is only that fraction of the variation in appetite which remains after allowing for variation in maintenance cost at constant weight and for variation in mature body size, which gives uncomplicated changes in food conversion efficiency; in ruminants, at least, this residual variation is not very great.

Variation in appetite is not, of course, the only determinant of food conversion efficiency. There seems to be considerable genetic variation in the way productive energy is partitioned between the tissues. Robertson (1963) has discussed the quite different ways in which different breeds of cattle lay down body fat or secrete different quantities and qualities of milk. Fowler (1966a) discussed the varying ability of breeds of pigs to grow lean tissue or lay down fat. If we require lean carcasses and higher food conversion efficiencies then we may be able to select for genotypes which can deal with high amounts of energy for lean tissue growth before they have to divert the excess to fat depots.

During the past 35 years we may have overestimated perhaps, because of the confusing effects of fat, the simplicity with which the breeder can change the growth curves of his animals towards earlier maturity and rapid early growth. The very high positive correlations found between lean tissue growth and size at different ages mean that an increase in early rate of growth will be accompanied almost inevitably by an increase in mature size and later maturity. Taylor's studies with cattle seem to show that the most important single variable is mature body size. This can be changed by using any one of a variety of selection criteria and it is then difficult to escape the other correlated responses, desirable or not. This is at once both a depressing but also a simplifying point of view.

One example of such correlated responses in the growth curve, produced by continued selection for small size in pigs, is provided by Dettmers, Rempel and Comstock (1965), who reported results from 11 years of selection for low weight at 140 days of age in a population derived from crosses between several small wild types. The mean weights during generations 1–3 may be compared with the corresponding weights in generations 9–11. Under full-feeding the estimated response was a reduction of 29 per cent in 140-day weight. The corresponding reduction in weight at 56 days was 23 per cent, and at birth 14 per cent. Unfortunately, no such comparison can be made for mature weights, though the 12 month weight of the selected animals in the last 3 years averaged only 177 lb. These results seem to be an example of fairly large correlated responses in weights other than the weight used as the single selection criterion.

Similarly, our conclusions now are that we probably cannot change the relative growth rates of different parts of the lean body nearly so easily as we had been accustomed to believe. When fat was excluded, Elsley, McDonald and Fowler (1964) and Fowler (*see* page 195) found little evidence of changed body proportions from the widely different nutritional treatments applied by Mc-Meekan (1940) to pigs and by Pálsson and Vergés (1952) to lambs. From a more theoretical viewpoint, Taylor's (1965b) calculations of the genetic correlations between size of one part of the body and the time taken by another part to mature, emphasize the expected difficulty of genetically changing relative growth rates without upsetting relative dimensions, although these detailed results have come from cattle and not from pigs. While it is certain that if we look hard enough, individuals will be found of the same mature size but with different shaped growth curves and with different lean body proportions, nevertheless, it would seem that the deliberate production of such deviants by genetic selection will not be a simple task. To use Taylor and Craig's (1965) phrase, the 'genetic flexibility' of the mean growth curve, and of the mean body proportions, is not very great.

VARIATION AMONG PIGS: BRITISH BREEDS

Genetic variation may be considered at two levels, within and between breeds. The immediate difficulty is that while there is

Table 1

Comparison of growth and conformation traits in different breeds under progeny test station conditions (From Johnson, 1962, by courtesy of The Pig Industry Development Authority, London)

Breed	Number tested	Average daily gain (lb)	Food conversion efficiency*	Average depth of back fat (mm)	Carcass length (mm)	'Eye' muscle area
Large White	4,880	1·47	4·25	34	805	26·6
Landrace	2,796	1·46	4·34	32	813	27·9
Welsh	332	1·46	4·38	34	796	27·6
Wessex	320	1·46	4·50	40	788	25·4
Essex	104	1·42	4·35	39	785	26·2
Gloucester Old Spot	56	1·37	4·73	42	765	25·7
Large Black	32	1·45	4·51	44	787	23·2
Tamworth	8	1·45	4·68	41	770	24·9

* lb food/lb carcass gain

evidence of variation, and genetic variation, in any trait related to growth and development, there are practically no studies available to help evaluate animals in the way in which the preliminary outline suggests is necessary. For example, how much of the observed variation in appetite between or within breeds of pigs is associated with differences in mature lean body size? It would be difficult even to tabulate the average mature lean sizes of common British breeds in 1967, and there do not seem to be any attempts to do this for individuals within a breed. Taylor presumably found similar difficulties some years ago with cattle but has deliberately set out to provide this information, using twins as a very economical way of estimating the genetic parameters.

With this limitation that, as yet, the data are not available in the required form, it is possible, nevertheless, to consider briefly some evidence of variation between breeds. Table 1 is taken from Johnson (1962) and gives data on the progeny of boars tested at the National testing stations up to June 1961. The number of pigs representing the minority breeds is clearly inadequate for reliable judgements. One can, nevertheless, observe that there is little variation between the first three (white) breeds but that they form a different group from the other (coloured) breeds. The latter are, in the broadest terms, slower growing, less efficient, shorter and fatter, under the particular conditions of feeding to appetite to the same live weight.

Table 2

Comparison of growth and conformation traits in the Large White and Landrace breeds. (From progeny test data, Smith *et al.*, 1962 and Smith and Ross, 1965, by courtesy of Oliver and Boyd)

Trait	Large White	Landrace
Daily gain (lb)	1·52	1·45
Food conversion efficiency		
(lb food/lb carcass gain)	4·30	4·34
Average depth of backfat (mm)	35·7	32·5
Carcass length (mm)	806	812
Leg length (mm)	605	596
Loin length (mm)	367	370
Wt. of head (lb)	12·2	11·1
Appetite (lb feed/day at 50 lb)	2·36	2·31
Appetite (lb feed/day at 125 lb)	6·38	5·47
Appetite (lb feed/day at 200 lb)	7·41	7·18

The two common white breeds have been studied in more detail by Smith and others (Smith, King and Gilbert, 1962; Smith and Ross, 1965) and the means for a number of traits are given in Table 2. These means are based on over 2,000 pigs by some 250 sires for each breed. In general, the Large White seems to eat slightly more, grow a little faster and is shorter and deeper than the Landrace, while their efficiency and backfat thickness may be very similar. But are these appetite and growth rate differences associated with different mature sizes and mature maintenance costs? It is important to know just how far our breeds really do differ in appetite, growth rate and efficiency when reduced to a common mature size. Otherwise we may lose some genes controlling useful deviations from the norm merely because they are at present associated with the smaller breeds with low overall performance. What applies to the evaluation of between-breed differences must also apply during selection within any one breed. We shall return to a consideration of selection procedures later.

NEW BREEDS

In addition to our established breeds, there has now been the opportunity to have a preliminary look at three breeds important in other areas of the world. These are the Hampshire, a saddleback

Table 3

Deviations of new breeds from average performance of white breeds of pig. A + sign means a deviation in the desirable direction; a − sign means the reverse

	Daily gain	Food conversion efficiency	'Eye' muscle area	Percentage lean	Back fat thickness	Carcass length
Lacombe	+ +	+ +	−	−	−	0
Hampshire	+	+ +	+	+	0	− −
Pietrain	−	0 to −	+ +	+ +	+	− −

type from the United States; the Lacombe, a white pig developed recently in Alberta; and the Pietrain, a spotted breed from Belgium. The performance of these three, and of their crosses with other breeds, is being evaluated in a series of trials by the Pig Industry Development Authority and the Animal Breeding Research Organization. The results of these are not yet fully available so rather than quote misleading figures based on small sets of data, these three new breeds have been characterized in a qualitative way in Table 3, based on King (1966, 1967). Their performance is indicated by comparison with the average performance of contemporary Large White and Landrace pigs.

The Lacombe seems to possess a large appetite, to have a high growth rate and good food conversion efficiency but to lay on more backfat than our white breeds, although less than our native coloured breeds. The Hampshire is only half-way through its crossing trials with A.B.R.O. but already it can be seen to grow quickly and efficiently and to produce large 'eye' muscles and a high percentage of lean meat. It is a short pig and, although it carries a smaller total percentage of fat, the thickness of backfat is greater than would be expected on a Large White carcass of similar total fat content.

An impression of the Pietrain may be gained from preliminary testing since 1964 at three different centres. The pure-bred animal is short and deep with apparently heavy shoulders and hams. It has a relatively slow growth rate, apparently associated with a small appetite, a normal food conversion efficiency and an extremely lean carcass; this combination of characteristics make it the most unusual of the three new breeds and raises the question of how the Pietrain fits into the framework outlined in the initial discussion. There are no precise figures available for the relative mature lean weights but the Belgian breed is almost certainly smaller than our British white breeds and would thus be expected to grow rather slowly. This implies that it maintains a high lean tissue growth rate relative to its appetite until nearly 200 lb live weight, which means a smaller amount of fat deposition. Whether or not its absolute level of lean tissue growth is remarkable, or merely its appetite control, needs further investigation. Preliminary data does suggest that the lean growth rate in the pure Pietrain is, in fact, higher than in other breeds, despite its lower live weight gains. Presumably its fairly normal food conversion efficiency, at least up to 140 lb live weight, results from two opposite processes; a decreased efficiency as a simple result of slow growth, and an increase owing to the energetic efficiency of laying down lean instead of fat.

The conclusion was drawn earlier that alteration of the conformation of the lean body will be difficult owing to the high inter-part genetic correlations. This is supported by dissection work from several species, including cattle from which Butterfield (1963) has given evidence of the constancy of various muscle groups as a percentage of total muscle in animals of widely different breeds with different live conformation. It is perhaps ironic, therefore, that when a pig breed is found with considerably different proportions in the trimmed side its origins and mode of development are obscure! The Pietrain breed looks different in shape from other breeds, but, having learnt to be sceptical from the beef cattle situation, we need accurate cut-out data. Again the preliminary evidence needs further support but it certainly suggests that lean tissue development has been increased in the ham and lower region of the foreleg at the expense of the streak. Such a changed conformation may be an advantage, especially where the Wiltshire trade does not dominate the market. The Pietrain is a breed which will repay much more study.

CROSS-BREEDING

So far consideration has been given entirely to differences between pure breeds. When these are crossed there is little evidence of anything other than intermediate inheritance of traits indicative of growth and development. Thus, the performance of crosses between two breeds is expected to be at least midway between the parental performances with perhaps some slight positive deviation for early growth rate and food conversion efficiency.

PHENOTYPIC EXPRESSION OF GENETIC MATURE SIZE

This survey of our established breeds, and of the three most recent imports, appears to have answered very few of the questions which must be asked before we attempt the comparatively simple task of between-breed selection. Before dealing with the more difficult topic of within-breed selection an interesting recent development is worth consideration. An important point in the arguments used earlier is that high growth rate and high efficiency over a given weight interval can be achieved most easily when using a genotype with a large mature size. This very size is a disadvantage in such a system, however, since it implies that the mature breeding females will have proportionately greater feed costs for maintenance. The assumption is, of course, that a breeding female with such a genotype will, in fact, achieve a large phenotype. This is inevitable if such animals are on an *ad lib.* feeding regime for at least part of their lives, which would presumably be the case for the normally grazing ruminant. Phenotypic size may also be desirable where the successful growth of the offspring depends upon a large maternal phenotype.

If, however, the female can have her feed intake controlled throughout her life, so that she fails to reach her genetically determined mature size, and if she can, at that reduced size, still reproduce effectively, then some of our previous fears about the consequences of selection for high early growth rate may be unfounded. It is known that feed restriction throughout life can limit

317

the phenotypic expression of size; it is known also that, from an early stage, the piglets' food intake may be supplemented with a creep feed which can compensate for deficiencies in the maternal postnatal environment. Experimental interest in recent years in controlling feed intake in the pregnant and lactating sow has created just this system. Lodge, Elsley and MacPherson (1966a, 1966b) have

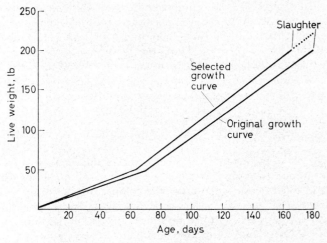

Figure 1. Simplified growth curves for pigs before and after selection causing increased mature size

shown that a sow's weight may be kept continuously 25 per cent below normal up to the end of the third lactation while she still farrows normal litters. Again, recently published figures from a larger trial indicate similar output from sows whose body weights differed by 40 per cent after three litters, where this weight difference was entirely a function of the reduced level of feed input (Elsley, Bannerman, Bathurst, Bracewell, Cunningham, Dodsworth, England, Forbes and Laird, 1967). It seems to be most important to learn precisely the relationship between feed input and litter output (measured in terms of litter size, survival and growth) which control performance from sows kept throughout their lives at different percentages below their genetic mature weight, and the effects of this on total lifetime performance. Not until we know this will we be in a position to balance the costs incurred per weaner from keeping a larger sow against the savings achieved from the more rapid and efficient growth to slaughter of her offspring. An approximate calculation may be made if some assumptions are accepted in place of established facts.

Let us suppose that pigs in the original unselected herd achieve a weight of 50 lb at 70 days of age and 200 lb at 180 days. Selection for growth rate and efficiency on test (50–200 lb) eventually results in a pig which is on average 10 per cent heavier at 180 days. This achievement may well involve a correlated response in mature size of the sow of, say, 10 per cent at a maximum. The sow will, therefore, have a higher maintenance cost of perhaps 7 per cent. What is this in terms of weight of food? The low-plane sows of Elsley *et al.* (1967) consumed about 2,000 lb per annum and their highest plane sows about 3,000 lb per annum, for maintenance, pregnancy and lactation; 7 per cent of 2,500 lb would amount to an extra 175 lb food per annum.

What are the savings involved? Each offspring grows more quickly and thus, by reducing its total maintenance requirement, more efficiently. The times taken to grow are summarized from the simplified growth curves given in *Figure 1*.

	Age at 50 lb (days)	Age at 200 lb (days)	Days on test
Original	70	180	110
Selected	64	167	103

A saving of 6 days up to 50 lb, and 7 days on test, is indicated. If we assume that the average maintenance requirement per day is 0·5 lb food before test and 1·25 lb while on test, we have a total saving of $6 \times 0.5 + 7 \times 1.25$ or 11·75 lb per pig. If a sow weans 16 pigs per year the total saving may be 11.75×16 or 188 lb food per year, against an estimated extra requirement in sow food of 175 lb.

There are inaccuracies on both sides of this calculation but the results may be sufficiently accurate to emphasize that an increased rate and efficiency of gain at the cost of increased mature size is not necessarily a useful result. The example may help to highlight the importance of keeping the sow's food consumption to a minimum and of understanding what this may do to her litter production. In passing it may be worth suggesting that there could be a bonus from genetically larger sows in the form of larger litters; there is evidence of this from mice where selection for faster gains produced larger mature mice having larger litters (Roberts, 1965).

Before discussing methods of genetic improvement of pigs, it is worth considering again the place of genetic improvement and, in particular, the phrase 'genetic potential'. Robertson (1963) has discussed this previously, but some statements by non-geneticists suggest that there is still a general misunderstanding surrounding the topic. The relevant section of a recent nutritional article concerned with improving pig performance is quoted below.

> The situation appears to be one in which there is always some limitation to the performance of the animal. If the environment is ideal in all respects then the limitation is genetic and further improvement can be achieved only by selection and breeding. If, on the other hand, the environment is not ideal, then the animal will not have reached its potential and much expenditure of time and effort on attempting to select superior breeding stock may be largely wasted because the best are not being challenged. It must be true that in the great majority of cases the pig is not being allowed to express its genetic worth because of defects in its environment.

Now I believe that this is an incorrect view. The assumption is made that genes basically capable of giving improved growth rate, for example, could not find expression at existing sub-optimum levels of management (this is diagrammatically shown in *Figure 2a* and *b*). Stated another way, this would imply that improvements from genetic and from environmental changes could not be combined simply in an additive fashion, but rather that the effect of changing the environment depends upon the particular genotype. Such an interaction between genotype and environment, if real, would indeed need to be considered in any improvement programme. But from theoretical reasoning (reviewed by Hull and Gowe, 1962) we should not expect to find serious interactions except where both genotypes and environments differ widely, by comparison with the variation within the environments. In addition, several experiments have been conducted on this very subject. Most of these have investigated the effects of feeding to appetite, compared with some form of restricted feeding, with two or more fairly different types of pigs. The results have been discussed by Hale and Coey (1963) and by King (1963), and, in general, they conclude that although interactions have been found they are not usually very important sources of variation. An exception may be the interaction of sex with nutrition. The more usual situation may, therefore, be that portrayed in Figure 2c.

These conclusions suggest, therefore, far from the conclusions given in the quotation, that at all times both genetic and environmental improvements should be sought simultaneously, but each only up to the economic limits. Thus a new feed additive, or

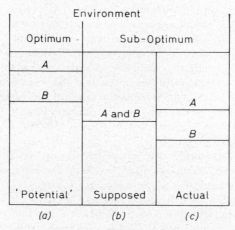

Figure 2. *Performance of two genetically different lines*

energy to protein ratio, or an improved ventilation system, will be worth while only if it increases profit, but it is likely to give a similar physical response whether the stock are good Large Whites or poor Large Whites. Equally, an expensive breeding programme can only be justified if it achieves results which can be passed on as higher profits from hundreds of individual pigs, but it is likely to produce useful improvement in herds fed high or low protein diets.

SELECTION FOR GREATER EFFICIENCY

The final aspect to consider briefly is that of selection procedures designed to improve the efficiency of conversion of resources into a saleable product. There could be considerable discussion on what the saleable product is, opinion ranging from the simplest definition, 'lean meat', which has considerable appeal as being a permanent definable trait, to a more sophisticated definition which recognizes that meat is produced in carcasses of particular shape and size, that particular muscles may have importance in their own right (the 'eye' muscle), that fat is quite acceptable and even desirable in certain forms and amounts, and that the consumer still buys joints with particular characteristics of size and shape. Personally, I think it unrealistic to consider that an economic activity like genetic improvement can be pursued in a democracy without taking account of trade and consumer preferences, even though these may be both somewhat irrational and rather transient. At present, therefore, it would not seem advisable for anyone to turn over entirely to selection for growth rate of lean tissues, or food cost per unit of lean tissue, though it is probably useful to record such data in order to gain the necessary information for a possible change towards these criteria.

We may confirm without repeating the arguments here (*see* P.I.D.A., 1965) that the selection procedure should be based on a performance test of the individual candidates for selection, perhaps supplemented by carcass data from sibs. What sort of feeding system should be employed? The currently accepted one at our National testing stations is to feed twice per day that amount of meal which the animal can clear up within 30 min, making weekly adjustments on the basis of observations by the stockman. The objection has been made that this permits animals with appetites above their lean tissue growth potential to become too fat. This is so, but such animals need not be automatically discarded on the basis of their fatness; their other attributes are considered too in their overall evaluation. An alternative proposal, to feed all pigs a fixed but increasing amount per day for a fixed test period, seems to fail to identify and utilize the large genetic variation in appetite which we accept to be available. A completely *ad lib.* feeding system could presumably still enable us to measure lean growth rate, and by using this, together with appetite, backfat and efficiency of conversion, we could attempt to select for a balance between them, while recognizing excellence in any one feature. Our own experience at Newcastle is that after an initial settling-down period it is quite possible to run a selection programme with *ad lib.* feeding without producing grossly fat carcasses.

It has been emphasized already that simple selection for fast growing, efficient pigs is likely to lead to an increase in genetic mature size and later maturity. Because improved food conversion can result both from faster growth and from the deposition of lean rather than fat tissue, it is possible that such selection might result in very fast growing but actually fatter pigs. Fowler (1966b) showed that the correlation between feed efficiency and carcass composition may be brought out only when the efficiency data are first corrected for the effect of growth rate on maintenance requirements. Such a selection result must be avoided by a joint consideration of carcass composition and growth traits, with appropriate weight given to each. If we insist that we want to identify those individuals which deviate positively from their expected growth curves (given their mature weight) then as Taylor and Craig (1965) have pointed out we are asking for something much more difficult. We could only expect to put useful selection pressure in such a direction by recording weight at frequent intervals, under closely specified nutritional conditions, and using a complex index taking account of all of the inter-age genetic correlations.

There may be, however, another way of avoiding some of the indirect consequences of selecting for faster growth rates. If pigmeat comes to be increasingly produced from three-way cross animals $(A \male \times (B \times C) \female)$, as seems likely, then the major selection for early growth rate can be confined to the breed or line destined to provide the final male parent (A). The cross-bred female parent $(B \times C)$ may perhaps be kept fairly small and thrifty, since her offspring will inherit rapid growth from their sire. This is not a new idea, it is the normal pattern in broiler production, but it will depend upon our ability to find at least three lines all with outstanding ability in at least one attribute. At present many people believe that there are only two established breeds in Britain worth incorporating in a cross, the Large White and the Landrace. This situation may change through the use of some of the new breeds already described, probably in combination with other native and imported stock. We can certainly afford to be less inhibited in our thinking about genetic methods for we should not accept that improvement will always be in the hands of normally intelligent farmers working with 50–100 sow units. If much larger breeding organizations evolve, as they give every sign of doing, then they will be able to take in their stride selection schemes which seem only wildly theoretical to the existing pedigree breeder.

SUMMARY

The main aim of the review has been to consider genetic aspects of growth and development in the pig in the context of current ideas on growth, appetite, food conversion efficiency and tissue composition, and their dependence upon mature genetic size. An outline of the consequences of selection for some of these attributes has been

323

attempted. A brief review of such traits in established British breeds is given besides a similar discussion of preliminary data on three recent imported breeds, Lacombe, Hampshire and Pietrain. The implications of larger sows are discussed in terms of extra maintenance costs but more efficient progeny, at the same time as the interesting consequences of restricting the mature size of sows through feed control. Finally, reference is made to the subjects of genetic potential, genotype-environment interactions, and methods of selection for improved efficiency.

REFERENCES

Butterfield, R. M. (1963). *Symp. Carcase Composition and Appraisal of Meat Animals.* Paper 7-1 (Ed. by D. E. Tribe). Melbourne, C.S.I.R.O.

Dettmers, A. E., Rempel, W. E. and Comstock, R. E. (1965). *J. Anim. Sci.* **24**, 216

Elsley, F. W. H., McDonald, I. and Fowler, V. R. (1964). *Anim. Prod.* **6**, 141

— Bannerman, M., Bathurst, E. U. J., Bracewell, A. G., Cunningham, J. M. M., Dodsworth, T. L., England, G. J., Forbes, T. J. and Laird, R. (1967). *Anim. Prod.* **9**, 270 (abstr.)

Fowler, V. R. (1966a). *Proc. Pig Industry Development Authority*, 2nd Conf. agric. Res. Workers and agric. Economists. p. 72

— (1966b). *Proc. 9th int. Congr. Anim. Prod.*, Edinburgh. p. 79 (abstr.)

Hale, R. W. and Coey, W. E. (1963). *J. agric. Sci., Camb.* **61**, 81

Hull, P. and Gowe, R. S. (1962). *Genetics* **47**, 143

Johnson, R. F. (1962). 'A study of the means of testing pigs in the United Kingdom.' *Memo. Pig Industry Development Authority, Lond.*

Joubert, D. M. (1963). *Anim. Breed. Abstr.* **31**, 295

King, J. W. B. (1963). *Anim. Prod.* **5**, 283

— (1966). *Proc. Conf. 'Breeding for Pig Improvement'.* Pig Industry Development Authority, Lond. p. 11

— (1967). *Pig Fmg.*, Jan. p. 75

Lodge, G. A., Elsley, F. W. H. and MacPherson, R. M. (1966a). *Anim. Prod.* **8**, 29

— — — (1966b). *Anim. Prod.* **8**, 499

McMeekan, C. P. (1940). *J. agric. Sci., Camb.* **30**, 276, 387 and 511

Pálsson, H. and Vergés, J. B. (1952). *J. agric. Sci., Camb.* **42**, 1

Pig Industry Development Authority (1965). *Rep. Study Group on Herd Improvement and Testing, P.I.D.A. Lond.*

Roberts, R. C. (1965). *Anim. Breed. Abstr.* **33**, 339

Robertson, A. (1963). *Proc. Wld. Conf. Anim. Prod.* **1**, 99

Smith, C., King, J. W. B. and Gilbert, N. (1962). *Anim. Prod.* **4**, 128

— and Ross, G. J. S. (1965). *Anim. Prod.* **7**, 291

Taylor, St. C. S. (1965a). *Anim. Prod.* **7**, 203
— (1965b). *A.B.R.O. Rep. Edinburgh.* p. 25
— and Craig, J. (1965). *Anim. Prod.* **7**, 83

DISCUSSION

PROFESSOR LUCAS (Bangor) asked DR. TAYLOR whether the results of his selection for mature size implied a linear response and might not the response in fact be curvilinear, due to biological limitations and the reduction of variability with time. A further point was whether it was equally meaningful to select for either relative growth rate or mature size. DR. TAYLOR replied that expressing results over 20 years was purely an artificial procedure to emphasize effects, not to show what would actually happen over that period. Previous work had demonstrated already the advantages to be derived in terms of weight gain; the object of the present exercise was to demonstrate that the greater part of the increase in weight gain was attributable to change in final size. He thought this indicated that size *per se* could not have any great economic advantage, otherwise breeders would have selected to the maximum for it and we would not have breeds like the Aberdeen Angus; the critical factor must be how fast the animal matures rather than maximum mature size.

MR. VIAL (Dublin) disagreed with this and suggested that the Aberdeen Angus had persisted not because of any outstanding genetic merit but because of the shrewdness of its breeders. DR. BLAXTER (Aberdeen) commented that if one took Dr. Taylor's arguments to their logical conclusion we should be producing the biggest possible mature animals to achieve maximum rates of gain, but DR. TAYLOR replied that this was the opposite of his contention; it was the present trend but no-one had yet shown that the large animal was more efficient than the small animal.

DR. PÁLSSON (Reykjavik) commented that we must have a selection index which enabled us to find animals growing relatively fast to mature size and asked whether there was any standard which related rapid early growth with mature size. DR. TAYLOR replied that it would not be difficult to produce a selection index which increased early growth rate without increasing mature size. PROFESSOR BOWMAN replied that, with sheep, it would also be possible to produce an index to select for a large animal in early life but this would depend on weaning the lamb early in order to increase heritability at the post-natal stage. He commented further that whereas arguments about sizes of meat animals were relatively unrestricted by retail trade requirements for cattle, this was not so for sheep; large cattle could be cut down easily into acceptable joints, whereas sheep could not.

MR. O'GRADY (Fermoy) asked whether it was not true that an animal with a large mature size would be leaner at any given weight than one with a smaller mature size, to which DR. TAYLOR replied that this was probably true and exemplified by sex differences, in that at birth bulls were only about 4 per cent of mature weight whereas heifers were about 6 per cent.

DR. McCARTHY (Dublin) asked for evidence that it *was* possible to select animals for rapid early growth rate without increasing mature size, and instanced the examples of the Friesian and Aberdeen Angus breeds. DR. TAYLOR replied that although there was little convincing evidence, there was latent genetic variation in most genetic characteristics and he believed it would prove possible to 'bend' growth curves. To a further question by Dr. McCarthy on optimum age for selection, DR. TAYLOR replied that any age prior to one year in cattle would have a slightly diminished effect but selection at or just beyond one year was about optimal for increasing early growth rate.

On the subject of bending growth curves, PROFESSOR BOWMAN commented that there was evidence that this could be done from poultry and turkeys. DR. TAYLOR added that 'double-muscled' animals showed this also, as they had much steeper growth curves in early life.

DR. BRAUDE (Reading) commented that there were two mature sizes of pig in the Large White breed and if one compared their growth rate and efficiency it was not related to ultimate size; similarly with miniature pigs. DR. BICHARD replied that it was difficult to believe that there was no relationship but it was possible that the degree of relationship was such that differences in performance were not predictable.

DR. MITCHELL (Reading) commented that it was very difficult to define feeding 'to appetite'; pigs in Progeny Test stations were not fed strictly to appetite but rather a certain amount twice daily; furthermore, pigs would eat more *wet* feed twice daily than they would *dry* feed *ad lib*. DR. BICHARD agreed that it was necessary to specify what was meant in any particular system.

DR. FOWLER (Aberdeen) asked whether selection for greater mature size and faster growth rate would result in a higher proportion of bone, because the pigs would be less mature at 200 lb. DR. BICHARD replied that he had assumed that slaughter would be at a constant weight but there was perhaps no need to always select an environment (which includes slaughter weight) and then select a genotype to fit it; the environment could be altered also and so the slaughter weight changed to avoid this problem.

DR. SALMON-LEGAGNEUR (Jouy-en-Josas) asked for evidence that increase in growth rate or mature size was positively related to reproductive efficiency, as the reverse seemed more likely to be true. DR. BICHARD replied that he had no direct evidence of this in the pig, because genetic variance for fertility was very low and therefore it was very difficult to estimate relationships between fertility and other characteristics, but in mice, selection for increased body weight had resulted in increased litter size.

PROFESSOR OSINSKA (Warsaw) commented on the statement that there were no genetic differences in digestibility of feed, as she was not aware of any digestibility data produced on a large enough scale to allow genetic conclusions to be drawn. DR. BLAXTER commented that the amount of phenotypic variation in digestibility coefficients was only about 9 per cent and the very limited evidence on genotypic variation indicated that it was very small.

DR. PRESCOTT (Newcastle) commented on the fact that genotype-environment interactions appeared to occur at least so far as sex differences were concerned, and perhaps one should accept that environmental effects might reduce genetic variation even though the ranking might remain the same. DR. BICHARD agreed that genetic/environmental interactions did occur and could be demonstrated if the procedures were sufficiently sophisticated; both environmental and genetic improvement should proceed simultaneously.

VI. NUTRITIONAL INFLUENCES

THE EFFECT OF THE DIETARY ENERGY
SUPPLY ON GROWTH

K. L. BLAXTER

The Rowett Research Institute, Bucksburn, Aberdeen

THAT THE amount of food and its quality are the most important factors determining rate of growth of an animal is well recognized. Perhaps the most spectacular demonstration of the effect of rate of food intake on rate of growth is the series of experiments by McCance and his colleagues (McCance, 1960; Lister, Cowen and McCance, 1966). By reducing the daily allowance of food for young poultry (*Figure 1*) and pigs, McCance stabilized their body weights at infantile values for long periods of time; in the case of pigs, at weights of about 7 kg for over a year (*see* year 224). When after a year these animals were given food *ad lib.*, weight gain was resumed and both the ultimate weight they attained and the rate of its attainment were not greatly different from values noted in controls fed normally. Females in particular were not affected to any appreciable extent by this nutritional stabilization of weight at infant values. There is some evidence (Fábry and Hrůza, 1956) that even in animals which do not attain full adult size after such treatment growth hormone administration enables them to do so.

Energy was without doubt the dietary component responsible for the variation in growth rate from zero to normal in these experiments. It can be argued that deficiencies of other nutrients also result in absence of gain and that restriction of food intake does not result in dietary deficit of energy alone. When a dietary deficiency of a specific nutrient is associated with a cessation of gain, however, food energy intake is also reduced. As shown by hundreds of paired feeding experiments, the induced reduction of the energy intake is a far more important factor in accounting for failure of growth than is any effect of the deficiency on the differential coefficient of weight on food energy intake dW/dM. The effect of a dietary deficiency of zinc on weight gain and food intake given in Dr. Mills' paper (page 368) illustrates the point that growth failure is associated with failure of voluntary intake of food.

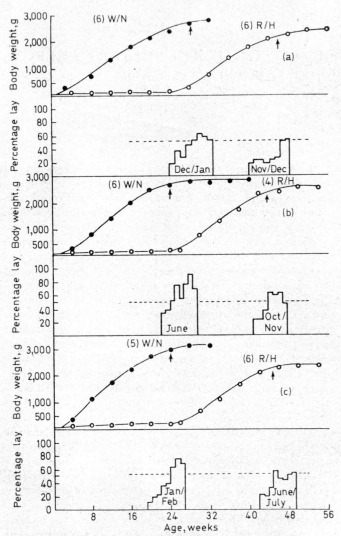

Figure 1. Growth curves and egg laying performance of normal (•——•) and rehabilitated (○——○) pullets. The numbers of birds in each group are given in parentheses. (a) Rhode Island Red birds: experimental (R/H) group hatched in February, control (W/N) group hatched in July; (b) Rhode Island Red birds: experimental (R/H) and control (W/N) groups hatched in January: (c) Light Sussex birds: experimental (R/H) and control (W/N) groups hatched in August. ↑ indicates age and body weight at which sexual maturity (50 per cent lay, shown by dotted lines) was attained (From Lister, Cowen and McCance, 1966)

By courtesy of Cambridge University Press

RATE OF FOOD ENERGY SUPPLY
AND RATE OF ENERGY RETENTION

The relationships between food energy intake and the protein and fat content of the body have been measured in many experiments. The general relationship is shown in *Figure 2* (Blaxter and Wainman, 1961), where daily rates of retention of energy as fat and protein

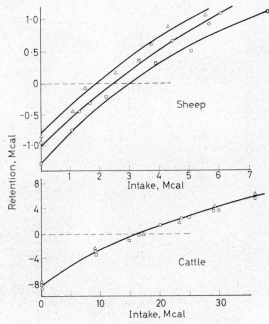

Figure 2. The relation between the heat of combustion of the food given and the energy retentions of 3 sheep and 3 cattle (From data of Blaxter and Wainman, 1961)

have been plotted against daily rates of food intake. The relationship is curvilinear, due partly to a slight fall in the digestive efficiency of the animal with increased feeding and partly to the fact that anabolic processes are less efficient than catabolic ones. The latter arises in part from thermodynamic considerations.

If the energy provided by food is expressed as metabolizable energy, thereby discounting changes which occur with increasing food intake in the proportional losses of energy from the body in the faeces, the urine, and as combustible gases, a good approximation to the general relationship is obtained by two straight lines intersecting at the point when energy retention is zero. This is shown in *Figure 3*. It should be noted that the rate of provision of food energy to maintain equilibrium of energy retention is a large proportion of the maximal rate of energy intake.

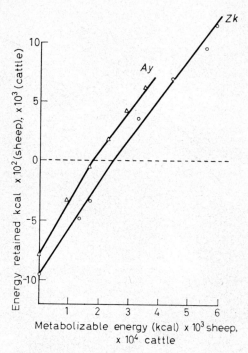

Figure 3. The relation between energy intake as metabolizable energy and energy retention for 1 of the sheep ($\mathcal{Z}k$) and 1 of the cattle (Ay) depicted in Figure 2

The rate at which energy is retained in the body in relation to the rate at which energy is provided in the diet can thus be expressed in two ways (Blaxter and Wainman, 1961).

When the animal receives more food than that required to maintain constant the energy content of the body

$$R = k_f \left(M - \frac{F}{k_m} \right) \qquad (1)$$

332

When the animal receives food in amounts that lead to no change or a loss of energy from the body

$$R = k_m M - F \qquad (2)$$

where

R = the retention of energy in calories/day.

M = the intake of metabolizable energy in calories/day.

F = the fasting metabolism in calories/day.

k_m = the efficiency of utilization of metabolizable energy for maintenance expressed as a decimal.

k_f = the efficiency of utilization of metabolizable energy for growth and fattening expressed as a decimal.

Table 1

The efficiency of utilization of the metabolizable energy of usual diets by different species

Species	Efficiency of utilization of metabolizable energy for maintenance (k_m)	Efficiency of utilization of metabolizable energy for growth and fattening (k_f)	Ratio k_f/k_m
Rat (1, 2)	95	84	0·88
Man (3)	92	?	
Chicken (4, 5)	81	75	0·88
Pig (6, 7)	80	70	0·88
Cattle (high roughage) (8)	68	40	0·60
and			
Sheep (low roughage) (8)	75	57	0·76
Horse (9)	71	?	
Dog (9)	89	75	0·84

(1) Hamilton, 1939
(2) Forbes and Swift, 1944
(3) Swift *et al.*, 1957
(4) Mitchell and Haines, 1927
(5) Hill and Anderson, 1958
(6) Breirem, 1938
(7) Nehring and Schiemann, 1966
(8) A.R.C., 1965
(9) Mitchell, 1964

The constants k_m and k_f have been measured in calorimetric experiments with a number of species and a number of diets. Table 1 summarizes values obtained for usual diets. The table is incomplete for some species but it appears that ruminants are unique in that they utilize metabolizable energy less efficiently than

333

non-ruminants. With ruminants the values of k_m and k_f vary more with diet than they do with omnivores.

The efficiency with which food energy is used to synthesize lipids does not differ to any large extent from that with which it is used to synthesize proteins. The synthesis of protein from amino acids involves a considerable expenditure of energy in the form of ATP, three moles ATP being required for each peptide bond formed. Synthesis of fat from carbohydrate or from smaller molecules is energetically efficient since it makes use of the reducing power of coenzymes, and H transfer appears generally to be a less wasteful process in terms of free energy transfer than does synthesis through phosphorylated intermediaries. Kielanowski's (1965) studies suggest that with mixed diets the energetic efficiency of protein synthesis in baby pigs is about 77 per cent and of fat synthesis about 81 per cent. Both values are higher than the average value for mixed diets given in Table 1, but they do not suggest large differences in the efficiencies of the two processes. With cattle we were unable to detect differences in overall efficiency (k_f) due to age of animal and hence to the proportions of fat to protein deposited in the body (Blaxter, Clapperton and Wainman, 1966b).

DERIVATION OF GROWTH EQUATIONS FROM ENERGY EQUATIONS

The relationship exemplified in equations 1 and 2 between rate of energy intake and rate of energy retention can be used to describe increase in body weight with time. The retention of energy and the fasting metabolism in equations 1 and 2 can be expressed in terms of body weight, W. Thus F, the rate of fasting metabolism, can be expressed generally as AW^n, the exponent having a value of about 0·7 as between species differing in size. Similarly R, the rate of energy retention can be expressed as $B\dfrac{dW}{dt}$, where B is the calorific value of gains.

Equation 1, describing the relationship between rate of energy retention and food intake above maintenance, can thus be written:

$$\frac{B\,dW}{k_f dt} + \frac{AW^n}{k_m} = M \tag{3}$$

and equation 2, describing energy utilization below maintenance, as:

$$\frac{B\,dW}{k_m dt} + \frac{AW^n}{k_m} = M \tag{4}$$

In these equations, M, the rate of food intake, has been regarded as constant but, in fact, during growth M varies with time. This aspect is considered later.

Equations 3 and 4 are both linear differential equations of precisely the same type which can be solved to describe the time course of growth. Their solution is not exact, the extent of the approximation being evident in Appendix 1.

Above maintenance the solution to equation 3 is

$$W - W_0 = k_m(M - M_0) \frac{k_m}{k_f} \frac{W_0^{1-n}}{An} \left(1 - \exp \left(- \frac{k_f An W_0^{n-1} t}{k_m} \frac{1}{B} \right) \right) \quad (5)$$

Below maintenance the solution to equation 4 is

$$W - W_0 = k_m(M - M_0) \frac{W_0^{1-n}}{An} \left(1 - \exp \left(- \frac{An W_0^{n-1} t}{B} \right) \right) \quad (6)$$

Equations 5 and 6 show that when the diet of an animal is changed from a maintenance value, M_0, to a new value, M, and then kept constant, body weight approaches a new value asymptotically. The exponents of the exponential terms in the two equations differ according to whether an increase or decrease in weight occurs. The efficiency of utilization of metabolizable energy for maintenance is higher than is that for growth and fattening. When a change in food intake results in an increase in weight, the exponent is smaller and the animal takes longer to reach weight equilibrium than when a change in food intake results in a decrease in weight. The exponent also increases, implying a more rapid approach to equilibrium the smaller the value of B, that is, the smaller the fat content and the higher the contents of protein and associated water.

EXPERIMENTAL VERIFICATION OF THE EQUATIONS

An experiment briefly reported by Clapperton and Blaxter (1965) was designed to test the validity of the equations relating food intake to the attainment of weight equilibrium. Sixteen 2-year old sheep were given a standard maintenance diet which was then increased or decreased by a constant amount and kept constant at that level for 10 months. Weight gain and energy metabolism were measured continuously in eight of the animals. The detailed analysis of the results is not complete but those summarized in Table 2 show that the rate of attainment of weight equilibrium following a change in diet was extremely slow, and that when the diet of a sheep was increased, only about 40 per cent of the ultimate response in weight was obtained in 10 months.

Table 2

Weights of sheep following changes in diet and the weights expected from equations 5 and 6

Group	Change in amount of metabolizable energy given*	Observed and expected †	Body weight (kg) in week		
			0‡	21	42
1	+1,406	observed	54·5 <	69·1	78·7
		expected		69·6	81·6
2	+703	observed	49·9 <	58·0	64·3
		expected		57·5	63·3
3	0	observed	48·6 <	50·8	50·6
		expected		48·6	48·6
4	−703	observed	43·8 <	37·2	33·2
		expected		33·0	25·7

* The initial allowance of metabolizable energy was 1,640 kcal/day
† The values of the constants in equations 5 and 6 were:
    (a) exponent of weight, n, 0·73
    (b) fasting heat loss/kg $W^{0.73}$, 57·5 kcal/day, determined by direct experiment (Blaxter, Clapperton and Wainman, 1966a)
    (c) k_f and k_m the efficiencies of utilization of metabolizable energy were computed from the metabolizable energy value of the diet (ARC, 1965) and were 0·50 and 0·72 respectively
    (d) the calorific value of weight changes was taken to be 6,000 kcal/kg
    (e) the initial body weights W_0 were taken to be those on the fourth day after the ration change
    (f) the changes in metabolizable energy intake were computed simply from the amounts of feed given and the metabolizable energy determined at the maintenance level
The resultant equations were:
    Group 1 $\Delta W = 71\cdot1\ (1 - e^{-0.00164t})$
    2 $\Delta W = 34\cdot6\ (1 - e^{-0.00168t})$
    3 $\Delta W = 0\cdot0\ (1 - e^{-0.00171t})$
    4 $\Delta W = 34\cdot6\ (1 - e^{-0.00254t})$
‡ This weight is that on the fourth day of the experiment: when diet is increased or decreased, the weight of the animal increases or decreases as a result of changes in its gut contents. When sheep given maintenance diets are fed, increases in body weight of up to 20 per cent can occur due to changes in gut contents. Conversely, fasting from the maintenance level results in losses of weight due to loss of gut contents of on average 8 per cent of body weight. (*See* Blaxter, Clapperton and Wainman, 1966a)

The table also shows that the sheep which were underfed lost less body weight than was expected from equation 6. This may well reflect adaptation to under-nutrition; indeed the sheep which were underfed were lethargic and not easily roused. The records of 24 h heat production do not suggest this adaptation was sequential, because after 3 weeks, heat production fell at the rate expected from the fall in body weight.

THE RELATION BETWEEN THE ENERGY EQUATIONS AND DESCRIPTIVE GROWTH EQUATIONS

It is of some interest that the exponents in equations 5 and 6, derived from considerations of the way in which a single increment of food is partitioned, have the same numerical magnitude as those in equations describing the so-called 'self-inhibitory' phase of growth. Brody (1945) has explored this phase of growth and has described the relationship by the empirical equation

$$W = A' \, (1 - e^{-k(t-t^*)}), \quad \text{or,} \quad W = A' \, (1 - e^{-1/\tau(t-t^*)}) \tag{7}$$

in which

A' = mature weight.

t^* = time origin of the equation.

$k = \dfrac{1}{\tau}$ is a constant.

Günther and Guerra (1955) and Taylor (1965) have examined the relationship between the exponent and the mature weight of animals of different mature size. Taylor's results were

$$\log \tau = 0.27 \log A + 2.0 \tag{8}$$

Not only is the form of equation 7, which describes the relation between weight and age in an individual species with no reference whatever to nutrition, similar to equations 5 and 6, which describe without reference to species the time dependent responses of weight to changes in nutrition, but Taylor's generalization of the descriptive growth equations of Brody yields an interspecies function for their time constants which is of precisely the same form as the time constants in the nutritional equations. Thus, time constants can be derived from equations 5 and 6 and compared with the value obtained by Taylor by substituting numerical values for the constants in the exponent in these two equations. The interspecies equation for the estimation of fasting metabolism from body weight is 70 $W^{0.73}$ (kcal/24 h), that is, in equations 5 and 6, n is 0.73 and A is 70.0 kcal/24 h. The ratio k_f/k_m is about 0.88 (*see* Table 2), and the mean calorific value of the bodies of animals is about 5,000 kcal/kg. Time constants so derived are:

337

	Value of time constant
From Taylor's statistical analysis of Brody's data	$100\cdot0\ W^{0\cdot27}$
From energetic considerations (equation 5 for gain of weight)	$111\cdot0\ W^{0\cdot27}$
From energetic considerations (equation 6 for loss of weight)	$97\cdot8\ W^{0\cdot27}$

The values for the time constants are virtually identical for a loss in weight; for a gain in weight from equation 5, the coefficient of $W^{0\cdot27}$ is increased by the ratio k_m/k_f, leading to higher values.

Values for the time constants obtained from Taylor's equation and from the energy equations are given in Table 3, while *Figure 4* compares the time constants estimated by Brody from records of weight and age of different animal species and that deduced above from energy considerations alone.

The concordance of the two sets of results suggest that the time constants of descriptive weight-age growth equations of the type computed by Brody have a deeper physiological meaning in terms of the bioenergetics of food utilization.

VARIATION IN FOOD ENERGY INTAKE AND GROWTH

The solutions to the differential equations apply to constant rates of feeding and simply describe the approach of weight to a new equilibrium when food intake is increased and then kept constant. When food intake varies with time then different solutions arise. If it is assumed, for example, that voluntary food intake is directly proportional to weight or to metabolic weight (W^n), the time course of weight is described by a positive exponential, weight increases throughout and no equilibrium is ever attained. It is possible that the accelerating phase of growth arises from the fact that in early life voluntary intake appears to increase with metabolic size. It is certain, however, that after the juvenile phase voluntary food intake expressed per kg metabolic size declines with increasing weight and that eventually a point is reached at which voluntary food intake is static though weight continues to rise. Oslage and Fliegel (1965) state that with pigs stability of food intake occurs at a weight of 120–130 kg, that is when the animal is probably about 50 per cent of its mature size. With steers of the Friesian breed given cereal diets throughout life, food intake tends to stabilize at 9·0 kg/day at weights of 400 kg; increases in feed intake from 300–400 kg being

Table 3

Numerical values for time constants in the growth equations of Brody (1945) as generalized by Taylor (1965) and in the energy equation describing the increase in weight of an animal following an increase in feed intake (equation 5)*

| Weight of animal (kg) | Time constants of exponential equations (days) | |
	Taylor's equation describing the self-inhibiting phase of growth	Equation 5 describing the response to a change in energy intake
1,000	746	717
500	536	594
100	346	384
50	287	318
10	186	206
1	100	111
0·1	54	60
0·01	29	32

* The time constants found by Brody (1945) were:

Animal	Weight kg	Days
Horse	1,000	806
Cow	460	560
Pig	200	451
Rabbit	3	112
Guinea-pig	1	100
Rat	0·3	78
Mouse	0·03	28

only about 0·5 kg/day compared with increases of about 2·0 kg/day in the weight range 100–200 kg (Kay, 1967). If it is assumed that voluntary food intake approaches a maximum value exponentially, the solution to the differential energy equation is a Gompertz growth equation in which there is a point of inflection and an equilibrium weight is reached.

What appears certain from such approaches is that the actual shape of growth curves reflects very much the time pattern of energy supply, and that growth curves of different types and obeying different 'laws' can be generated simply from a knowledge of the energy relationships involved, of which the most important is the amount of food consumed. Perhaps the most useful form in which

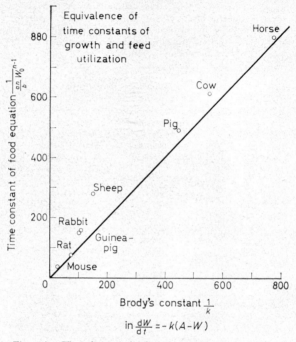

Figure 4. The relation (solid line) between the time constants of growth curves for animals of different mature body size as predicted from a statistical analysis of these constants by Taylor (1965) and those predicted from the energy relationships given in the text. The points for different species are the values actually found by Brody (abscissa) plotted against the values predicted from equation 5 (ordinate)

equation 5 can be generalized is as an integral equation in which the weight at any time t, depends on the past pattern of rate of food intake $M(\theta)$, θ being a time antecedent to time t

$$W(t) = W(t_0) + \int_{t_0}^{t} M(\theta)\alpha(\theta)(1 - e^{-\beta(\theta)(t - \theta)})\,d\theta \qquad (9)$$

The constants $\alpha(\theta)$ and $\beta(\theta)$ are the complex constants of equation 5, and they also vary with the conditions appertaining at time θ

when changes in M occur.   Equation 9 in its discrete form can be solved numerically.

ENERGY RELATIONSHIPS IN THE GROWTH OF PARTS OF THE BODY

The algebraic description of the response of energy retention and of body weight to food intake ignores the fact that different tissues

Table 4

Mean daily retention of energy, fat and flesh (protein + water) and growth of wool, by sheep during a 10-month period of constant feeding (Clapperton and Blaxter (1965), unpublished)

Group	Change in metabolizable energy intake	Composition of gain or loss/day			
		Energy value (kcal)	Wool including skin secretions (g)	Body fat (g)	Body flesh including water (g)
1	+1,406	645	7·3	65·0	35·9
2	+703	438	6·2	43·6	34·2
3	0	56	4·6	3·8	−8·0
4	−703	−61	2·7	−7·6	−17·6

respond differently to changes in the rate of food energy provision. Protein deposition and fat deposition occur at different rates depending on the proportion of mature size the animal has attained and the rate of energy provision at the time.

Table 4, which gives mean results from the experiment of Clapperton and Blaxter referred to in Table 2, shows this.   The mean daily rate of wool growth (including all skin products) increased from 4·6 to 7·3 g/day from maintenance to the highest feeding level, an increase of 58 per cent.   Energy retention increased, however, over elevenfold and the proportion of fat in the energy gains to the body increased with increase in feeding level.

The factors which determine differential growth, which can be expressed as a partition of the energy of diets adequate in amino acids and other dietary essentials as between tissue components, is not known.   The generalization that rates of protein deposition

341

have maxima with respect to rates of food consumption whereas rates of fat deposition have not, merely describes the whole phenomenon rather than explains it. An understanding of the biochemical nature of the control of the rate of protein synthesis in animals and why it apparently depends on the maturity of the animal rather than its age or, within limits, the amount of food it is given, is necessary to achieve progress in this field.

APPENDIX 1

Solution of the differential equation

$$\frac{B \, dW}{k_f dt} + \frac{AW^n}{k_m} = M$$

This is of the form

$$\frac{b \, dW}{dt} + aW^n = M$$

Let

$$W = (W_0 + \Delta W)$$

and let

$$aW_0^n = M_0$$

Then

$$aW^n = aW_0^n \left(1 + \frac{\Delta W}{W_0}\right)^n$$

and from the binomial expansion

$$aW^n \simeq aW_0^n \left(1 + \frac{n\Delta W}{W_0}\right)$$

$$aW^n \simeq aW_0^n + anW_0^{n-1}\Delta W$$

$$aW^n \simeq M_0 + anW_0^{n-1}\Delta W$$

Insert this expression in the differential equation

$$\frac{d\Delta W}{dt} + \frac{an}{b} W_0^{n-1} \cdot \Delta W = \frac{1}{b}(M - M_0)$$

Let

$$\frac{anW_0^{n-1}}{b} = \alpha$$

$$\frac{d\Delta W}{dt} + \alpha\Delta W = \frac{1}{b}(M - M_0)$$

$$e^{\alpha t}\Delta W = \int \frac{1}{b} (M - M_0)e^{\alpha t}$$

$$\Delta W = \frac{1}{\alpha b} (M - M_0) + Ce^{-\alpha t}$$

when

$$t \to 0, \; e^{-\alpha t} \to 1 \cdot 0 \quad \text{and} \quad \Delta W \to 0$$

therefore

$$C = - \frac{1}{\alpha b} (M - M_0)$$

$$W - W_0 = \frac{1}{\alpha b} (M - M_0)(1 - e^{-\alpha t})$$

Substituting for α

$$W - W_0 = \frac{1}{an} (M - M_0) W_0^{1-n} \left(1 - \exp \left(-\frac{an}{b} W_0^{n-1} t \right) \right)$$

REFERENCES

Agricultural Research Council (1965). *Nutrient Requirements of Farm Live-stock. 2. Ruminants.* London; ARC
Blaxter, K. L., Clapperton, J. L. and Wainman, F. W. (1966a). *Br. J. Nutr.* **20**, 283
——— (1966b). *J. agric. Sci., Camb.* **67**, 67
— and Wainman, F. W. (1961). *J. agric. Sci., Camb.* **57**, 419
Breirem, K. (1939). *Biedermanns Zentr. B. Tierernährung* **11**, 487
Brody, S. (1945). *Bioenergetics and Growth.* New York; Rheinhold Pub. Co.
Clapperton, J. L. and Blaxter, K. L. (1965). *Proc. Nutr. Soc.* **24**, xxxiii
Fábry, P. and Hrůza, Z. (1956). *Physiologia. bohemoslov.* 5. Suppl., p. 10
Forbes, E. B. and Swift, R. W. (1944). *J. Nutr.* **27**, 453
Günther, B. and Guerra, E. (1955). *Acta physiol. latinoam.* **5**, 169
Hamilton, T. S. (1939). *J. Nutr.* **17**, 583
Hill, F. W. and Anderson, D. L. (1958). *J. Nutr.* **64**, 587
Kay, M. (1967). Private communication
Kielanowski, J. (1965). In: *Energy Metabolism.* (Ed. by K. L. Blaxter.) p. 13. London; Academic Press
Lister, D., Cowen, T. and McCance, R. A. (1966). *Br. J. Nutr.* **20**, 633
McCance, R. A. (1960). *Br. J. Nutr.* **14**, 59
Mitchell, H. H. (1964). *Comparative Nutrition of Man and Domestic Animals.* Vol. II. New York and London; Academic Press
— and Haines, W. T. (1927). *J. agric. Res.* **34**, 927
Nehring, K. and Schiemann, R. (1966). In: *Vergleichende Ernährungslehre des Menschen und Seiner Haustiere.* Jena; Gustav Fischer

Oslage, H. J. and Fliegel, H. (1965). In: *Energy Metabolism.* (Ed. by K. L. Blaxter.) p. 297. London; Academic Press

Swift, R. W., Barron, G. P., Fisher, K. H., Magruder, N. D., Black, A., Bratzler, J. W., French, C. E., Hartsook, E. W., Heishberger, T. V., Keck, E. and Stiles, F. P. (1957). *Bull. Pa. agric. Exp. Stn.* 618

Taylor, St. C. S. (1965). *Anim. Prod.* **7**, 203

GROWTH AND DIETARY AMINO ACID BALANCE

D. LEWIS and J. P. F. D'MELLO

*University of Nottingham School of Agriculture,
Sutton Bonington, Loughborough*

SYNTHESIS of protein, that is so inescapably a characteristic of growth, requires that there should be available a supply of amino acids balanced according to the need. Provided that there is such a source of amino acids it is difficult to visualize any barrier to synthesis other than by the actual apparatus of synthesis itself, probably through a limitation in the amount of messenger RNA. It has been shown in many ways that a lack of balance in amino acid supply or a deficiency of a particular amino acid will lead to a halt in synthesis. It must be emphasized, however, that the ultimate site at which a particular balance is needed is within the cell carrying out the anabolic process. A different relative supply of amino acids within the diet is needed to meet this demand; there are differentials introduced in amino acid absorption and in involvement in metabolism other than protein synthesis.

There is considerable data available on the relative rates of absorption of amino acids (*see* Wiseman, 1964). The data of Adibi, Gray and Menden (1967), presented in Table 1 and referring to the human subject, show a considerable differential, a twofold range, in the rates of absorption of essential amino acids. It is of particular interest that the two amino acids which usually receive most attention are at each end of the scale. One can also consider these two amino acids for emphasizing differences in metabolism. Lysine enters into few metabolic pathways other than protein synthesis and is not readily degraded in the body, whereas methionine is involved in many other transformations and is easily catabolized.

It is possible to emphasize the effects of these differences by comparing the composition of tissue protein or whole body proteins with standard dietary amino acid requirements (Table 2). Tissue composition and requirements are given both for the chick and the pig. In the case of the chick a few whole body values are included, in parenthesis, emphasizing the significance of including feathers in the assessment. It is particularly illuminating again to compare lysine and methionine, where the requirements for the pig are respectively about 60 per cent and 90 per cent of the relative figures for tissue composition.

A more detailed examination of the subject of growth and amino acid balance can be divided into two parts. It is possible to con-

Table 1

Relative rates of intestinal amino acid absorption
(Human subject; adapted from Adibi, Gray and Menden, 1967)

Methionine	100	Cystine	64
Leucine	98	Tryptophan	63
Isoleucine	98	Glycine	61
Valine	91	Threonine	54
Arginine	86	Histidine	48
Phenylalanine	83	Lysine	47
Tyrosine	78		

sider the establishment of dietary amino acid requirements by what is essentially an empirical approach and also to remark upon the consequences of lack of ideal amino acid balance based upon a more functional outlook.

AMINO ACID REQUIREMENTS

In defining allowances an empirical viewpoint is probably essential; involving the recording of growth performance in relation to graded levels of a particular nutrient. It is doubtful whether there is sufficient knowledge available to justify what has been called a factorial approach; a procedure of adding together the necessary contributions of a particular nutrient to meet the component parts of its overall biological function. This situation does not imply that a casual experimental programme can be devised. There are numerous pitfalls that may be encountered in developing a satisfactory experimental situation. It is essential to ensure that a series of potential hazards are recognized; for example the objective and, in particular, the chosen growth parameter must be clearly defined; safeguards are necessary to ensure that there is overall dietary amino acid balance; it must be acknowledged that adding graded levels of an amino acid to a basal diet merely serves to define the barrier of the next limiting nutrient; due attention must be given to an appropriate allocation within the protein component of materials that can contribute to the formation of amino acids which are individually dispensable; the problem of differentials in amino acid availability must be recognized; and the relative values of D-amino acids must be recognized if racemic mixtures are used in experimental studies.

Table 2

Tissue composition and amino acid requirements*

g /16 g N

	Chick† muscle	Chick requirements	Pig muscle	Pig requirements
Glycine	5·7(9·2)†	3·0	5·0	—
Arginine	6·7	4·25	6·6	—
Histidine	2·0	2·0	2·8	1·6
Lysine	7·7(6·9)	4·8	8·5	5·9
Phenylalanine	4·1	3·5	4·5	3·1
Tyrosine	2·7	3·0	3·1	2·4
Tryptophan	1·0	0·9	1·1	0·9
Methionine	2·4	2·25	2·5	2·2
Cystine	1·0(2·1)	1·75	1·4	1·7
Threonine	4·0	3·0	4·6	2·8
Leucine	8·2	6·5	8·0	4·4
Isoleucine	4·2	3·5	4·7	3·4
Valine	4·1(5·5)	4·0	5·5	3·1

* Data for muscle based upon ion-exchange resin chromatographic analysis; requirements proposed from an appraisal of published information and the findings of Cooke (1968) and Hewitt (1968)
† Values in brackets are for whole body analyses

The importance of defining the objectives can be exemplified by considering the case of establishing an amino acid allowance for the bacon pig. Graded supplements of lysine were added to a basal bacon pig ration with 16 per cent crude protein (N × 6·25) and 3,150 kcal digestible energy per kg (Crehan, Lewis and Lodge, 1965), growth performance was recorded and the lean content of the carcass determined at slaughter. The results are presented in *Figure 1* and suggest that there was no advantage in terms of growth in increasing the lysine content beyond 0·88 per cent of the diet, whereas the proportion of lean in the carcass increased up to a lysine level of 0·95 per cent of the diet. These relative values may represent a consistent relationship though the appropriate absolute allowances may be defined by other nutrients, possibly the supply of energy-yielding nutrients.

It is possible to emphasize these issues further by considering an experimental programme designed to establish the tryptophan

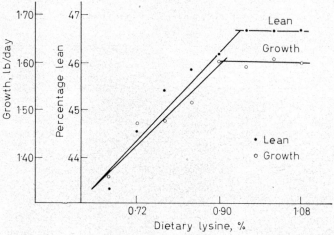

Figure 1. Effect of dietary lysine on live weight gain and carcass lean content (Crehan, Lewis and Lodge, 1965)

requirement of the growing pig (Cooke, 1968). A basal diet was used (Table 3) containing only 0·08 per cent of tryptophan, and synthetic lysine, methionine and isoleucine were added in order to achieve the amino acid levels indicated. The experimental arrangement allowed graded supplements of tryptophan to be added so that tryptophan content constituted eight treatments ranging from 0·08 to 0·20 per cent of the diet. It was considered that DL-tryptophan was equivalent to two-thirds of the same amount of L-tryptophan (Firth and Johnson, 1956). The diets were fed on a restricted scale to groups of 8 gilts and records taken of growth rate, efficiency of feed conversion and carcass quality. The results given in *Figure 2* show clearly that the best performance was obtained with a diet supplying 0·14 per cent tryptophan. It must be emphasized that the total tryptophan (including D-isomer) in this case was around 0·17 per cent of the diet. It is also necessary to point out that these findings refer to a situation wherein there is a relative excess of dietary nicotinic acid. The depression in performance at higher tryptophan levels serves to emphasize the importance of a correct balance of amino acids within the dietary protein. It is difficult, however, to identify a metabolic situation that might respond so markedly to such a marginal surplus of tryptophan.

Table 3

Diet used in determination of tryptophan allowance*

Diet (percentage)		Amino acids (percentage)	
Ground wheat	30	Lysine	0·95
Maize meal	50	Isoleucine	0·59
Maize gluten meal	8	Methionine	0·39
White fish meal	3	Cystine	0·28
Minerals and vitamins supplement	3·5	Threonine	0·45
Supplementary amino acids	5·5	Tryptophan	0·08

* Basal diet of 16 per cent protein (N × 6.25) and 3,200 kcal digestible energy per kg (Cooke, Lewis, Lodge and Pappas, 1967)

Several allusions have been made to the importance of overall dietary nutrient balance and in this respect the overwhelming importance of the energy-yielding nutrients must be stressed. Though many empirical studies have emphasized the importance of nutrient balance it has not been possible in all such cases to feel sure that findings were not influenced by a specific limiting nutrient. A research programme at the University of Nottingham concerned with the bacon pig has aimed for some years at identifying satisfactory dietary energy—protein balance (see Robinson, 1962). In an extension of this work, recent studies have accommodated a more careful definition of dietary amino acid adequacy. Furthermore, the potential for high nutrient concentration and a sensitive definition of optimum balance has been emphasized by the slaughter of pigs at lower weights (for example 120 lb), by using gilts, by using a strain of pig better selected for lean meat production (Large White × Landrace crossbred) and by carrying out the trial under circumstances more suitable for limiting the inhibitory effects of disease. With these principles in mind a trial has been conducted recently (Cooke, Lewis, Lodge and Pappas, 1967) in which the dietary concentration of protein and of the energy yielding fraction were varied in a factorial design. Every effort was made to ensure adequacy in all diets of vitamins, minerals and amino acids and care was taken to avoid possible harmful surpluses. A series of sixteen diets was used with a range in protein from 15 to 24 per cent of the diet and digestible energy values from 2,850 to 3,525 kcal/kg. Raising the supply of energy-yielding nutrients (Table 4) appears to encourage in all instances a more rapid rate of liveweight gain. With increases in protein level, growth rate reached a maximum at 18–21 per cent dietary protein and a fall at the higher protein level. There

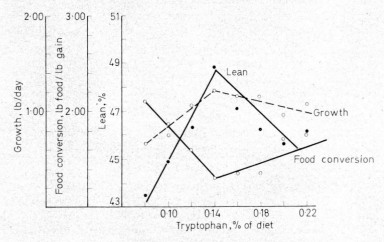

Figure 2. Live weight gain (lb/day), *efficiency of food conversion* (lb food/lb gain) *and carcass lean* (%) *in relation to dietary tryptophan (Cooke, 1968)*

appears to be a tendency for the protein content for maximum growth rate to be greater at the higher energy levels, emphasizing for this parameter the importance of dietary energy-protein balance.

An appraisal of carcass lean content as an index of quality shows a somewhat different pattern (Table 4). An increase in dietary protein at all levels of energy-yielding nutrients leads to an improvement in carcass quality whereas raising the energy level at all points of dietary protein supply results in a reduction in quality. A more realistic impression of the implication of these findings may be obtained in terms of the daily production of lean, an approximate index of which can be obtained from the product of daily live weight gain and carcass lean content. The supply of energy-yielding nutrients appears to have little effect upon the quantity of lean produced. On the other hand, raising the dietary protein level from

15 to 18 per cent significantly increased the deposition of lean: further increase in dietary protein did lead to more lean production. These findings do not appear to identify an overall plateau of

Table 4

Growth performance of the growing pig (20–50 kg)*
offered diets varying in energy-protein balance

kg/day *Live Weight Gain*

% N × 6·25	2,850	3,075	3,300	3,525
	kcal *Digestible Energy*/kg			
15	0·59	0·65	0·65	0·67
18	0·63	0·68	0·68	0·70
21	0·61	0·65	0·71	0·71
24	0·60	0·65	0·65	0·65

Percentage Carcass Lean

% N × 6·25	2,850	3,075	3,300	3,525
	kcal *Digestible Energy*/kg			
15	45·6	44·1	42·5	41·1
18	50·0	48·4	46·9	45·7
21	50·4	47·1	48·6	45·2
24	51·4	50·6	48·6	46·6

kg/day *Lean Gain*

% N × 6·25	2,850	3,075	3,300	3,525
	kcal *Digestible Energy*/kg			
15	0·21	0·23	0·20	0·21
18	0·26	0·27	0·25	0·25
21	0·26	0·24	0·27	0·25
24	0·26	0·27	0·25	0·25

* Experiment conducted in a factorial design using 2 groups of 64 pigs. Lean gain per day approximately assessed as product of live weight gain and final carcass lean content (Cooke, Lewis, Lodge and Pappas, 1967)

response but probably serve to emphasize the importance of the point beyond which increase in nutrient supply results in an inadequate response in economic terms.

AMINO ACID BALANCE AND SURPLUS

Since amino acids in general enter into a variety of metabolic pathways, it has been widely assumed that any surplus ingested and not subsequently used for protein synthesis exerts no adverse effects. It is now acknowledged that in most instances a surplus of a particular amino acid will impose a limitation upon efficiency of nutrient utilization commensurate with the magnitude of the deviation from

Table 5

Amino acid imbalance

Effect of tryptophan on growth of rats fed diets containing 8 per cent casein and gelatin or threonine. Adapted from Morrison, Reynolds and Harper (1960)

	Weight gain g/2 week
8 per cent casein diet	15
+6 per cent gelatin	5
+gelatin+0·1 per cent tryptophan	32
+0·36 per cent threonine	4
+0·36 per cent threonine+0·1 per cent tryptophan	32

a perfect balance. There are now recognized, however, certain instances when a dietary excess of an amino acid or a mixture is known to precipitate an ill-effect that is totally disproportionate to the degree of imbalance. It was shown, for example, by Morrison, Reynolds and Harper (1960) that the growth of rats was much reduced (*see* Table 5) by adding gelatin or threonine to a basal casein diet. This depression in growth was prevented by supplements of either nicotinic acid or tryptophan. It has subsequently been shown clearly that a specific relationship exists between threonine and tryptophan; in some cases addition of as little as 0·1 per cent threonine caused a severe growth retardation whereas additions of other amino acids were without effect.

These findings have led to a considerable study of other purified proteins, for example zein, gluten or casein. It has become generally recognized that besides the inherent simple deficiencies of some amino acids, certain proteins also contain large excesses of other amino acids which accentuate the penalty already imposed by the initial deficiencies.

Other instances of disproportionate effects of amino acids in excess have been observed in young chicks (*see* Lewis, 1965). In particular, the effect of a relatively small excess of lysine results in a marked growth depression which is prevented if supplementary arginine is also present (Table 6). Subsequent experimental work showed that the ill-effects of excess lysine could be reversed only by

Table 6

Lysine–arginine interaction in the chick*

Treatment	g/day
Basal only	13·2
+0·2 per cent L-arginine	13·3
+0·6 per cent L-lysine	11·9
+lysine+arginine	13·5
	S.E. ±0·25

* Mean values (g/day) for 6 replicates of 10 birds for the period from 7 to 14 days; groups of 15 birds placed in cages at day-old, reduced to 10 at 7 days and experimental diets given. (D'Mello, 1967)

Table 7

Specificity of lysine–arginine interaction

Effect of lysine addition and supplementation with first and second limiting amino acids on growth and plasma amino acid levels. (D'Mello, 1967)

	Growth g/day	Plasma levels (μ moles/100 ml.					
		Lysine	Arginine	Histidine	Threonine	Isoleucine	Leucine
Basal	12·9	83	16	23	27	16	22
+Arginine (Arg)	12·9	90	20	17	23	12	22
+Threonine (Thr)	19·5	69	15	19	82	12	19
+Arg+Thr	20·6	80	25	19	94	15	24
+Lysine (Lys)	7·8	120	7	17	34	12	19
+Lys+Arg	12·8	169	14	17	40	13	24
+Lys+Thr	9·8	115	6	16	133	11	19
+Lys+Arg+Thr	16·7	120	16	19	80	9	18

arginine and that this interaction would occur even if another amino acid, for example threonine, appeared to be first limiting. The results in Table 7 demonstrate the specificity in terms of two parameters, growth rate and free amino acid levels in peripheral plasma. Excess lysine produced the predictably severe growth retardation which was corrected only by arginine. Lysine also depressed the level of arginine in plasma without in any way altering the level of threonine. The simultaneous supply of arginine did not succeed appreciably in reducing the plasma levels of lysine when the latter was in excess in the diet, suggesting indirectly that the interaction is not quantitatively reversible.

The inter-relationships between leucine, isoleucine and valine have been studied in some detail (Spolter and Harper, 1961; Rogers,

Table 8

Leucine–isoleucine interaction

Effect of leucine and isoleucine supplementation of a basal diet deficient in isoleucine, on growth and plasma amino acid levels (D'Mello, 1967)

	Growth g/day	Plasma levels (μ moles/100 ml)				
		Leucine	Isoleucine	Valine	Arginine	Tyrosine
Basal	12·8	15	4	19	15	14
+ Isoleucine (0·2 per cent)	16·1	12	6	17	17	14
+ Leucine (1·5 per cent)	7·8	20	2	10	14	11
+ Leucine + iso-leucine	15·9	19	4	8	16	13

Spolter and Harper, 1962) in the young rat. Similar responses for these interactions have been observed in the young chick. In Table 8 the data illustrate the existence of a leucine–isoleucine interaction. Excess leucine causes a growth depression and isoleucine counteracts it, though only partially. The plasma amino acid data particularly demonstrate the effect of excess leucine on the plasma valine levels. The valine values are depressed to a greater extent than are the levels of isoleucine. It can be assumed, with good reason, that the branched-chain amino acids are involved inextricably with each other in terms of their effect upon the growth of the young animal.

INTERDEPENDENCE IN NUTRITION

Under conditions of optimum dietary amino acid balance it can be considered that the requirements for individual essential amino acids are minimal. As the balance deviates from the ideal pattern, the requirement for some is increased proportionately. Almquist (1954) observed that free amino acids in plasma tend to accumulate when an indispensable amino acid is suboptimal. This is due probably to a consequential decrease in protein synthesis. The plasma amino acid data in Tables 7 and 8 also show that under conditions of relative excess, there is an accumulation of free amino acids in plasma. Efficient utilization is therefore reflected in minimal total amino-nitrogen in the plasma and other tissues.

Table 9

Quantitative assessment of the interaction between lysine and arginine

Effect of graded supplements of lysine and arginine on live weight gain of chicks. Each figure in the body of the table represents a mean value of four replicate groups of 4 chicks during the period 7–21 days)

Arginine (percentage of diet)	Lysine (percentage of diet)			
	1·10	1·35	1·60	1·85
0·85	17·7	16·1	13·0	7·2
1·00	19·6	19·5	17·1	14·0
1·15	19·8	20·3	18·4	15·6
1·30	19·8	19·3	19·1	17·2
1·45	19·9	19·4	19·4	19·0
1·60	19·5	19·4	19·4	19·8

It can be considered that in these instances of interaction the consequence of the presence of a surplus of one amino acid is to increase the requirement for another. This interdependence in amino acid requirements has been demonstrated in the chick for several categories of interaction (D'Mello, Hewitt and Lewis, 1967). The results in Table 9 demonstrate a clear relationship between dietary lysine levels and the chick's requirement for arginine. A progressive growth depression occurs with increasing dosage of lysine; arginine supplementation completely counteracts these effects at all increments of lysine showing convincingly that the effect of lysine is to alter the metabolic fate of arginine alone. This is further reflected in the plasma levels of lysine and arginine (Table 10). As the dietary levels of lysine increase, there is a progressive depression in the plasma arginine status confirming the existence of a metabolic phase in the involvement between lysine and arginine.

The significance of the lysine–arginine interaction and the clear demonstration of interdependence in amino acid requirements can be recognized in *Figure 3*, prepared from the data in Table 9. The straight line correlation indicates that even small, seemingly innocuous, additions, of lysine above the requirement are likely to

Table 10

Lysine–arginine interaction

Effect of graded supplements of lysine and arginine in the diet on their respective levels in plasma
(μ moles/100 ml.) Lys. = lysine; Arg. = arginine (D'Mello, 1967)

Arginine (percentage of diet)	Lysine (percentage of diet)							
	1·10		1·35		1·60		1·85	
	Plasma arginine and lysine (μ moles/100 ml)							
	Arg.	Lys.	Arg.	Lys.	Arg.	Lys.	Arg.	Lys.
0·85	9·1	40·0	7·7	47·5	6·1	64·0	5·4	94·4
1·00	10·7	41·8	6·6	31·0	8·0	48·6	7·2	81·8
1·15	16·1	32·8	10·3	31·7	17·3	56·8	8·5	85·0
1·30	24·9	36·5	9·5	28·7	19·2	57·0	17·2	86·0
1·45	24·0	36·6	13·4	27·0	21·5	55·1	20·3	73·1
1·60	20·0	22·6	25·6	30·3	24·8	56·8	25·6	72·3

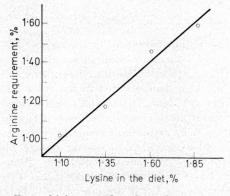

Figure 3. Effect of lysine on requirement for arginine (D'Mello, 1967)

affect chick growth adversely, unless a simultaneous supply of arginine is also available. It follows, therefore, that the requirement of the young chick for arginine cannot be estimated without considering also the level of lysine in the diet. This probably explains why there is considerable uncertainty regarding the arginine requirement of the chick.

A similar approach has been extended to the study of the inter-action between threonine and tryptophan. Although its existence has been widely documented, precise quantitative data on the extent of dependence of the tryptophan requirement on dietary threonine

Table 11

Quantitative assessment of the interaction between
threonine and tryptophan

Effect of graded supplements of threonine and tryptophan on live weight gain
of chicks.   Each figure in the body of the table represents a mean value of four
replicate groups of 4 chicks during the period 7–21 days (D'Mello, 1967)

Tryptophan (percentage of diet)	Threonine (percentage of diet)			
	0·80	1·30	1·80	2·30
0·12	11·0	9·7	9·7	9·0
0·17	17·8	16·5	16·3	15·7
0·22	18·0	18·2	18·4	17·0
0·27	19·0	18·8	18·0	17·5
0·32	18·2	17·9	18·4	18·0
0·37	18·5	18·7	18·1	18·4

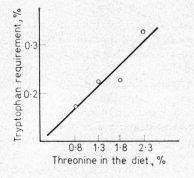

Figure 4.   Effect of threonine on require-ment for tryptophan (D'Mello, 1967)

levels have been lacking.   The results in Table 11 produce such quantitative data.   Increments of threonine in the diet result in significant growth depressions; tryptophan completely reverses these effects.   These growth data closely resemble the ones presented for the lysine–arginine relationship.   In both experiments, within the ranges studied, reversibility is not readily apparent.   If the data in Table 10 are used to examine quantitatively the relationship between dietary threonine and tryptophan requirement, a straight line correlation is again found to exist (*Figure 4*).

An entirely similar situation exists in the case of another well-established interaction, that between leucine, isoleucine and valine. In the first instance leucine and valine were examined keeping the isoleucine content of the diet adequate. The next stage was the investigation of the relationship between leucine and isoleucine in quantitative terms, again ensuring adequacy of the third branched-

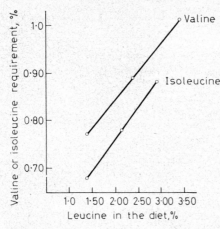

Figure 5. *Effect of leucine on the requirement for valine or isoleucine* (*D'Mello, 1967*)

chain amino acid. In both cases, as the dietary leucine level increased there was a linear increase in the requirement of the young chick for valine and isoleucine respectively. These patterns of interdependence are illustrated in *Figure 5*.

These findings emphasize the point that, for at least some indispensable amino acids, requirement values are variable entities. Complete reliance, therefore, on minimum requirements for these amino acids is not to be recommended and tables of amino acid requirements have value only as a guide in obtaining an optimum dietary pattern of amino acids. It follows also that the technique of adding graded supplements of a single amino acid to determine its requirement not only serves merely to define the relative level of the next limiting amino acid but may also serve to confuse the point at which its limiting influence becomes operative.

MECHANISMS OF AMINO ACID INTERACTIONS

In spite of the considerable attention which has been given to the study of amino acid imbalance, no mechanism has been proposed which can account for all the observations. Furthermore, attempts at classifying these adverse effects of excessive amino acid intake are not based on a knowledge of the mechanisms involved. Elvehjem (1956) separated these effects into three categories; imbalances, in which the effect of the added amino acid is reversed by supplementing the diet with the first limiting amino acid; antagonisms, in which the effect of the added amino acid is alleviated by a structurally related amino acid; and toxic effects in which no simple combination of amino acid supplementation counteracts the growth depression. Harper (1964) encouraged this approach and suggested that its use be continued until evidence of a common basis for these effects be discovered. However, in considering explanations for imbalances, antagonisms and toxicities, it was concluded that similar characteristics could be identified for these adverse effects, suggesting that the tripartite classification was in fact based on differences in the underlying causes. Its continued use, therefore, cannot be adequately defended or justified.

Lewis (1965) suggested that the disproportionate consequences that accrued from addition of amino acids in excess could best be considered as interactions between pairs of amino acids. One of these in a pair was defined as the 'agent', the addition of which precipitated the ill-effects on growth. The 'target' constituted the second amino acid of the pair which had the ability to alleviate the growth retardation. Such a hypothesis allows a systematic study to be made of the scope, specificity and reversibility of such reactions. The findings give a reasonable foundation for a functional approach in identifying the aetiology of the ill-effect. Within such interacting pairs it is possible to accommodate the groups that have been considered; namely, threonine and tryptophan, lysine and arginine, and leucine, valine and isoleucine. This concept was proposed as an alternative to the Elvehjem (1956) classification primarily due to the lack of evidence to support disparate mechanisms for the three classes of imbalance. The concept of interacting pairs is also consistent with much of the evidence available at present and particularly with the data already presented. A virtue of this outlook is that it attempts to establish a basis for recognizing mechanisms involved in amino acid imbalance.

After ingestion of protein relatively high levels of amino acids are attained in the portal blood but the venous systemic circulation shows much smaller increments. This difference is due largely to interception by the liver and other tissues, chiefly muscle. These tissues absorb amino acids into their interior, the liver disposing of its accumulation rapidly while muscle tissues tend to retain their surfeit of amino acids for a relatively longer period. The animal is thus capable of dealing with a seemingly infinite quantity of amino acids. The inevitable question then arises of how amino acids ingested beyond the need exert deleterious effects on overall growth of the young animal. The immediate influence of excess amino acids upon growth led early investigators to postulate a direct effect in terms of a low palatability of the imbalanced diets. Since the effects are reversed by further supplementation of certain amino acids to the imbalancing mixture, low palatability has been discounted as a major cause. This possibility cannot be completely precluded, however, in imbalances caused by the addition of amino acid mixtures constituting as much as 10 or 15 per cent of the diet.

Competition between amino acids during absorption from the intestine has been considered also as a possible mechanism leading to the exclusion, and therefore loss, of the limiting or target amino acid. However, there are several reasons that argue against this proposition. The large absorptive capacity of the intestine in relation to the amount of amino acids released during digestion and presented for absorption makes it unlikely that competition between amino acids is of substantial influence. A strongly competitive amino acid would be rapidly absorbed in the initial part of the intestine, so that other amino acids of lower transport affinity would be absorbed lower in the intestine. Furthermore, Tannous, Rogers and Harper (1966) in their studies on the intact rat, observed no accumulation of isoleucine or valine either in the intestinal wall or in its contents when excess leucine was consumed. This would have occurred if competition between leucine, isoleucine and valine were operative in inducing the growth depression. Another reason is seen in the observations presented in Table 7. Supplemental arginine relieves the effect of excess lysine after the latter has been absorbed, since arginine does not reduce the high plasma levels of lysine although it succeeds in restoring optimum growth rate.

It has been recorded consistently that following the consumption of an imbalanced diet there is both a reduction in the rate of growth and in food intake. There has been, in fact, considerable controversy as to which of these features is the primary consequence of the consumption of the imbalanced diet. Convincing information was produced by Harper and Rogers (1965) in which rats showed a very marked depression in intake after only 3–6 h (*Figure 6*). The results imply that the depression in food intake can be regarded as a primary effect and responsible for the ensuing retardation in growth. A considerable body of evidence exists to support this premise. Appetite in animals consuming an imbalanced diet can be maintained by insulin injections, by adjustment of the protein:calorie ratio (Fisher and Shapiro, 1961), and by exposure of the animals to a cold environment (Klain, Vaughan and Vaughan, 1962). All these manipulations result in improved growth relative to the control animals.

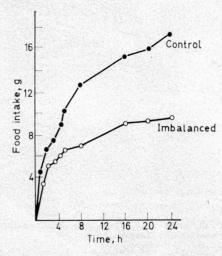

Figure 6. Food intake patterns over 24 h of rats fed on control (6 per cent casein), or imbalanced diets (control + EAA mix — threonine) (Harper and Rogers, 1965)

In an attempt to account for the reduction in food intake, Harper and Rogers (1965) arrived at the conclusion, in the absence of any other metabolic change resulting from the ingestion of an imbalanced diet, that the altered plasma amino acid pattern was instrumental in producing the adverse effect on appetite. It is, necessary, however, to explain how the altered plasma amino acid pattern develops in the initial phases of the phenomenon. In particular, the mechanism of the depression in the plasma levels of the limiting or target amino acid remains to be accounted for, as indeed does the mechanism of the subsequent influence on the appetite-regulating centres of the hypothalamus.

The effect upon food intake does not appear to be a direct influence of an enhanced plasma amino acid level upon the satiety centre of the hypothalamus, for when the effect is corrected by the addition of a 'target' amino acid the plasma level of the 'agent' bringing about the ill-effect remains high. It is thus necessary to look to a response to a depressed plasma level of the 'target' and in particular to a mechanism that will give rise to this depression. Two hypotheses, which seem at first glance to be somewhat conflicting, can be put forward to account for this depressed plasma level; they are, in brief, that there is either an increased catabolism (or excretion) or an increase in anabolism as a response to the consumption of the imbalanced diet.

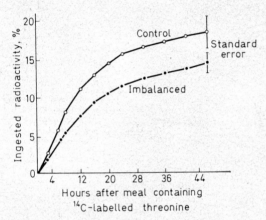

Figure 7.   Cumulative percentage of $^{14}CO_2$ expired by rats fed on control or imbalanced diets containing uniformly ^{14}C-labelled threonine (Harper and Rogers, 1965)

In support of the latter hypothesis Harper and Rogers (1965) have demonstrated that an imbalancing mixture lacking one amino acid causes a depression in the oxidation of the limiting amino acid which is reflected in low $^{14}CO_2$ respiratory output relative to control animals (*Figure 7*). They interpreted this to imply that an imbalancing mixture conserved and utilized more effectively the small quantity of limiting or target amino acid that was present. On the basis of these results it was concluded that the surplus of the indispensable amino acids in the liver stimulated synthesis or suppressed breakdown of protein so that more of the limiting amino acid was retained in the liver in the imbalanced than in the control group. The supply of the limiting or target amino acid for peripheral tissues was thereby reduced, so that the free amino acid patterns of muscle and plasma became severely imbalanced, food intake being depressed as a consequence.

Loss of the amino acid in shortest supply is postulated often also as an alternative mechanism leading to the derangement in plasma amino acid pattern. Competition between the amino acids added in excess and the first limiting or target amino acids at certain post-intestinal sites is suggested to occur and result in net loss of the target amino acid. The kidney may well be the site most likely to offer conditions for such competition. Sauberlich and Salmon (1955) observed that the amount of tryptophan excreted by rats fed a tryptophan-deficient diet increased when gelatin was added to induce a growth depression.

Studies at the University of Nottingham, on the other hand, have emphasized the potential significance of a catabolic route in lowering the effective levels of the amino acid in least supply. Lewis (1965) advanced the hypothesis that the extent to which an amino acid in excess causes a detrimental effect when present in excess is related to the ease with which it enters into pathways of catabolism. One may, for instance, contrast the behaviour of lysine and arginine; the former is relatively inert metabolically and is a potent agent of inter-action, whereas the reverse is true with arginine. The viewpoint now adopted is that the agent amino acid present in excess enhances oxidative catabolism and inadvertently encourages the con-comitant loss of the target amino acid. It is proposed that the interactions between lysine and arginine, threonine and tryptophan, and leucine, valine and isoleucine are examples in which enhanced catabolism of the appropriate target amino acid may occur. An avian liver L-amino acid oxidase studied by Struck and Sizer (1960) possesses limited specificity and its activity is to some extent adaptive in relation to dietary intake (Shinwari and Lewis, 1968). It is also possible that the necessary change in effective activity in response to the imbalanced input may be a function of a naturally occurring process of inhibition or activation of the enzyme (Shinwari and Falconer, 1967). The catabolic activity of such enzymes leads to a depression in circulating levels of the 'target' amino acid and could result directly in a retardation in growth through inhibition of protein synthesis. Alternatively, the altered plasma amino acid pattern could be monitored by the appetite centres of the hypo-thalamus, with the resultant onset of anorexia.

Restriction of the rate of protein synthesis through inadequate supply of the target amino acid is an attractive theory, especially when considered in the light of recent findings by Wunner, Bell and Munro (1966). Rats force-fed on amino acid mixture deficient in tryptophan had a considerably higher proportion of small ribosomal fractions (oligosomes) in the polysome fraction than the control group fed a complete amino acid mixture. This difference was detected and was of greatest magnitude 2–6 h post-prandial. The rate of incorporation of DL-(1-^{14}C) leucine and L-(Me-^{14}C) tryptophan was also measured in cell-free systems of polysomes and oligosomes from rats previously fed the complete or deficient amino acid mixtures. It was observed that polysomes derived from rats fed the tryptophan deficient mixture incorporated significantly less leucine and tryptophan than polysomes from the control animals. There is also evidence to sustain the premise that appetite is governed by levels of circulating amino acids. The initial studies of Mellinkoff, Frankland, Boyle and Greipel (1956) encouraged the view that a relationship exists between plasma amino acid concentration and fluctuations in appetite. Sanahuja and Harper (1963) concluded that a hypothalamic control system regulated appetite and was sensitive to the pattern of amino acids in the plasma.

Although the two hypotheses regarding the manner in which an altered plasma amino acid pattern is produced appear incompatible, it is now becoming clear that this may be due to the fact that they represent two distinct mechanisms, anabolic and catabolic. It is envisaged that interactions of the anabolic type are characterized by a homeostatic response to conserve and limit the loss of the target amino acid, eventually leading to an enzymic adaptation and some recovery in the pattern of growth (*Figure 8*). On the other hand, in interactions of the catabolic type the predisposing feature is the continued loss of the target or limiting amino acid through oxidative pathways. No adaptive change to correct for this can be detected and there is no recovery in terms of the deranged growth pattern (*Figure 9*). Both situations demand that when an imbalanced diet is consumed there is, subsequent to absorption from the intestine, an obligatory phase of intense metabolism before the ultimate depression in growth is observed. In some instances anabolic functions are enhanced; in others catabolic pathways preponderate. Homeostatic mechanisms are overwhelmed by the acute alteration in plasma amino acid patterns which also leads to the depression in the voluntary intake of food and resultant retardation of growth.

It is tempting, from the limited evidence available, to suggest another system of classification of adverse effects of excessive amino acid intake based on an understanding of the mechanisms involved. Two categories may be recognized:

(1) *Anabolic interactions*: exemplified by those instances in which deleterious effects are produced by mixtures lacking one amino acid,

and by doses of single amino acids like tyrosine, histidine and methionine.

(2) *Catabolic interactions*: comprising specific groups and to which

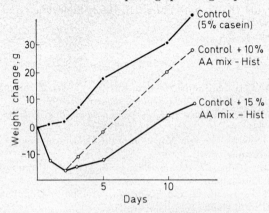

Figure 8. Depression of growth of rats by imbalanced diets (Harper and Rogers, 1965)

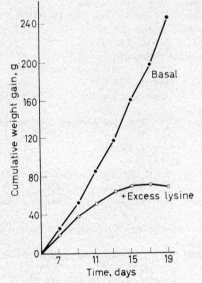

Figure 9. Effect of excess lysine (1·5 per cent) on growth of chicks (D'Mello, 1967)

no adaptation occurs. Small additions cause disproportionate effects on growth. It is suggested that the inter-relationships between threonine and tryptophan, leucine, isoleucine and valine, and

365

lysine and arginine are instances which fall into this category. It is possible also that methionine should be included in this category since there is evidence to suggest that a relationship exists between it and glycine (Klain and Vaughan, 1962; Benevenga and Harper, 1967).

REFERENCES

Adibi, S. A., Gray, S. J. and Menden, E. (1967). *Am. J. clin. Nutr.* **20**, 24

Almquist, H. J. (1954). *Archs Biochem. Biophys.* **52**, 197

Benevenga, N. J. and Harper, A. E. (1967). *Fedn Proc. Fedn Am. Socs exp. Biol.* **26**, 522

Cooke, R. (1968). 'Protein Nutrition of the Growing Pig.' *Ph.D. Thesis*, Univ. of Nottingham

— Lewis, D., Lodge, G. A. and Pappas, S. (1967). Unpublished data, Univ. of Nottingham

Crehan, M. P., Lewis, D. and Lodge, G. A. (1965). Unpublished data, Univ. of Nottingham

D'Mello, J. P. F. (1967). 'Amino Acid Interaction in Chick Nutrition.' *Ph.D. Thesis*, Univ. of Nottingham

— Hewitt, D. and Lewis, D. (1967). *Proc. Nutr. Soc.* **26**, vii

Elvehjem, D. A. (1956). *Fedn Proc. Fedn Am. Socs exp. Biol.* **15**, 965

Firth, J. and Johnson, B. C. (1956). *J. Nutr.* **59**, 223

Fisher, H. and Shapiro, R. (1961). *J. Nutr.* **75**, 395

Harper, A. E. (1964). In: *Mammalian Protein Metabolism*, Vol. II. (Ed. by H. N. Munro and J. B. Allison.) Ch. 13. New York; Academic Press

— and Rogers, Q. R. (1965). *Proc. Nutr. Soc.* **24**, 173

Hewitt, D. (1968). 'Amino Acid Allowances for the Growing Chick.' *Ph.D. Thesis*, Univ. of Nottingham

Klain, G. J. and Vaughan, D. A. (1962). *Fedn Proc. Fedn Am. Socs exp. Biol.* **21**, 7

— — and Vaughan, L. N. (1962). *J. Nutr.* **78**, 359

Lewis, D. (1965). *Proc. Nutr. Soc.* **24**, 196

Mellinkoff, S. M., Frankland, M., Boyle, D. and Greipel, M. (1956). *J. appl. Physiol.* **8**, 535

Morrison, M. A., Reynolds, M. S. and Harper, A. E. (1960). *J. Nutr.* **72**, 302

Robinson, D. W. (1962). 'Protein and Energy Nutrition of the Bacon Pig.' *Ph.D. Thesis*, Univ. of Nottingham

Rogers, Q. R., Spolter, P. D. and Harper, A. E. (1962). *Archs Biochem. Biophys.* **97**, 497

Sanahuja, J. C. and Harper, A. E. (1963). *Am. J. Physiol.* **204**, 686

Sauberlich, H. E. and Salmon, W. D. (1955). *J. biol. Chem.* **214**, 463

Shinwari, M. A. and Falconer, I. R. (1967). *Biochem. J.* **104**, 53

— and Lewis, D. (1968). *Proc. Nutr. Soc.* (In press)

Spolter, P. D. and Harper, A. E. (1961). *Am. J. Physiol.* **200**, 513

Struck, J. and Sizer, I. W. (1960). *Archs Biochem. Biophys.* **90**, 22

Tannous, R. I., Rogers, Q. R. and Harper, A. E. (1966). *Archs Biochem. Biophys.* **113**, 356

Wiseman, G. (1964). *Absorption from the Intestine*, Ch. 5. New York; Academic Press

Wunner, W. H., Bell, J. and Munro, H. N. (1966). *Biochem. J.* **101**, 417

THE ROLES OF MINERAL ELEMENTS IN TISSUE GROWTH AND DEVELOPMENT

C. F. MILLS

Rowett Research Institute, Bucksburn, Aberdeen

THE SKELETAL tissues of mammalian organisms account for the greater part of the mineral components of the body. The effects of suboptimal mineral supply on the growth and development of these tissues have long been known and recent studies in this field are being described by another contributor to this symposium.

Less well appreciated is the fact that many mineral elements play vital roles in the growth of other tissues and that changes in mineral status of the organism have a marked and often dramatic influence on the rate of cellular development, on cell division and on the utilization of non-mineral nutrients for growth. This review will be concerned with these particular aspects and will illustrate some of the more important functional roles of metals in the biochemical and physiological development of those tissues which have a relatively low mineral content in comparison with skeletal tissue.

For this purpose it is convenient to consider separately the following processes upon which tissue growth ultimately depends:

(1) nutrient ingestion.

(2) the modification of these nutrients by digestive processes to facilitate absorption.

(3) the transport of these modified nutrients to the developing cells and their subsequent entry into those cells.

(4) the synthesis of new cellular components to permit growth and, ultimately, cell division.

Since particular stages of processes (1) to (4) are dependent upon the efficient utilization of energy without which cellular proliferation will not take place, some attention will also be paid to the roles of mineral elements in the utilization of nutrient energy for biosynthesis.

GROWTH FAILURE ACCOMPANYING ANOREXIA IN MINERAL DEFICIENCY SYNDROMES

Failure of appetite is an early feature of several mineral element deficiency syndromes and the adverse effects of these deficiencies on growth is in some cases largely accounted for by the fall in nutrient intake. This situation is particularly pronounced in animals maintained on low intakes of sodium, phosphorus or zinc. There is as yet no clear indication of the mechanism involved in any of these conditions. Sodium depletion gradually leads to a pronounced decrease in the production of parotid saliva (Blair-West, Coghlan, Denton, Goding and Wright, 1964) and it is possible that this may in turn lead to mechanical difficulties in mastication. In other cases it appears probable that mineral depletion may lead to anorexia as a consequence of a direct or indirect involvement of the appetite regulating centres of the brain. For example, the nutrient intake of the young rat falls abruptly only 4 days after withdrawing zinc from the diet and long before clinical lesions of zinc deficiency are apparent. From an analysis of recent results it appears possible that this fall in food intake may be immediately preceded by the sudden growth arrest that is characteristic of zinc deficiency. This suggests that food intake may have fallen as the consequence of a 'feed-back' from a developing metabolic lesion, but the nature of the regulatory mechanism is unknown and warrants closer study. This is particularly so in view of the fact that the inhibitory effects of zinc deficiency on appetite are abolished within 4 hours of feeding a zinc supplemented diet.

Anorexia preceding a reduced rate of growth has been reported in many studies of the effects of phosphorus and cobalt deficiencies (for example Preston and Pfander, 1964; Marston, Allen and Smith, 1961) and in animals maintained on moderately low intakes of magnesium (Greenberg, Lucia and Tufts, 1938). Again, the nature of the physiological and biochemical changes which lead to appetite repression are unknown. Since phosphorus and magnesium are so intimately concerned in the processes of energy utilization for biosynthesis in all mammalian species and cobalt (as a component of enzymes containing vitamin B_{12} as a co-factor) is concerned in particular steps of the same process in ruminants (*see* later) it is tempting to speculate that, as with zinc, the fall in nutrient intake may be an indirect consequence of the failure of specific processes involved in tissue growth and proliferation rather than a direct effect of these mineral nutrients upon the appetite regulating centres.

The plausibility of this suggestion is supported, in the case of zinc deficiency, by the observations of Harper and Rogers (1965) that changes in the composition of tissue amino acid pools may be associated with changes in appetite, and our own recent finding that similar changes in plasma amino acid composition arise at a very early stage of zinc deficiency as appetite fails and are rapidly reversed after zinc supplements are given.

THE ROLE OF MINERAL ELEMENTS IN NUTRIENT HYDROLYSIS DURING DIGESTION

The elements, calcium, magnesium, zinc, cobalt and manganese are integral components or are activators of many of the enzymes concerned with hydrolysis during digestion of lipids, proteins, peptides and polysaccharides judging from *in vitro* studies on purified preparations (for detailed review see Dixon and Webb, 1964). Despite this wealth of knowledge of the *in vitro* behaviour of these systems there have been very few studies of the effects of specific mineral element deficiencies on the activity of these enzymes *in vivo* and on the relationship of such possible effects to changes in nutrient absorption and growth of the intact animal.

One study of this type in which the nature of the lesions leading to growth failure in the zinc deficient rat were investigated showed that the activity of pancreatic carboxypeptidase (a zinc-containing enzyme responsible for the release of single amino acid units during digestion of proteins and peptides) fell after 5 days of zinc depletion to a level that was only 40 per cent of that in pair-fed control rats and rose equally rapidly when zinc was reintroduced. During later experiments in which the fate of ^{14}C-Chlorella protein fed to zinc deficient and pair-fed control rats was investigated it was found, however, that this marked decline in carboxypeptidase activity was insufficient to limit the rate of protein degradation and amino acid absorption and was thus not primarily responsible for the growth failure of zinc deficient animals (Mills, Quarterman, Williams, Dalgarno and Panic, 1967).

It is perhaps surprising that the influence of mineral nutrient supply on the activity of the enzymes responsible for nutrient hydrolysis within the digestive tract has not been more widely investigated, particularly when so many studies have indicated that suboptimal mineral intake has adverse effects on the apparent digestibility of the major nutrients and when the activity of so many digestive enzymes appears to be greatly influenced by the inorganic composition of incubation media used in assay techniques.

The most serious obstacle to a clearer understanding of these possible relationships lies in our lack of knowledge of the extent to which changes in mineral element intake may influence the mineral status of secretory tissues associated with the function of the digestive tract. Although many studies have been made of sodium and potassium homeostasis in the digestive tract, the factors influencing the active concentration of those elements that apparently have a more important effect upon the activity of many digestive enzymes have not so far been investigated.

POSSIBLE ROLES FOR METALS IN THE CELLULAR UPTAKE OF AMINO ACIDS

Early work on the *in vitro* culture of mammalian tissues first led to the suspicion that certain metal ions might be involved in the transport of amino acids into cells. This belief received support from studies in the field of plant nutrition in which it was repeatedly found that mineral nutrient deficiencies, particularly deficiencies of the transition elements, led to extracellular accumulation of amino acids and ultimately to a failure of protein synthesis. Since this time many attempts have been made to define clearly those systems which are involved in the process of intracellular accumulation of amino acids against concentration gradients.

There is as yet no universally acceptable hypothesis which can account for observations on the dynamics of transport of all the biologically important amino acids. However, with some amino acids there is good evidence from studies with intact rats (Ueda, Adeko and Suda, 1960), from work with intestinal preparations (Fridhandler and Quastel, 1955) and from isolated Ehrlich ascites tumour cells that the entry of the amino acid may involve the formation of a Schiff base between pyridoxal phosphate and the amino acid.

Observations that showed that amino acid uptake was inhibited both in bacteria and tumour cells when metal-binding agents were added to the medium, coupled with the knowledge that the formation of pyridoxylidene amino acid complexes may involve chelate formation with metals (for example, *see* Blake, Siegal, Katz and Kilpatrick, 1963) led Christensen and his colleagues to investigate the possible role of pyridoxal phosphate/metal/amino acid complexes in amino acid transport (Pal and Christensen, 1959; Christensen, 1960). It was found that the accumulation of amino acids by preparations of isolated cells was accompanied by an uptake of manganese (but not of copper, iron or zinc) and the rates of both amino acid and manganese absorption were increased by the addition of pyridoxal phosphate to the medium. Although such results could arise either if manganese happened to be a component of the amino acid carrier system or if it was merely accumulated separately in the form of a manganese chelate, the former possibility seems the more likely from observations on the inhibitory effects of transition metal chelating agents on amino acid accumulation that have been reported by others.

The activation of an amino acid carrier system by the sodium ion is the postulate advanced by both Vidaver (1964) and by Kipnis (1965) to account for the direct relationship which exists between the rate of cellular uptake of amino acids by a variety of tissues and the extracellular Na^+ concentration. There is clear evidence from these studies that the acceleration of amino acid uptake is not dependent upon the simultaneous uptake of Na^+ from extracellular fluid, since the accelerated rate of amino acid influx in high-Na^+ media remains unaffected by stimulation or inhibition of Na^+ transport. This stimulatory effect is not produced by other alkali metal ions, indeed one study suggests that K^+ may competitively inhibit the action of Na^+, nor is the effect produced by Ca^{2+} or Mg^{2+} ions.

It is not yet possible to assess the significance of these findings in terms of their effects in the intact animal. For example, we know far too little about the influence of mineral deficiency upon the concentrations of mineral at active sites on cell membranes through which substrates must penetrate, to be able to conclude that a dietary deficiency of manganese or of sodium will rapidly lead to failure of amino acid accumulation by growing tissues. We are well aware that depletion of tissue mineral reserves does not imply a uniform rate of loss of mineral element from all sites at which that mineral is stored or is functioning. Without evidence obtained at the histochemical level which indicates metal depletion at the cell membranes, the significance of *in vitro* experiments which suggest possible roles for mineral elements in the cellular uptake of other nutrients is difficult to assess.

THE ROLE OF METALS IN PROTEIN BIOSYNTHESIS

Attempts to achieve protein synthesis in cell free systems have repeatedly demonstrated that certain metals have pronounced stimulatory effects on the incorporation of labelled amino acids into protein (for example Korner, 1961). It now appears probable that this effect is exerted at two levels in the chain of events leading to the synthesis of protein from individual amino acids. The first step in this sequence is mediated by amino acid activating enzymes which, with free amino acids and adenosine triphosphate (ATP) as substrates, yield amino acid–adenylate complexes and pyrophosphate anions. The activity of this system is strongly dependent upon the magnesium ion concentration of the medium, concentrations of about 5 mM Mg^{2+} being optimal (Korner, 1961). The precise role of magnesium in this step has not yet been elucidated but since magnesium pyrophosphate is the preferred substrate of the enzyme pyrophosphatase, it is feasible that it may be concerned in the removal of pyrophosphate from amino acid activation sites and thus in the establishment of an equilibrium favouring synthesis of activated amino acids. There is good evidence that the potassium ion also stimulates amino acid incorporation in micro-organisms, probably at the stage where the activated amino acid becomes incorporated into the developing peptide chain (Schlesinger, 1964), but the intervention of this element in the mammalian system is less clearly established.

A further point of intervention of a metal in the sequence of reactions leading to the formation of the protein molecule is in the later organization of the functional complex between soluble (s)RNA carrying activated amino acids and ribosomal RNA within which the peptide or protein is assembled. Ultracentrifuge studies of the sedimentation characteristics of mixed species of RNA in this system have shown that both the nature of the associations formed between these species of RNA and their activity in promoting incorporation of isotopically labelled amino acids into protein is influenced by the magnesium concentration of the environment. It is envisaged by Moldave (1965) that magnesium is involved in the processes of orientation of messenger RNA and ribosomal RNA.

Both RNA and deoxyribonucleic acid (DNA) have a very strong affinity for many divalent metals (Wacker and Vallee, 1959). Studies with other metallo-organic complexes show that the elements, copper, zinc, iron and cobalt form complexes of much greater stability than the comparable magnesium complex. In view of this it appears probable that complexes of the former metals and polynucleotides may exist under physiological conditions. Several of these metal–polynucleotide complexes are resistant to attack and degradation by polynucleotidases (for example Wojnar and Roth, 1964) and it is possible that these metals may ultimately be found to confer protection upon the RNA/protein synthesizing system in the intact cell, or to have an even more direct role in the mechanics of protein synthesis. These possibilities are being actively investigated in the present studies in which the influence of mineral element deficiency on polynucleotide turnover rate and the rates of synthesis of specific proteins are being examined.

POLYNUCLEOTIDE BIOSYNTHESIS
IN ZINC DEFICIENCY

Our knowledge of the involvement of polynucleotides in protein biosynthesis, cell growth and cell replication, coupled with an appreciation of the great affinity of polynucleotides for the trace metals, has recently stimulated interest in the possibility of functional associations between these components and their possible roles in the processes of tissue growth and development.

In 1962 Wacker published the results of a study of the effects of zinc deficiency on the growth of the micro-organism *Euglena gracilis*. Omission of zinc from the incubation medium decreased the cell content of RNA and protein and retarded cell division and growth.

While this work was in progress the effects of zinc deficiency in rats, calves and lambs were being investigated at the Rowett Institute. In these species growth arrest takes place within a very short time after zinc supplements are withdrawn from the diet; 4–5 days in the young rat and 7–10 days in calves and lambs. One part of the programme of work to determine the nature of the biochemical lesion responsible for this effect was to examine the effects of zinc deficiency on tissue polynucleotide concentration and possible effects on RNA and DNA labelling after ^{32}P-orthophosphate injection. In rats sacrificed shortly after the onset of zinc deficiency (judged by growth arrest) the RNA content of both liver and pancreas was lower than that of pair-fed control rats. Liver DNA was also lower but not pancreatic DNA. In experiments in which the rates of synthesis and 'turnover' of liver RNA and DNA were followed by studying the incorporation of radioactive phosphorus (^{32}P) into these polynucleotides, it was found that the specific activities of liver RNA and DNA was appreciably lower in animals at a very early stage of zinc deficiency than in pair-fed controls. Investigation of the ^{32}P activity of the component nucleotides of liver RNA after alkaline hydrolysis indicated that the labelling pattern in the RNA from zinc deficient animals was different from that in controls (Williams, Mills, Quarterman and Dalgarno, 1965). In these experiments animals were sacrificed 2 h after ^{32}P-orthophosphate injection and it is not possible to determine from these results whether the low relative specific activity of RNA-^{32}P and DNA-^{32}P arises through differences in precursor labelling (which may be expected to be at a maximum 15–20 min after injection) or to a higher rate of polynucleotide turnover in the presence of a precursor pool that is declining in specific activity. Since the turnover rates of RNA and DNA are so very different and yet the decrease in labelling was of the same order in both polynucleotide species in the zinc deficient tissue, the effect is more likely to be due to a decrease in labelling of common mononucleotide precursors than to differences in polynucleotide turnover rate. These possibilities are being investigated. In connection with these speculations on the effects of zinc deficiency on RNA and DNA synthesis, it is of interest that we have recently obtained evidence that early in zinc deficiency, defects develop in the major pathway leading to the production of ribose 5'-phosphate, an obligatory precursor of both purine and pyrimidine ribo- and deoxyribonucleotides. This evidence is insufficient for it to be claimed either that this defect limits the rate of RNA or DNA synthesis or even that it is a primary lesion in zinc deficiency, since similar changes in the activity of enzymes of this metabolic pathway in rat liver have resulted from the imposition of changes in the feeding pattern (Fábry and Brown, 1967).

These observations on RNA metabolism in zinc deficiency are of particular interest in relation to the findings that there is both a lowered efficiency of incorporation of isotopically labelled amino acids into protein (Mills, Quarterman, Williams, Dalgarno and Panic, 1967) and a decrease in the total protein content of mitochondria, ribosomes and cell sap in the zinc deficient rat. Since these differences cannot be attributed to a decrease in the efficiency of intestinal proteolysis and intestinal absorption of amino acids (Mills et al., 1967) they may possibly arise as a consequence of the defects in RNA synthesis just described.

Studies of the sequence of events preceding DNA synthesis and cell division in cultured rabbit kidney cortex cells led Leiberman and Ove (1962) and Lieberman, Abrams, Hunt and Ove (1963) to the discovery that the addition of the chelating agent ethylene diamine tetra-acetic acid (EDTA) to the medium, shortly after initiation of the culture, strongly inhibited DNA synthesis as judged by the incorporation of [3]H- or ^{14}C-thymidine into DNA. This effect was obtained when EDTA was added to the culture at the initiation of growth or within a period of 20 h thereafter. EDTA prevented the rise in thymidylate kinase and DNA polymerase that normally precedes DNA synthesis later in the development of the cell. Later studies showed firstly, that the addition of EDTA later than 20 h after the initiation of the culture had no effect on the activities of these enzymes and did not inhibit DNA synthesis and, secondly, that the effects of EDTA added during early cell growth were abolished if zinc was simultaneously added. Fujioka and Lieberman (1964) later showed that DNA synthesis in the liver of partially hepatectomized rats was inhibited by perfusion with EDTA unless the tissue was also perfused with Zn^{2+} and they demonstrated that this effect of Zn^{2+} was not shared by Fe^{2+}, Ni^{2+}, Mn^{2+}, Co^{2+}, Cu^{2+} or MoO_4^{2-} ions.

The mechanisms underlying these effects of Zn^{2+} on DNA synthesis *in vitro* and *in vivo* are as yet obscure. Lieberman and his colleagues have suggested that a zinc protein may be involved in the sequence of events preceding DNA synthesis. An alternative and equally plausible hypothesis is that zinc is in some way involved in the production of nuclear RNA, since the period of maximum sensitivity to EDTA and zinc corresponds closely to the period at which nuclear RNA synthesis is proceeding with greatest rapidity in the young cell. This postulate is particularly attractive when one considers the similarity of the effects of *low* concentrations of the inhibitor actinomycin D and the metal chelating agent EDTA on DNA synthesis. Both inhibit the synthesis of the rapidly labelled fraction of RNA present in the nucleolus without influencing the rate of synthesis of other RNA species (it is emphasized that this specific effect of actinomycin D is only obtained with low concentrations; higher concentrations have a more general effect on RNA and protein synthesis) and both lead ultimately to a failure of DNA synthesis and inhibit cell division.

If the action of zinc and the inhibitory effects of EDTA are at the site of synthesis of rapidly labelled RNA in the nucleolus, it is perhaps significant that Fuji (1958) has shown by histochemical techniques that almost all of the zinc of the cell nucleus is centred in the nucleolus. Isolation and analysis of nucleoli revealed that the zinc concentration was within the range 1·3–3·7 mg/g; values that are approximately fifty to one hundred times greater than those found in most other tissue structures.

This detailed description of work suggesting functional roles for zinc in polynucleotide metabolism has been given because of its possible relevance to the dramatic growth arrest that supervenes when an animal becomes zinc deficient. It should be emphasized, however, that many of these studies are at a very early stage of development. We do not know, for example, how widely these effects on cell division may account for the failure of growth of individual tissues or whether the marked stimulatory effects of zinc on tissue regeneration after trauma are related to the above processes.

The clear resolution of the role of zinc in tissue growth and development is hindered by the fact that restriction of zinc intake rapidly leads to appetite failure and thus to a decreased supply of substrates and energy for synthetic processes. However, as is mentioned in another section of this paper, it is probable that the decline in growth rate may immediately precede the fall in appetite and that the primary lesion is not at the site of appetite regulation. This view is supported by the results of several years work with pair-fed rats in which food intake was equated to that of zinc deficient littermates. Restriction of food intake, although severely limiting growth, did not prevent it completely as in zinc deficient animals.

THE ROLES OF MINERAL ELEMENTS IN THE UTILIZATION OF ENERGY FOR GROWTH

The elaboration of all tissue components for cellular growth and replication is accompanied by the utilization of energy. It follows that if energy supply is limited, either through a low calorie intake or through inefficient utilization, this will ultimately limit the rate of tissue growth.

The influence of energy supply on growth is considered in greater detail elsewhere in this symposium; the object of the present discussion is to briefly indicate the points in the processes of energy utilization in which mineral elements have an important role. In this context it is important to appreciate that the metabolic changes described do not solely affect the process of tissue growth; they must also influence the resting metabolism of cells and the process of tissue repair and replacement during normal resting metabolic turnover and wastage.

Most biosynthetic processes involve at a multiplicity of steps the utilization of phosphate bond energy derived from the nucleotide triphosphates. The 'trapping' of nutrient energy in this form is achieved by coupling the oxidation of substrates to the generation of nucleotide triphosphates both during the early stages of degradation of energy sources and at the terminal steps of the oxidative sequence during the reoxidation of reduced co-enzymes.

The overall efficiency of energy trapping during the terminal oxidation may be as high as 45 per cent when optimal coupling between oxidation and phosphorylation is achieved. The efficiency

377

of coupling of oxidation to phosphorylation is, however, partly determined by the intracellular Mg^{2+} ion concentration and *in vitro* studies have shown that at low concentrations of Mg^{2+} the efficiency of coupling and consequently the transfer of energy from cell nutrient to the nucleotide triphosphates is greatly reduced (Vitale, Nakamura and Hegsted, 1957).

The importance of this metabolic defect as the primary cause of growth failure in magnesium deficiency is not known. It is apparent, however, that species differences exist in the effects of chronic marginal magnesium deficiency on growth. For example, in the fowl (Bird, 1949) and the rat (Leroy, 1926) magnesium deficiency has been found repeatedly to decrease the rate of growth and efficiency of food utilization. In the pig and in ruminants there is no clear evidence of such an effect. This may be due merely, however, to the fact that the nervous sequelae of magnesium deficiency are so striking and appear with such regularity in these species at an early stage of depletion that effects on growth are of relatively minor importance.

This point of intervention of Mg^{2+} is only one of many at which this and other elements are involved in the processes of energy utilization in biological systems. Thus, the haem enzymes of the terminal respiratory chain, cytochromes b, c, a and a_3 (cytochrome oxidase) and the enzyme succinic dehydrogenase all contain iron as a component. A decline in the activity of cytochrome c and succinic dehydrogenase occurs in severe iron deficiency (Beutler and Blaisdel, 1958, 1960). Cytochrome oxidase contains both copper and iron and a decline in the activity of this enzyme is the earliest known indication of copper deficiency in mammalian tissues (Gallagher, Judah and Rees, 1956; Mills, Williams and Poole, 1963).

Although these changes must inevitably decrease the rate of production of energy-rich nucleotide triphosphates and thus the availability of energy for synthesis of tissue components, there have been no quantitative studies of the *in vivo* effects of deficiencies of magnesium, iron and copper on tissue nucleotide triphosphate concentrations.

Considering the magnitude of effects of deficiencies of these elements involved in energy transfer on the growth of the intact animal the following generalizations may be made:

(1) in most species a decline in growth rate is an early feature of copper deficiency even in the absence of other clinical lesions: low

378

activities of the copper enzyme cytochrome oxidase have been found in the tissues:

(2) growth failure is a comparatively late feature of iron deficiency syndromes; similarly, the tissue concentration of the iron-containing enzymes of the terminal respiratory chain do not decrease markedly until tissue iron reserves have been depleted by prolonged iron deprivation:

(3) the most characteristic features of magnesium deficiency in all mammalian species are hyperirritability and tetany; a decrease in rate of growth is usually apparent only when the magnesium intake is sufficient to prevent dysfunction of the neuromuscular system. The possible relationship of growth failure to a decreased efficiency of oxidative phosphorylation has not so far been examined.

The above discussion has centred on the role of mineral elements in energy trapping through oxidative phosphorylation; there is ample evidence, however, that metals are involved in many metabolic transformations that are essential preliminaries to this process. Thus many oxidative steps are mediated by dehydrogenase enzymes which contain zinc as a co-factor. A further example of the obligatory intervention of a metal at this level of substrate transformation is the role of cobalt as a component of cyanocobalamin which acts as a co-factor in the enzymic transformation of propionate to succinate in ruminant species. Cobalt deficiency in ruminants leads to a decline in growth rate and it appears probable that a decreased efficiency of utilization of energy from propionate may be an important factor contributing to this situation. Whether this is the most important site of action of cobalt in tissue growth must now be questioned, however, in the light of recent findings that in micro-organisms the synthesis of deoxyribonucleotides for DNA production and cell division is a cobalamin-dependent process.

INTEGRATING DISCUSSION

The preceding sections serve to illustrate the many diverse sites of action of the mineral elements in the processes that are obligatory before cell growth and replication can proceed. Which of these processes fail during mineral depletion is determined largely by the extent to which the concentration of each element falls at the local site of action of the mineral or at the site of its incorporation into an enzyme complex. We know that in some instances the organism responds to depletion in such a way as to conserve mineral element concentrations at the active site. Sodium depletion is a typical example of this process in which the immediate response to depletion is the efflux of extracellular water to maintain the sodium concentration at cell interfaces. In other cases depletion is accompanied by the mobilization of tissue stores of the element to maintain the metabolic *status quo*. Copper depletion brings about a steady but immediate withdrawal of liver copper stores which prevents a rapid decline in the activity of those copper-dependent processes that are involved in the maintenance of tissue growth. The effectiveness of compensatory responses to depletion differ, however, from organ to organ within the body in the sense that one metal-dependent process may fail while another dependent upon the same metal may remain at least temporarily unaffected. A good illustration of this situation and of the nature of species difference in the response to mineral depletion is provided by copper depletion. In most mammalian species clinical copper deficiency is accompanied by growth failure and is preceded by a fall in the cytochrome oxidase activity of liver and muscle tissue. The cytochrome oxidase activity of nervous tissue remains unaffected and development of this tissue proceeds normally. In contrast, growth is not affected in most cases of copper deficiency in the lamb and liver cytochrome oxidase is normal but oxidative defects and degenerative lesions arise in the central nervous system.

Generalized effects on tissue growth occur with the greatest rapidity and most dramatically when the organism is unable to maintain the concentration of the essential element at important functional sites through the process of redistribution of the element from tissue stores. This situation probably exists in zinc deficiency. Plasma zinc concentrations may fall to one-quarter or one-fifth of normal within 24 h of feeding a zinc deficient diet and there appear to be no significant tissue depots of zinc that can be drawn upon to maintain even temporarily the normal zinc concentration of tissue fluids. Gross analysis of tissues fails to reveal any significant redistribution of zinc in the brief interval between the decline in plasma zinc and growth failure and we are forced to the conclusion that in this instance a local depletion of zinc from its sites of involvement in the growth process must occur. Good evidence that zinc is not withdrawn at a uniform rate from all its sites of action in other processes is provided by our own observations and those of several other groups of workers that the zinc dependent dehydrogenases so far investigated fall in activity only at the terminal stages of zinc deficiency.

The above examples serve to illustrate the diversity of points at which mineral elements intervene in the processes of tissue growth and development. It is regrettable that in many of the situations described it is feasible to present only a tempting view of possibilities rather than to clearly relate *in vitro* studies to observations made in the intact animal suffering effects on growth and development through a specific mineral deficiency. Our frequent inability to close this gap is a reflection of what we have all known for many years, that the biochemist, the nutritionist and the physiologist meet all too rarely in their attempts to consider the animal as an intact organism!

REFERENCES

Beutler, E. and Blaisdel, R. K. (1958). *J. Lab. clin. Med.* **52**, 694
—— (1960). *Blood* **15**, 30
Bird, F. H. (1949). *J. Nutr.* **39**, 13
Blair-West, J. R., Coghlan, J. P., Denton, D. A., Goding, J. R. and Wright, R. D. (1964). In: *Salivary Glands and their Secretions* (Ed. by L. M. Sreeby and J. Meyer). London; Pergamon
Blake, M. I., Siegal, F. P., Katz, J. J. and Kilpatrick, M. (1963). *J. Am. chem. Soc.* **85**, 294
Christensen, H. N. (1960). *Adv. Protein Chem.* **15**, 239
Dixon, M. and Webb, E. C. (1964). *Enzymes.* London; Longmans
Fábry, P. and Brown, P. (1967). *Proc. Nutr. Soc.* **26**, 144
Fridhandler, L. and Quastel, J. H. (1955). *Arch. Biochem. Biophys.* **56**, 424
Fuji, T. (1958). *Proc. Fac. Sci. Univ. Tokyo* **7**, 313
Fujioka, M. and Lieberman, I. (1964). *J. biol. Chem.* **239**, 1164
Gallagher, C. H., Judah, J. D. and Rees, K. R. (1956). *Proc. R. Soc.* (B) **145**, 134
Greenberg, D. M., Lucia, S. P. and Tufts, E. B. (1938). *Am. J. Physiol.* **121**, 424
Harper, A. E. and Rogers, Q. R. (1965). *Proc. Nutr. Soc.* **24**, 173
Kipnis, D. M. (1965). In: *Control of Energy Metabolism* (Ed. by B. Chance, R. W. Estabrook and J. R. Williamson). New York; Academic Press
Korner, A. (1961). *Biochem. J.* **81**, 168
Leroy, J. (1926). *C.r. hebd. Séanc. Acad. Sci., Paris* **94**, 431
Lieberman, I., Abrams, R., Hunt, N. and Ove, P. (1963). *J. biol. Chem.* **238**, 3955
— and Ove, P. (1962). *J. biol. Chem.* **237**, 1634
Marston, H. R., Allen, S. H. and Smith, R. M. (1961). *Nature, Lond.* **190**, 1085
Mills, C. F., Quarterman, J., Williams, R. B., Dalgarno, A. C. and Panic, B. (1967). *Biochem. J.* **102**, 712
— Williams, R. B. and Poole, D. B. R. (1963). *Biochem. J.* **87**, 10P

Moldave, K. (1965). *Ann. Rev. Biochem.* **34**, 419
Pal, P. R. and Christensen, H. N. (1959). *J. biol. Chem.* **234**, 613
Preston, R. L. and Pfander, W. H. (1964). *J. Nutr.* **83**, 369
Schlesinger, D. (1964). *Biochim. biophys. Acta* **80**, 473
Ueda, K., Adeko, H. and Suda, M. (1960). *J. Biochem., Tokyo* **48**, 584
Vidaver, G. A. (1964). *Biochemistry, N.Y.* **3**, 662
Vitale, J. J., Nakamura, M. and Hegsted, D. M. (1957). *J. biol. Chem.* **228**, 573
Wacker, W. E. C. (1962). *Biochemistry, N.Y.* **1**, 859
— and Vallee, B. L. (1959). *J. biol. Chem.* **234**, 3257
Williams, R. B., Mills, C. F., Quarterman, J. and Dalgarno, A. C. (1965). *Biochem. J.* **95**, 29P
Wojnar, R. S. and Roth, J. S. (1964). *Biochim. biophys. Acta* **87**, 17

DISCUSSION

MR. LITTLE (B.O.C.M. Ltd.) enquired whether continued growth of wool on underfed sheep was part of an adaptive mechanism to reduce heat loss, to which DR. BLAXTER replied that, in fact, they grew less wool than overfed ones and their temperature regulation became grossly upset, so that they had to be kept very much warmer than was necessary early in the experiment.

DR. TOPPS (Aberdeen) asked why sheep fed to 'maintenance' gained body fat and lost flesh, to which DR. BLAXTER replied he didn't know but the daily gain of calories as fat was very small (35) and the loss of calories as protein was even smaller (about 6).

DR. BRAUDE (Reading) expressed surprise that the two approaches DR. BLAXTER had presented agreed so closely and wondered how this finding could be applied. DR. BLAXTER agreed that the conformity was surprising and thought that the main value of the result was in putting things into perspective by showing that food was the main determinant of growth; it expressed shapes of growth curves in terms of food supply. Numerical solutions could be produced and most of the constants were built in; the only constant needed to solve the equation was the change in the proportion of fat to protein in the gains.

DR. TIMON (Dublin) asked whether the equation relating the maintenance requirement of the animal to body weight 0·734 was equally applicable at all stages of growth regardless of body composition, or would a linear skeletal measurement be better. DR. BLAXTER replied that the so-called 'metabolic body size' equation was useful but only in the context in which it was first promulgated, namely, as a description of the fact that as species increase in size so their metabolism increases but at a smaller proportional rate. There was evidence that in the very young animal the energy cost of maintenance was very high relative to the older animal or an older animal of a different species but the same size as itself; i.e. the metabolic body size concept was not generally applicable over the whole age relationship of animals. Within a single species a linear equation with an intercept was probably just as accurate as a power relationship.

DR. FALCONER (Nottingham) stated that from the figures shown for predicted values for body weight against observed values, it appeared that the only major discrepancy was where the animal was on a sub-maintenance diet. If in these circumstances the animal was replacing body protein or fat with extracellular fluid, this discrepancy would be expected. DR. BLAXTER replied that although there was no oedema in these animals this could be part of the reason for the discrepancy with weight loss; mean calorific value of gains had been used in the computation of the equation.

Dr. Rerat (Jouy-en-Josas) enquired whether any difficulty had been experienced in obtaining the maintenance level because of adaptation by animals to the lower feeding levels, as occurred with rats on low protein diets. Dr. Blaxter replied that he had not looked at adaptation to protein undernutrition; the energy retentions measured in maintenance-fed sheep showed more positive values in the first few weeks of the experiment than in the later stages. Dr. Armstrong (Newcastle upon Tyne) observed that Dr. Widdowson had mentioned very watery muscle in several restricted animals and referred to Kielonowski's data showing high efficiency for fat deposition and wondered whether this might have been due to high fat levels in the diet resulting in direct incorporation of dietary fat into body fat. Dr. Blaxter replied that Kielonowski should really put another term in his equation to allow for direct incorporation.

Dr. Armstrong also asked what the Kf values of Dr. Blaxter's animals had been during their growth period, to which Dr. Blaxter replied that the net availability of metabolizable energy had been 0·50 on a dried grass diet.

Dr. Taylor (Edinburgh) agreed with Dr. Blaxter that when the range in values was only about twofold, the difference between using the linear relationship and the two-thirds power was imperceptible. He went on to observe that he had fitted exponential curves to data on animals fed a constant amount and they fitted to within ± 3 per cent, which was excellent confirmation of the validity of Dr. Blaxter's similar equation. In relation to Dr. Braude's question on the use of such an equation, he stated that this was exactly what geneticists had been looking for as the Kf and Km values would be genetic characteristics of the animals which could be a basis for selection; animals with high Kf/Km ratio would belong to an early-maturing type. However, as the genetic variation in these values for individuals was only likely to be 2–3 per cent, such an equation would have to be accurate to within 0·5–1 per cent before it could be used in animal breeding. Dr. Blaxter replied that it was a slow and laborious business to determine Kf and Km values, and thought that in determining shapes of growth curves the factors of importance, whether of genetic or environmental origin, were those which determined patterns of intake of foods of widely different qualities.

Dr. Shaw (Merck, Sharp and Dohme Ltd.) asked for further information on the measurement of direct incorporation of fat, to which Dr. Blaxter replied that this could be done isotopically; an indirect way was by measuring efficiencies, which revealed that it was biochemically impossible for the energy retentions to have been as high as they were with fatty acids if the fat had been first degraded.

Dr. Rerat queried the expressing of amino acid requirements as a percentage of the diet and thought it was better to express the needs of the pig in relation to energy intake. Professor Lewis agreed that if a reference point was needed then energy was better for this than dry matter but when the circumstances were less clearly defined then data was best accumulated by referring to the diet as a whole; the concentration of nutrients should be graded in relation to the energy-yielding fraction because it was the only factor which could be supplied at the same concentration but occupying different fractions of the diet, depending, for example, on the proportion of fat present.

Dr. Jones (Aberdeen) asked how the differential absorption rates for the different amino acids were measured and would this vary with the proportion of synthetic amino acid in the diet, to which Professor Lewis replied that differences in rates of absorption had been demonstrated in the human subject but the data were inadequate to provide quantitative values for use in calculation of dietary requirements.

Dr. RERAT commented on the absorption of free lysine in the pig by saying that in experiments with wheat and synthetic lysine rates of absorption had been the same from both sources.

Dr. SHAW asked for further opinion on methods of arriving at amino acid requirements of ruminants; PROFESSOR LEWIS replied that he and his colleagues were currently initiating work of this nature, using the technique of assessing plasma amino acid levels and extrapolating conclusions previously obtained on the relationship between relative amino acid balance and allowances; once a deviation from the ideal amino acid was produced this resulted in an increase in the surplus amino acid in the plasma, together with certain others due to a restriction in protein synthesis. The technique, therefore, was to feed to animals with a duodenal loop a basic diet of known amino acid balance, to add selected supplements of amino acids and then to measure the response in terms of plasma amino acid level; this would indicate whether an amino acid had been added which was previously limiting.

Dr. SALMON-LEGAGNEUR (Jouy-en-Josas) asked how excess of tryptophan caused a decrease in lean content, to which PROFESSOR LEWIS replied that it may be the result of an interaction between amino acids, in the same way as occurred with lysine/arginine in the chick; the relative levels may be significant also, in that the range of tryptophan levels was very wide, from 0·08–0·22 per cent (33 per cent range).

Dr. MITCHELL (Reading) wondered whether upsetting the balance of amino acids by adding synthetic forms might influence absorption, to which PROFESSOR LEWIS replied that there was considerable evidence that there was no competition effect for absorption at the alimentary level; evidence at the cellular level was less conclusive.

Mr. PIKE (Leeds) asked whether there had been much between-animal variation in the effect of amino acid imbalance, to which PROFESSOR LEWIS replied that there had been little and MR. D'MELLO confirmed this.

To a question by DR. SHAW on what the effect of zinc deficiency would be if the depressing effect on feed intake were overcome by force feeding, DR. MILLS replied that there were the two problems, one of the deficiency and the other of food intake, but even if the food was ingested it was not properly utilized and did not promote growth; food intake was important but not of overriding importance.

Mr. LITTLE asked what was the effect of low mineral intake on the mature animal, such as the cow; to which DR. MILLS replied that effects were generally less dramatic and much slower to appear; with zinc, for example, there was a slow decline in live weight and a slow appearance of skin lesions. In the lactating animal milk yield declined rapidly with deficiency of both zinc and copper; claims had been made that sterility occurred with these also. There were both age differences and species differences in reaction to specific mineral deficiencies; for example, copper deficiency in mature sheep was unknown in Britain but well known as affecting the brain of the lamb, whereas in calves it affected not the brain but liver and muscle.

PROFESSOR LEWIS (Nottingham) asked for further comment on the relationship between mineral intake and feed intake, as to whether there was a common 'feedback' mechanism or response to individual levels. DR. MILLS thought that routes must converge on the regulatory centre; initial lesions giving rise to appetite failure may be different because of differences in time of response, for example very rapid with zinc and slower with copper. If zinc deficient animals were offered two feeds, one containing zinc and one not, they started to select that with zinc within 2 h and by 4 h the difference in consumption was significant. Evidence on the relationship between this and plasma amino acid levels was very sparse but some [results suggest that an abnormal pattern of amino acids may begin to change back to normal within half an hour of giving zinc; there were also changes in plasma free fatty acids.

MR. O'GRADY (Fermoy) commented on the varying response to mineral deficiencies in spite of the very basic function of these in the animal's metabolism. Dr. MILLS thought that much of this variation, as in cytochrome oxidase values, was due to imperfection in the technique; furthermore, some measures of response, such as liver copper or blood copper, were not sufficiently well related to the direct effect of the deficiency and, therefore, gave poor correlations.

PROFESSOR LUCAS (Bangor) remarked that manganese deficiency in pigs had been found to produce not only bone lesions but excess fat deposition in animals gaining at the normal rate and asked whether this was a species difference, in that the pig did not suffer appetite depression, or a peculiarity of manganese. DR. MILLS replied that as the same effect had been reported in rats it must be a peculiarity of manganese rather than a species effect, the mechanism of which he did not know.

DR. ADLER (Jerusalem), on the question of minerals and feed-back mechanism on appetite, asked whether cobalt acted on appetite maintenance in the ruminant through vitamin B_{12} production. DR. MILLS replied that B_{12} was required for conversion of propionate to succinate before the energy of propionate could be used and that this had been proposed as the explanation of emaciation in cobalt deprived animals, but an alternative explanation came from recent work with bacteria, namely, that the enzyme responsible for production of deoxyribose required for DNA production contained B_{12} as a co-factor; this could result in an immediate interference with growth.

DR. MESSAGE (Cambridge) remarked that a recent report had stated that in the rat brain the localization of zinc was in the fore-brain, i.e. where it could influence appreciably the hypothalamus. He commented further, in relation to protein synthesis and B_{12}, that long-standing B_{12} deficiency in man gave rise to carcinoma of the stomach and this implied rapid cellular proliferation and massive DNA synthesis. He went on to ask for comment on the fact that it was very difficult to label muscle RNA in the growing animal but very easy in the older animal. DR. MILLS replied that this may depend upon what the labelling material was; ortho-phosphate and ^{14}C ribose were well retained.

MR. EVERETT (Walls Ltd.) asked how long it took for experimental animals to show clinical signs of mineral deficiency, particularly scouring, to which DR. MILLS replied that scouring did not normally occur with zinc deficiency but with copper it occurred in the terminal stages, both with sheep and cattle, except where molybdenum was high in which case severe scouring occurred in the early stages.

DR. EVERITT (Ruakura) asked for comment on the relationship between selenium and muscular tissue, but DR. MILLS replied that little was known about the function of selenium.

DR. BLAXTER (Aberdeen) commented on the appetite effect and re-emphasized that while *most* deficiencies resulted in a change in intake, vitamin E deficiency in calves was *not* associated with immediate decline in intake and the rate of protein synthesis in these animals was extremely high.

VII. CARCASS QUALITY AND ASSESSMENT

BEEF CARCASS EVALUATION:
FAT, LEAN AND BONE

M. A. CARROLL and D. CONNIFFE

The Agricultural Institute, Castleknock, Co. Dublin

THE PROPORTION of fat, lean and bone in a carcass and the distribution of these between the cheap and expensive joints are not the only factors to be taken into consideration in beef carcass evaluation, also important are tenderness and the water-holding capacity of the lean meat, together with the colour of the fat and lean. However, limitation of time and space necessitates that the present discussion be limited to the measurement of fat, lean and bone in the carcass, the composition of the carcass, the distribution of lean and the prediction of total carcass 'lean' from 'lean' in a sample joint.

THE MEASUREMENT OF FAT, LEAN AND BONE

A carcass or joint may be dissected into separable fatty tissue (both subcutaneous and intermuscular), separable muscle, separable connective tissue and bone, as set out in Table 1, or the subcutaneous fatty tissue can be removed, the carcass or joint boned out and the subcutaneous fatty tissue and boneless joint separately minced and a sample of each chemically analysed for fat, moisture, protein and ash (Table 2). In the first of these methods for measuring carcass composition, the amount of fat and connective tissue in the separable muscle can be responsible for variation between operators, between seasons and, particularly, between animals of different degrees of fatness. Separable muscle always contains the intramuscular fat and this can vary by as much as from 1·3 to 14·5 per cent (Callow, 1962b). This variation in the amount of fat in separable muscle could be quite a serious source of interaction between treatment and method of measurement when assessing the proportion of total carcass separable muscle occurring in a particular joint or joints. It is the main objection to using 'trimmed wholesale cut' as a measurement in carcass work.

Basing the calculation of carcass composition on chemical analysis eliminates the problem of a variable amount of fatty tissue occurring in the separable muscle. However 'lean', which is the sum of the moisture, protein and ash in the boneless joint, contains the moisture, protein and ash of the subcutaneous, intermuscular and intramuscular fatty tissue together with that occurring in the connective

Table 1

Dissection

Separable fatty tissue	includes	Subcutaneous Intermuscular
Separable muscle	includes	Intramuscular fat Muscle
Separable connective tissue	includes	Tendon Ligament Fascia Periosteum
Bone	includes	Bone Cartilage

Table 2

Dissection and chemical analysis

Subcutaneous fatty tissue	Fat Moisture Protein Ash	Fat 'Lean'
Intermuscular fatty tissue Intramuscular fatty tissue Muscle Connective tissue	Fat Moisture Protein Ash	Fat 'Lean'
Bone		Bone

tissue. Subcutaneous fatty tissue contains about 80 per cent fat, so about 20 per cent of this fatty tissue is attributed to 'lean'. Do these limitations invalidate the use of 'lean' as representing muscle in the carcass when comparing the muscle content of carcasses or joints? We do not think so, and regard 'lean' as a useful measurement. However, if 'lean' is used as a basis for economic calculations, allowance must be made for the proportion of 'lean' occurring in fatty and connective tissue and, undoubtedly, the best parameter to use would be the fat-free separable muscle. John (1960) has presented a useful consideration of this aspect of carcass appraisal.

CARCASS COMPOSITION

Throughout the rest of this paper frequent reference will be made to Harte's (1966) data. He compared the feed efficiency and carcass composition of Friesian, Hereford × Shorthorn and Aberdeen Angus × Shorthorn steers. In his first experiment, the cattle were

Table 3

Comparison of three beef 'breeds' (Harte, 1966)

	Friesian	Hereford × Shorthorn	Aberdeen Angus × Shorthorn
Experiment I			
Number of animals	10	8	7
Age (days)	723	695	659
Final live wt. (lb)	$1,181 \pm 4\cdot4$	$1,096 \pm 4\cdot8$	$990 \pm 5\cdot0$
Hot carcass wt. (lb)	$681 \pm 6\cdot6$	$624 \pm 7\cdot0$	$573 \pm 7\cdot5$
Experiment II			
Number of animals	9	10	10
Age (days)	737	737	737
Final live wt. (lb)	$1,039 \pm 20\cdot5$	$1,047 \pm 19\cdot4$	$973 \pm 19\cdot4$
Hot carcass wt. (lb)	$588 \pm 13\cdot2$	$616 \pm 12\cdot5$	$566 \pm 12\cdot5$

fed concentrates as a constant proportion of live weight and roughage *ad lib.* and were killed at a predetermined live weight; Friesians at 1,200 lb, Hereford × Shorthorns at 1,100 lb and Aberdeen Angus × Shorthorn at 1,000 lb. In his second experiment, the steers were fed *ad lib.* (throughout most of their life) a constant proportion of concentrates to roughage and were killed at a constant age of 737 days (Table 3).

Lean is the most valuable part of a carcass. It is most important, therefore, for the carcass to contain no more fat than the consumer requires, as any increase in fat means less lean. From Tables 4 and 5, it can be seen that the Friesian carcasses contain more lean than either of the other two breeds, despite the fact that on a fat-free basis the Friesian carcasses contain about 1·5 per cent more bone.

It is important, in this connection to remember that management factors, such as plane of nutrition and live weight at slaughter, have a marked influence on the fat composition of a carcass.

Table 4

Carcass composition of three beef 'breeds' (Harte, 1966)

	Experiment	Friesian	Hereford × Shorthorn	Aberdeen Angus × Shorthorn	Level of significance
Fat per-centage	I	$15\cdot3 \pm 1\cdot17$	$22\cdot8 \pm 1\cdot35$	$21\cdot1 \pm 1\cdot25$	**
	II	$14\cdot1 \pm 1\cdot26$	$19\cdot4 \pm 1\cdot04$	$23\cdot4 \pm 1\cdot10$	***
'Lean' per-centage	I	$71\cdot1 \pm 0\cdot92$	$65\cdot8 \pm 1\cdot06$	$67\cdot0 \pm 0\cdot99$	***
	II	$70\cdot7 \pm 1\cdot16$	$67\cdot7 \pm 0\cdot94$	$65\cdot3 \pm 1\cdot00$	**
Bone per-centage	I	$13\cdot6 \pm 0\cdot26$	$11\cdot5 \pm 0\cdot31$	$11\cdot9 \pm 0\cdot31$	***
	II	$15\cdot0 \pm 0\cdot34$	$12\cdot7 \pm 0\cdot32$	$11\cdot6 \pm 0\cdot32$	***

** = P>0·01; *** = P<0·001

Table 5

Carcass composition on a fat-free basis of three beef 'breeds' (calculated from Table 4)

	Experiment	Friesian	Hereford × Shorthorn	Aberdeen Angus × Shorthorn
Lean percentage	I	83·9	85·1	84·9
	II	82·5	84·2	84·9
Bone percentage	I	16·0	14·9	15·1
	II	17·5	15·8	15·1
'Lean':Bone	I	5·2	5·7	5·6
	II	4·7	5·3	5·6

The 'lean' to bone ratio is also a useful measurement in comparing carcasses as it eliminates the complications of a third variable, fat. It is a measurement which geneticists rightly think well worth considering in animal breeding plans.

MUSCLE DISTRIBUTION

An estimate of the distribution of expensive and cheap joints in a carcass is a basis on which live animals and carcasses are often judged. Because the muscle in a joint is of far greater value than the fat or bone, it can be misleading to compare carcasses on the basis of joint weights; it is far better to compare carcasses on the basis of muscle distribution. Harte (1966) dissected his carcasses into 12 joints which were comparable to the London and Home Counties cuts. In Table 6 are given the data for the 'lean' in each cut as a percentage of total carcass 'lean'. When compared in this way, there seems to be very little difference between what can be regarded as three very different types of cattle. This is not surprising since Willey, Butler, Riggs, Jones and Lyerly (1951) and Stonaker, Hazaleus and Wheeler (1952) found only marginal differences in the distribution of wholesale joints when they compared 'Compact' and 'Regular' type Herefords, and Butler, Warwick and Cartwright (1956) found little difference between Brahman × Herefords and Herefords. Dumont, Le Guelte and Arnoux (1961) report little variation in distribution of muscle within 29 Charolais and Butterfield (1963) found little difference in muscle distribution between Herefords, Aberdeen Angus × Brahmans and unimproved Shorthorns.

Butler (1957) concluded that 'The animal breeder has considerable latitude in selecting animals of different shapes without encountering great changes in the proportion of wholesale cuts'. Two animals could have the same weight of lean in the carcass and the same distribution of lean between cuts but have different lengths of carcass or lengths of hind leg. This means that Yeates' (1959) gross fleshing index (total weight of carcass 'lean' divided by carcass length) would be different for the two animals. Likewise, the same type of index could be applied to the leg; total 'lean' in leg over leg length. Fleshing is more than likely one of the attributes being selected for in the show ring and in carcass competitions although assessment of the characteristic by eye judgment is made difficult by fat cover. However, selection for this characteristic does not mean that muscle distribution is changed.

THE PREDICTION OF CARCASS 'LEAN'
FROM 'LEAN' IN SAMPLE JOINTS

In order to save time and expense, the composition of sample joints

Table 6

Lean meat in each cut as a percentage of total carcass lean in three beef 'breeds' (Harte, 1966)

London and Home County type cuts	Experiment I				Experiment II			
	Friesian	Hereford × Shorthorn	Angus × Shorthorn	Level of significance	Friesian	Hereford × Shorthorn	Angus × Shorthorn	Level of significance
Clod and sticking and fore and hind shin	21·9±0·4	22·3±0·4	21·6±0·4	N.S.				
Clod and sticking					13·4±0·2	14·7±0·2	14·1±0·2	**
Fore shin					2·7±0·1	2·7±0·1	2·6±0·1	N.S.
Hind shin					5·0±0·1	4·8±0·1	4·6±0·1	*
Chuck	12·6±0·3	12·6±0·4	12·0±0·3	N.S.	14·6±0·4	14·4±0·3	14·0±0·3	N.S.
Brisket	6·5±0·3	6·3±0·3	6·3±0·3	N.S.	6·4±0·2	6·3±0·2	6·2±0·2	N.S.
5th, 6th and 7th ribs	5·7±0·1	5·5±0·2	5·6±0·1	N.S.	5·4±0·1	5·1±0·1	5·3±0·1	N.S.
8th and 9th ribs	3·7±0·1	3·9±0·1	3·8±0·1	N.S.	3·6±0·1	3·6±0·1	3·6±0·1	N.S.
10th rib	1·8±0·1	1·9±0·1	1·6±0·1	N.S.	1·7±0·1	1·8±0·1	1·6±0·1	N.S.
Plate	1·3±0·1	1·4±0·1	1·3±0·1	N.S.	1·5±0·1	1·4±0·1	1·5±0·1	N.S.
Loin	6·4±0·1	6·4±0·1	6·6±0·1	N.S.	6·4±0·2	6·5±0·1	6·6±0·1	N.S.
Flank	7·5±0·1	8·7±0·1	8·4±0·1	***	7·6±0·2	8·0±0·2	9·1±0·2	***
Round	32·7±0·4	31·0±0·5	32·9±0·5	*	31·7±0·4	30·8±0·4	30·7±0·4	N.S.

N.S. = non-significant; * = P<0·05; ** = P<0·01; *** = P<0·001 is often used to predict total carcass composition. Relationships established between a sample joint and total carcass composition in some previous work are often taken as justification for using the sample joint, whereby it is assumed that the relationship established

Table 7

Simple and multiple correlations between indirect measurements and total carcass lean

	N	Carcass weight (lb)	Correlations between total carcass lean and		
			Carcass weight	3 rib lean	1 rib lean
Hopper (1944) (Hereford steers)	56	501 ± 322		0·94	
Hankins and Howe (1946) (Mixed breed and sex)	120	284–629		0·85	
Mason (1951) (Mixed breed and sex)	31				0·88†
Crown and Damon (1960) (Mixed breed and sex)	24	424		0·94†	0·82†
Cole et al. (1960) (Mixed breed and sex)	99		0·77	0·74 0·85*	
Ledger and Hutchison (1962) (Steers)	32				0·84†
Martin and Torreele (1962) (Mixed sex)	17	$323 \pm 15·4$		0·95†	

* Multiple correlation including carcass weight
† Correlation of percentage muscle in carcass with percentage muscle in joint

in some previous experiment holds good for the present one. Unfortunately, most of the reported work is based on miscellaneous collections of cattle and interactions between treatment and the part/whole relationships have not been measured.

Harrington (1963) has discussed some statistical problems encountered in establishing part/whole relationships. He pointed out that if there is wide variation in the carcass weight of cattle in the sample used, a predictor may be well correlated with total 'lean' simply because it is closely related to carcass weight. In fact, it is the correlation with total 'lean', having eliminated the correlation with carcass weight, that is of interest. In many papers on this topic, variation in carcass weight or live weight is not reported. In Tables 7 and 8 are summarized most of the reports in the literature where the correlations in cattle between total carcass 'lean' and carcass weight and/or sample joints have been computed.

Table 8

Simple and multiple correlations between indirect measurements and total carcass lean

	N	Carcass weight (lb)	Correlations between indirect measurement and total carcass lean on		
			Carcass weight	Shin lean	Biceps femoris
Orme et al. (1960)	43	558 (range 382–810)	0·81		0·96
(Hereford cows)					
Cole et al. (1960)	99		0·77	0·81 0·89*	0·96*
(Mixed breed and sex)					
Callow (1962a)	24			0·90	
(Mixed breeds)					
Butterfield (1962)	35			0·98	0·99
(Mixed breeds)					
Butterfield (1965)	29	451 (range 206–737)	0·97	0·95	0·98
(Mixed breeds)					

* Multiple correlation including carcass weight

The pooled within-breed correlations for Harte's (1966) data have been computed and are presented in Table 9. From this, it can be seen how misleading simple correlation coefficients can be when considering the usefulness of sample joint dissection. The simple correlation between total carcass 'lean' and 'lean' in 8–9th rib is 0·82 and that between total carcass 'lean' and carcass weight is 0·81, but the multiple correlation coefficient on both these variables is only 0·89 showing that much of the correlation between total carcass 'lean' and 'lean' in 8–9th rib could be attributed to carcass weight.

It is of interest to note the considerable increase in the value of multiple correlation from 0·89 to 0·98 when the weight of the 8–9th rib joint was included in the multiple correlation. Harrington and King (1963) found a similar effect on residual standard errors when they included joint weight in a multiple regression.

We also analysed the data overall and included experiment and breed in multiple correlations. The results are given in Table 10. When breed was included in the multiple correlation, there was an increase in its value from 0·922 to 0·954, which indicates a breed

Table 9

Pooled within-breed correlations calculated from Harte's (1966) data

Total carcass 'lean' on	Correlation
Carcass weight	0·81
'Lean' in 8–9th rib joint	0·82
'Lean' in 8–9th rib joint and Carcass weight	0·89
'Lean' in 8–9th rib joint Carcass weight Weight of 8–9th rib joint	0·98

Table 10

Multiple correlations for 8–9th rib (Harte's 1966 data)

Total carcass 'lean' on		Total carcass 'lean' on	
Experiment Breed Carcass weight 'Lean' in cut	0·954	Experiment Breed Carcass weight 'Lean' in cut Cut weight	0·985
Breed Carcass weight 'Lean' in cut	0·954	Breed Carcass weight 'Lean' in cut Cut weight	0·985
Carcass weight 'Lean' in cut	0·922	Carcass weight 'Lean' in cut Cut weight	0·985

effect. However, if cut weight was included then taking out breed did not reduce the correlation, indicating that whatever breed effect remained was taken out by cut weight. Harrington (1963) suggested that comparison of the predicted values obtained by

Table 11

Standard error of estimate

	N	Carcass (lb)	Predictors lean on			
			3 rib lean	1 rib lean	Shin lean	Biceps femoris lean
Hopper (1944)	56	501 ± 322	1·12 % (3·6 lb)			
Hankins and Howe (1946)	120	range 284–629	2·51 %			
Ledger and Hutchison (1962)	32			2·14 %		
Callow (1962a)	24			18·0 lb	12·19 lb	
Harrington and King (1963)	24			12·3 lb*	8·1 lb*	
Cole et al. (1960)	99		9·8 lb		8·5 lb	
Orme et al. (1960)	43	558 (range 382–810)				5·5 lb¶
Butterfield (1962)	35				8·2 lb	5·1 lb¶
Butterfield (1965)	50	SD 113§			5·0 lb†	
Carroll and Conniffe (Unpublished, from Harte's 1966 data)	54	SD 13·2 ‖	3·8 lb‡			

* Multiple regression including carcass weight
† Multiple regression including carcass weight and depth of fat over eye muscle
‡ Multiple regression 8–9th rib lean, 8–9th rib weight and carcass weight
§ Pooled within-breed standard deviation taken from Berg and Butterfield (1966); it is only an approximation as it includes 12 extra cattle in addition to the above 50
‖ Greatest within-breed standard deviation
¶ See Harrington (1963)

different workers should be made by examining the size of the standard error of estimate. Very few reports contain this information or measures of the variability of the dependent variable. However, most of those given for beef cattle are included in Table 11. Standard errors or ranges for carcass weight are given as a measure of the variability of the dependent variable total 'lean'.

REFERENCES

Berg, R. T. and Butterfield, R. M. (1966). *Anim. Prod.* **8**, 1

Butler, O. D. (1957). *J. Anim. Sci.* **16**, 227

— Warwick, B. L. and Cartwright, T. C. (1956). *J. Anim. Sci.* **15**, 93

Butterfield, R. M. (1962). *Nature, Lond.* **195**, 193

— (1963). *Symp. Carcass Composition and Appraisal of Meat Animals.* Paper 7-1. (Ed. by D. E. Tribe.) Melbourne, C.S.I.R.O.

— (1965). *Res. vet. Sci.* **6**, 24

Callow, E. H. (1962a). *Anim. Prod.* **4**, 37

— (1962b). *J. agric. Sci., Camb.* **58**, 295

Cole, J. W., Orme, L. E. and Kincaid, C. M. (1960). *J. Anim. Sci.* **19**, 89

Crown, R. M. and Damon, R. A. (1960). *J. Anim. Sci.* **19**, 109

Dumont, B. L., Le Guelte, P. and Arnoux, J. (1961). *Annls. Zootech.* **10**, 321

Hankins, O. G. and Howe, P. E. (1946). *Tech. Bull.,* U.S. Dep. Agric. **926**

Harrington, G. (1963). *Ann. N.Y. Acad. Sci.* **110**, 642

— and King, J. W. B. (1963). *Anim. Prod.* **5**, 327

Harte, F. J. (1966). *Ph.D. Thesis,* Dubl. Univ.

Hopper, T. H. (1944). *J. agric. Res.* **68**, 239

John, M. G. (1960). *6th Meet. Eur. Meat Res. Wkrs,* Utrecht

Ledger, H. P. and Hutchison, H. G. (1962). *J. agric. Sci., Camb.* **58**, 81

Martin, J. and Torreele, G. (1962). *Annls. Zootech.* **11**, 217

Mason, I. L. (1951). *Anim. Breed. Abstr.* **19**, 1

Orme, L. E., Cole, J. W., Kincaid, C. M. and Cooper, R. J. (1960). *J. Anim. Sci.* **19**, 726

Stonaker, H. H., Hazaleus, M. H. and Wheeler, S. S. (1952). *J. Anim. Sci.* **11**, 17

Willey, N. B., Butler, O. D., Riggs, J. K., Jones, J. H. and Lyerly, P. J. (1951). *J. Anim. Sci.* **10**, 195

Yeates, N. T. M. (1959). *J. Aust. Inst. agric. Sci.* **25**, 301

GENETIC STUDIES OF GROWTH AND CARCASS COMPOSITION IN SHEEP

VIVIAN M. TIMON

The Agricultural Institute, Castleknock, Co. Dublin

PLANNED CHANGES in the meat potential of our sheep breeds can only be effective when we have a clear picture of the direction in which we wish to change meat production. These objectives must be expressed in terms of the relative importance of the components of meat quality on the consumer market. As yet, consumer preference in meat has not been clearly defined, possibly because finite consumer preferences do not exist except on a very general basis. In the broad sense, however, it is generally accepted that the consumer requires lean meat. This immediately pinpoints the importance of one aspect of carcass quality, namely carcass composition. The question then arises to what extent we can modify the expression of this trait by changing the animal environment or, in the long term, by changing the underlying genetic basis for body and tissue growth. Hammond's (1932) work provides us with a broad description of the fundamental principles of growth at tissue level. Later work by his colleagues (Pálsson, 1939; Pálsson and Vergés, 1952) and the recent re-analysis of this work (Elsley, McDonald and Fowler, 1964) has clearly demonstrated the importance of the environment in determining rate of tissue growth and, consequently, body and carcass composition.

On the other hand, very little information is available on the importance of genetic influences on meat traits in sheep. The difficulty of measuring carcass composition on adequate numbers of animals has been responsible for this deficiency. Indeed, this difficulty has been so widely recognized that much of the research on carcass composition over the past 20 years has been orientated towards evaluating techniques of measurement rather than in studies of carcass composition *per se*. The principal aim in these studies has been to identify simple practical measurements that give accurate and precise estimates of carcass composition. In this context it is important to emphasize the criteria on which carcass

400

measuring techniques must be assessed. Ease and speed of measurement are obvious requirements. The major criteria of assessment, however, must be the precision and accuracy (unbiased) with which the index will reflect true variation in the tissue it is describing. Precision of estimate is not described adequately by the widely quoted correlation coefficient as this parameter is a direct function of the variation in the data upon which it is based: $x = [1 - \sigma_y^2 \cdot x / \sigma_y^2]^{\frac{1}{2}}$. Consideration of the residual variance, on the other hand, provides more meaningful information. This criterion determines the predictive efficiency of an index in discriminating between progeny, treatment or commercial groups of animals in the particular set of conditions for which its use is intended. It is at this point that one must consider the possible effects of treatment and genetic differences on the prediction parameters. As yet there is very little information available on this aspect of carcass predictions.

A further aspect of carcass studies which has received very little attention is the relation between tissue growth and the overall body growth curves. This problem is particularly relevant in selection programmes which are aimed at changing the growth pattern of our meat breeds. Will changes in early postnatal growth result in correlated changes in the shape of the growth curve and in the mature size of the adult animal? This last point assumes major importance in determining productive efficiency in intensive sheep farming systems.

The major questions raised above may be summarized as follows:

1. What are the predictive values of various indices of lamb carcass composition?

2. To what extent is variation in growth rate and carcass composition inherited?

3. What are the relationships between growth rate, carcass composition and the overall growth curve?

Some recent analyses of lamb and mouse data provide some answers to these questions.

INDICES OF LAMB CARCASS COMPOSITION

There have been many attempts to evaluate the predictive value of various indices of lamb carcass composition following Hammond's study of sample joints in 1932. The principal workers in this field (Pálsson, 1939; Hankins, 1947; Walker and McMeekan, 1944; Barton and Kirton, 1958; Kirton and Barton, 1962) each presented evidence of the usefulness of a particular set of measurements as carcass predictors. The various indices studied may be classified conveniently as follows: sample joint composition, specific gravity determinations, dimensional measurements on the intact and jointed carcass and visual assessments. A common feature of all these studies has been the very wide variation in the data analysed. This

Table 1

Correlation coefficients between carcass fat, muscle and bone percentages and various carcass predictors

Predictor	Carcass percentage		
	Fat	Muscle	Bone
Leg+loin dissection	0·98	0·96	0·93
Loin dissection	0·96	0·93	0·84
'Best-end' dissection	0·94	0·92	0·75
Carcass specific gravity	−0·93	0·84	0·56
Carcass measurements*			
internal and external	0·86	0·78	0·78
external	0·83	0·71	0·76
Subjective score	0·71	−0·55	−0·43

* Multiple equations involving the measurements carcass weight, caul fat weight, gigot width, cannon bone weight, back flesh depth and 4 feet weight, recorded on the uncut carcass (external), together with measurements of eye muscle area at the 6–7th and 12–13th rib interfaces (internal)

wide variation precludes any confident extrapolation of the correlation analyses to less variable situations. Consequently, a comparative study of the different indices in less variable data was undertaken by the present author (Timon and Bichard, 1965a, b and c). The analyses were carried out on 83 Clun Forest lambs slaughtered at approximately 80 lb live weight. Differences in gut fill and in killing-out percentage, however, resulted in some variation in carcass weight (S.D. = 2·5 lb).

The more important results of this investigation are summarized in Tables 1, 2 and 3. The relative precision of the different indices studied is apparent in Table 1. Sample joint composition provides more reliable estimates than specific gravity measurements and these in turn are better predictors than carcass measurements or subjective assessments. It is clear from the sample dissections that the bigger the joint the more precise estimate of composition it provides. However, the 'best-end' joint is much more efficient as a predictor in that it provides reliable predictions and requires minimum dissection time (2 h). Specific gravity determinations also estimate fat and muscle quite well but provide very poor predictions of carcass bone. Several carcass measurements were evaluated, both singly and in multiple combinations (multiple regression analysis); none provided worth-while predictions singly. The following best external measurement combinations do give some reasonable estimates of carcass composition:

$$Y_1 = 56\cdot22 + 3\cdot61\ X_1 - 1\cdot66\ X_2 - 0\cdot43\ X_3 + 0\cdot51\ X_4 \pm 2\cdot22$$
$$(R = 0\cdot83)$$
$$Y_2 = 65\cdot27 - 2\cdot63\ X_1 + 0\cdot44\ X_3 - 0\cdot39\ X_4 - 6\cdot20\ X_7 \pm 2\cdot25$$
$$(R = 0\cdot71)$$
$$Y_3 = 0\cdot12 - 1\cdot28\ X_1 + 0\cdot79\ X_2 - 1\cdot58\ X_5 + 3\cdot80\ X_7 \pm 1\cdot03$$
$$(R = 0\cdot76)$$

where Y_1, Y_2 and Y_3 denote carcass fat, muscle and bone percentages respectively, and X_1 = caul fat weight, X_2 = gigot width, X_3 = cannon bone weight, X_4 = carcass weight, X_5 = back flesh depth (probe), and X_7 = weight of 4 feet.

Further marginal improvements in these predictions were found when internal carcass measurements were included in the equations, as follows:

$$Y_1 = 46\cdot88 + 3\cdot63\ X_1 - 1\cdot21\ X_2 - 0\cdot33\ X_3 + 0\cdot66\ X_4 - 0\cdot61\ X_9 \pm$$
$$2\cdot01\ (R = 0\cdot86)$$
$$Y_2 = 65\cdot60 - 2\cdot39\ X_1 + 0\cdot36\ X_3 - 0\cdot56\ X_4 - 5\cdot59\ X_7 + 0\cdot83\ X_8 \pm$$
$$2\cdot04\ (R = 0\cdot78)$$
$$Y_3 = 4\cdot46 - 1\cdot22\ X_1 + 0\cdot56\ X_2 - 2\cdot28\ X_5 + 3\cdot45\ X_7 + 0\cdot26\ X_9 \pm$$
$$1\cdot01\ (R = 0\cdot78)$$

where the additional variables X_8 and X_9 denote eye muscle areas at the 6–7th rib and 12–13th rib interfaces respectively. A disturbing feature of these prediction equations, however, was the variation in the relative sizes of the partial coefficients in the two samples of data (1960 and 1961). This would suggest that these measurements would be meaningful only if their equations were established in a subsample of animals in a particular experiment.

The subjective scores were made by an experienced meat trade judge. However, the size of the correlations and the unexpected negative relationship between the muscle scores and carcass muscle would preclude their inclusion as useful predictors.

As pointed out earlier, however, correlation analyses only provide a relative ranking of the different indices. The question may be asked; how useful are carcass indices (X) in describing treatment differences in carcass composition (Y)? Before we consider this question it is necessary to make a distinction between the accuracy or bias of prediction estimates and the precision of the estimates. If treatment differences affect the parameters 'a' and/or 'b' in the equation, $\hat{Y} = a + b\,X$, then the predicted values $(\hat{Y})$ will be biased and will not reflect the true treatment differences in Y. Biased estimations of carcass fat were found among sire comparisons in the

Table 2

Confidence limits on individual predictions (s_Y) and group predictions $(S_{\bar{Y}})$* based on loin dissection

Percentage	$t_{0.05}\,S_Y$	$t_{0.05}\,s_{\bar{Y}}$
Fat	$\pm 2{\cdot}22$	$\pm 0{\cdot}63$
Muscle	$\pm 2{\cdot}44$	$\pm 0{\cdot}71$
Bone	$\pm 1{\cdot}62$	$\pm 0{\cdot}47$

* These estimates are based on a group size of 12 animals

Clun Forest data (Timon, 1965). Cunniffe and Moran (1968) have recently discussed this aspect of carcass predictions and have presented evidence of bias in predictions of lean in beef carcasses from different breeds. They also point out that the statistical tests to establish the significance of treatment bias (in predictions) are not very sensitive when X (carcass index) and Y (carcass composition) are poorly correlated. It is clear therefore that much greater emphasis must be placed on assessing the importance of treatment bias in carcass predictions. Indeed prediction estimates have no future relevance in treatment comparisons if the absence of bias cannot be established.

In comparisons where treatment differences can be shown not to influence the prediction parameters it is then meaningful to evaluate the precision of prediction estimates. Precision of estimate may be expressed in terms of confidence limits on the prediction estimates. Confidence limits $(P = 0{\cdot}9)$ on individual and group $(n = 12)$ mean prediction estimates of carcass composition based on loin dissections are shown in Table 2. It is clear that the magnitude of the confidence limits on individual predictions $(t_{0.05}\,s_Y)$ preclude their usefulness except in describing very wide animal differences. On the

404

other hand, the limits on the group mean estimates would suggest that the loin index could be used to reflect moderate differences in carcass composition. The question then arises: can these prediction errors be offset by increasing treatment group size? In an earlier paper (Timon and Bichard, 1965c) the author presented some estimates of the increases in group size necessary to offset the prediction errors associated with different indices. These examples were based on the model

$$P = y + e$$

and

$$\sigma_p^2 = \sigma_y^2 + \sigma_e^2 + 2\sigma ye$$

where P = predicted record, y = true record and e = error of prediction. On the assumption that $\sigma ye = 0$, $\sigma^2 p$ was estimated as the sum of the true record variance $(\sigma^2 y)$ and the prediction error variance $(\sigma^2 e)$. Cunniffe and Moran (1968) have rightly pointed out that this assumption is not justified. Indeed the variance among predicted values $(\sigma^2 p)$ will be less than the variance of the true values $(\sigma^2 y)$.

In this context fewer animals per group are needed to describe treatment differences among predicted estimates $(\hat{Y})$ of carcass composition. This may seem paradoxical until it is realized that a predicted estimate $\hat{Y} = a + bX$ is an average estimate of the possible 'Y' values associated with a particular X. Cunniffe and Moran (1968) have suggested inverse estimation as an interesting alternate approach which overcomes this situation. In this case the regression of the carcass index (X) on carcass composition (Y) is evaluated $(X = c + d Y + e)$ and the individual Y values (Y') are estimated as $Y' = (\hat{X} - c)/d$. The variance of the predicted estimates $(\sigma^2 y')$ is then easily shown to be greater than among the actual values of Y; $\sigma^2 Y' = \sigma^2 X/d^2 = \sigma^2 Y/r^2$ where $d = r \cdot \sigma X/\sigma Y$ and r = correlation coefficient.

In the examples of the earlier paper (loc. cit.) the estimated treatment group size (n') would now be derived as $n' = n/r^2$, where n = required treatment group size when whole carcass dissections were used. Some examples are shown in Table 3. These results suggest that if the index (X) is highly correlated with carcass composition (Y) then relatively few additional animals per treatment group will offset the increased variation. This is in agreement with the earlier examples. When the correlation coefficient is low (for example subjective assessment), however, treatment group size must be increased to a much greater extent than the earlier examples suggest.

In summary it must again be stressed that the possibility of biased predictions may preclude indirect estimation of carcass composition in treatment comparison. Future carcass evaluation studies should establish the extent of bias in different indices and in different treatment comparisons.

Table 3

Number of animals per treatment group required to detect a significant* difference (δ') of different orders of magnitude between two treatments in carcass fat, muscle and bone percentages when carcass dissections and predictions of carcass composition are available

Size of difference (δ')	Index of composition	Percentage		
		Fat	Muscle	Bone
	Carcass dissection	11	11	11
	Leg and loin dissection	11	12	13
	Loin dissection	12	13	16
$1\cdot0\sigma$	Best neck dissection	12	13	26
	Specific gravity	13	16	35
	Carcass measurements†	15	18	18
	Subjective assessment	22	36	60
	Carcass dissection	37	37	37
	Leg and loin dissection	39	40	43
	Loin dissection	40	43	52
$0\cdot5\sigma$	Best neck dissection	42	44	87
	Specific gravity	43	52	118
	Carcass measurements†	50	61	61
	Subjective assessment	73	122	200

* Probability of detecting a real difference, $\beta = 0\cdot75$. Probability of a chance difference as real, $\alpha = 0\cdot04$
† Includes all measurements which showed a significant effect ($P < 0\cdot05$) in multiple regression equations

THE INHERITANCE OF GROWTH AND CARCASS COMPOSITION IN LAMBS

The inheritance of growth in sheep has been widely studied and has been reviewed recently by Bowman (1966). Apart from some estimates of the heritability of cannon bone measurement by Taneja (1958) and Purser (1958), and body conformation scores by

American workers (reviewed by Terrill, 1958), there has been no large-scale analysis of the inheritance of carcass traits in sheep. Some preliminary investigations of the inheritance of carcass traits in Clun Forest lambs have been published by the present author (Timon, 1963, 1965). These estimates suggest that most of the major carcass traits are highly inherited.

A study of carcass traits in Galway sheep was initiated in 1964 at the Agricultural Institute, Dublin. The results of some preliminary analyses of 196 wether lambs, the progeny of 26 sires on test in 1965, will be summarized below. These lambs were slaughtered on a constant age basis. The data analysed include general size traits, slaughter and carcass weight, six skeletal size measurements, weight distribution as measured by joint cut-out weights and yields and estimated carcass composition. The carcass composition estimates are based on a 6–12th rib saddle (best-end) which after boning was analysed chemically for protein, ether extract, moisture and ash. Several types of analyses were carried out by adding further co-variates to the following basic least squares model.

$$Y_{ijkl} = \mu + \alpha_i + \beta_j + \gamma_k + b_l X_{ijkl} + e_{ijkl}$$

where

Y_{ijkl} = an observation on the lth animal in the ijkth subclass,

μ = an effect common to all lambs

α_i = an effect common to all lambs in the ith birth/rearing class;
$$i = 1 \ldots 3$$

β_j = an effect common to all lambs in the jth age of ewe class;
$$j = 1 \ldots 4$$

γ_k = an effect common to all lambs in the kth sire group;
$$k = 1 \ldots 26$$

b_l = the pooled within subclass regression of Y_{ijkl} on ewe body weight at lambing, X_{ijkl}

e_{ijkl} = a random (error) effect peculiar to each animal assumed N.I.D. $(0, \sigma^2 e)$.

The main classes were analysed as deviations from their subclass means on the assumption that

$$\sum \alpha_i = \sum \beta_j = \sum \gamma_k = 0$$

The additional covariates included in the analyses were carcass weight and carcass fat percentage. The results of these analyses show the inheritance of carcass traits on a weight constant and fat constant basis.

The results in Table 4 summarize the heritabilities of some of the general carcass traits analysed. All traits have high heritabilities and indicate that selection for these traits should be successful in this particular population. It is necessary to mention at the outset that these heritabilities are only preliminary estimates (note the size of the standard errors) and to emphasize that these genetic parameters are only descriptive of the inheritance of these traits in this particular population at the time of measurement. The heritabilities of 'eye'

Table 4

Heritabilities and least square sire differences in general carcass traits

	Range in sire means	$h^2 \pm$ S.E.
Slaughter weight (lb)/age	0·20–0·27	0·57 ± 0·26
Carcass weight (lb)	26·1–40·4	0·53 ± 0·25
'Eye' muscle area (cm²)	9·3–13·8	0·56 ± 0·26
Middle weight (lb)/age	0·02–0·04	0·53 ± 0·25
Skeletal size measurements:		
Carcass length (cm)	88·0–94·5	0·28 ± 0·21
Tibia length (cm)	19·9–21·7	0·75 ± 0·28
Leg length (cm)	24·8–26·6	0·24 ± 0·28
Cannon bone weight (g)	33·0–44·3	0·49 ± 0·25
Cannon bone length (cm)	12·4–14·7	0·68 ± 0·28
Cannon bone circumference (cm)	4·2–4·8	0·46 ± 0·25

muscle area and of variation in midback weight-for-age (midback = loin + best end) indicate the scope for increasing the size of these more valuable parts of the carcass. The range in differences between sires (least squares means) indicates considerable scope for selection on these traits.

Similarly the sire differences and heritability of skeletal measurements indicate a real possibility to change these traits. The estimates in Table 4 are in good agreement with the earlier estimates on the Clun Forest data (Timon, 1963, 1965). Together these studies would suggest that selection should be effective in changing the overall size of the lamb carcass.

This finding is also supported by the heritabilities of joint weights in Table 5. The estimates in the first column refer to the absolute weights of the joints. However, when joint weights are expressed as percentages of carcass weight, as a measure of weight distribution, they show greatly reduced heritabilities and especially when variation due to carcass fat is removed.   However, if we consider Yeates' (1952) index of blockiness, i.e. the ratio of carcass length to carcass weight, these data suggest that this trait is highly inherited.   If, however, we correct for differences in carcass fat, much of the sire variation in this ratio is removed.   These data suggest, therefore, that weight distribution at a constant level of fatness is not highly inherited.

Table 5

Heritabilities of weight distribution among the carcass joints

	$h^2 \pm$ S.E.
Leg weight (lb)	0.41 ± 0.24
Leg weight (% carcass weight)	0.02 ± 0.16
Leg weight (fat corrected) *	Neg.
Loin weight (lb)	0.41 ± 0.24
Loin weight (% carcass weight)	0.15 ± 0.20
Loin weight (fat corrected) *	0.13 ± 0.19
Carcass length/carcass weight	0.61 ± 0.27
Carcass length/carcass weight (fat corrected) *	0.17 ± 0.20

* Based on covariance analysis of trait with the variance associated with carcass fat removed

If we now consider sample joint composition we see a very similar trend.   Table 6 would suggest that carcass composition traits are highly inherited and especially the protein and ether extract fractions.   These heritabilities compare very closely with the estimates for the dissectable tissues, fat, muscle and bone in the Clun Forest data (loc. cit).   However, when these fractions are analysed on a fat-free basis it would appear that they do not show any genetic variation.   Indeed the overall variation in these traits (fat-free protein percentage, fat-free moisture percentage) is very small, as is shown by the size of the coefficients of variation.   In fact these traits were found to have negative sire variance in these data.

The possibility of changing the chemical composition of the lean body mass is therefore not clear.   However, in terms of consumer preference for lean meat, the lean/bone ratio (lean = total meat–ether extract) is possibly the most important carcass characteristic.   These data suggest that this trait is moderately inherited even in animals of constant fatness.

RELATIONSHIPS BETWEEN GROWTH CURVE PARAMETERS AND CARCASS COMPOSITION

Growth curves in farm animals have been widely used by Brody (1945) in describing the exponential nature of body growth in mammals. Brody studied mammalian growth in two distinct equations, namely, an increasing curve in the pre-pubertal phase

Table 6

Coefficients of variation (C.V. percentage) and heritability estimates ($h \pm$ S.E.) for sample joint composition

Basis for assessment	C.V. percentage	$h^2 \pm$ S.E.
Total joint		
Percentage		
ether extract	21·4	0·50 ± 0·26
protein	9·1	0·51 ± 0·26
bone	11·4	0·32 ± 0·22
Meat/bone	13·4	0·32 ± 0·22
Fat-free joint		
Percentage		
protein (fat-free)	3·8	Neg.
moisture (fat-free)	1·1	Neg.
protein/moisture	4·9	Neg.
lean/bone	9·3	0·34 ± 0·23
lean/bone (fat corrected)	9·3	0·36 ± 0·23

and a decreasing curve in post-pubertal growth. His general conclusion was that there is very little variation in the exponential growth rates and he termed his growth curve parameters Genetic Growth Constants. Recently Taylor (1965), extended Brody's work by relating Brody's specific growth parameters to mature size. There is, however, very little information in the literature on the extent to which the animal growth curve can be altered. Abplanalp, Ogasawara and Asmundson (1963) have changed the growth curve of turkeys in a restricted index selection by selecting for increased 6-week weight holding 24-week weight constant. More recently Laird and Howard (1967) have shown that inbreeding caused differences in the growth curves of different strains of mice.

The present investigation is planned to ascertain appreciable variation in the growth curve parameters within an outbred strain of mice and to measure the inheritance of this variation and its covariation with carcass composition. As this investigation was started very recently, the following results are presented merely as preliminary estimates.

The first problem was in choosing an appropriate growth curve that would furnish an accurate description of mouse growth. It was thought desirable to choose a curve which would reflect the point of growth inflection, as this parameter marks a point of physiological

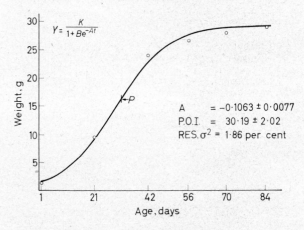

Figure 1. Logistic growth curve (Q15 mice)

equivalence on the growth curve. This automatically excluded Brody's (1945) self-inhibiting and self-accelerating equations as these curves begin and end respectively at the inflection point. Two sigmoid curves have been tested, namely, an orthogonal polynomial regression of weight on age and the logistic curve

$$Y = \frac{k}{1 + be^{-at}}$$

where Y = weight at a given age, t; k = mature size; b = an integration constant; a = the exponential rate of change in growth and e = the natural logarithmic base. The growth constant 'a' is more clearly understood as the regression of $\log_e (K - Y)/Y$ on age.

The overall growth curve as fitted to a sample of 244 mice is shown in *Figure 1*. The growth data ranged from birth to 84 days (0, 21, 42, 56, 70, 84) and fitted well; the average residual variation

411

in the individual mouse was 1·86 per cent of the overall variation in body weight. However, the curve tended to underestimate weight gain prior to 6 weeks and to overestimate subsequent weights. Considerable variation was found in age at the point of inflection

Table 7

Least squares estimates of sire and sex differences in growth curve parameters

		Growth constant (a)	Point of inflection (days)
Sex	Males	−0·1085 ⎤ *** −0·1036 ⎦	30·47 ⎤ ** 29·73 ⎦
	Females		
Sire	Maximum	−0·1134 ⎤ N.S. −0·0939 ⎦	30·10 ⎤ ** 28·13 ⎦
	Minimum		
Heritability		0·07 ± 0·18	0·51 ± 0·27

**Denotes $P<0·01$
***Denotes $P<0·001$

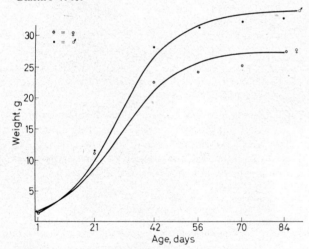

Figure 2.   Logistic growth curve (Q15 mice)

(C.V. = 6·7 per cent) and in the exponential growth parameter 'a' (C.V. = 7·3 per cent) in these mice, which were the progeny of 32 sires.

An analysis of this variation (Table 7) shows that sex differences in these parameters were quite pronounced ($P < 0.01$). This is also clearly evident in *Figure 2*. However, the between sire variation as estimated in half-sib analysis (32 half-sib groups) was found to be almost negligible ($h^2 = 0.07 \pm 0.18$) in the growth parameter 'a', but quite pronounced in age at point of inflection ($h^2 = 0.51 \pm 0.27$).

Table 8

Correlation coefficients between carcass tissues and growth curve parameters based on 46 mice (24 ♂♂, 22 ♀♀)

	Growth constant (a)			Point of inflection		
	Males	Females	Total	Males	Females	Total
Ether extract %	0·31	0·38	0·34*	0·13	0·34	0·27
Protein %	−0·13	−0·10	−0·36*	−0·03	−0·21	−0·18
Protein % (fat-free)	0·13	0·13	−0·17	0·07	−0·02	−0·03
Protein/ether extract	−0·34	−0·46*	−0·39**	−0·10	−0·31	−0·24

* denotes $P < 0.05$
** denotes $P < 0.01$

Table 9

Correlation coefficients between body weights at 21, 42, 56, 70 and 84 days and the growth constant (a) based on 46 mice (24 ♂♂, 22 ♀♀)

Weight at (days)	Males	Females	Total
21	0·38	0·25	0·11
42	0·13	−0·01	−0·46**
56	−0·14	−0·17	−0·58**
70	−0·30	−0·36	−0·67**
84	−0·10	−0·09	−0·50**

** Denotes $P < 0.01$

Correlation analyses relating these growth parameters to carcass composition were carried out on a sample of 46 mice which had been slaughtered at 12 weeks and which were analysed for protein, ether extract, water and ash. The results of these analyses (Tables 8, 9) do not show any close relationships between the growth parameters and carcass composition. The only significant relationships were those between ether extract and the growth parameter 'a'. The correlations between specific growth and body weights at 3, 6, 8, 10 and 12 weeks while significant ($P < 0.05$) overall, were not significant within the sexes. This might suggest considerable flexibility to change carcass composition without altering the growth curve. Further analyses are being made to investigate the generality of this result. Similar analyses will be undertaken in lambs.

The general conclusion of this paper may be summarized as follows:

1. Part carcass dissections may provide useful predictions of whole carcass composition, if the absence of biased estimates in treatment comparisons can be established.

2. The inheritance of general size traits in lambs would appear to be high, whereas the inheritance of the components of the fat-free body was not evident.

3. The logistic curve gave a useful description of post-natal mouse growth. Genetic variation in the point of inflection indicated differences in rates of physiological maturity. Sex difference in growth curve parameters were highly significant. There were no strong relationships between body composition and the growth parameters.

REFERENCES

Abplanalp, H., Ogasawara, F. X. and Asmundson, V. S. (1963). *Br. Poult. Sci.* **4**, 71

Bowman, J. C. (1966). *Anim. Breed. Abstr.* **34**, 1

Barton, R. A. and Kirton, A. H. (1958). *J. agric. Sci., Camb.* **50**, 331

Brody, S. (1945). *Bioenergetics and Growth.* New York; Reinhold

Cunniffe, D. and Moran, A. (1968). *Anim. Prod.* (in press)

Elsley, F. W. H., McDonald, I. and Fowler, V. R. (1964). *Anim. Prod.* **6**, 141

Hammond, J. (1932). *Growth and Development of Mutton Qualities in Sheep.* Edinburgh; Oliver & Boyd

Hankins, O. G. (1947). *Tech. Bull. U.S. Dep. Agric.* 944

Kirton, A. H. and Barton, R. A. (1962). *J. Anim. Sci.* **21**, 553

Laird, A. K. and Howard, A. (1967). *Nature, Lond.* **213** 786

Pálsson, H. (1939). *J. agric. Sci., Camb.* **29**, 544

— and Vergés, J. B. (1952). *J. agric. Sci., Camb.* **42**, 1

Purser, A. F. (1958). *Proc. 10th Int. Conf. Genet.*

Taneja, G. C. (1958). *J. Genet.* **56**, 103

Taylor, St. C. S. (1965). *Anim. Prod.* **7**, 203

Terrill, C. E. (1958). *J. Anim. Sci.* **17**, 944

Timon, V. M. (1963). *Ph.D. Thesis*, Univ. Newcastle upon Tyne
— (1965). *Proc. Eur. Ass. Anim. Prod.*, The Hague
— and Bichard, M. (1965a). *Anim. Prod.* **7**, 173
— — (1965b). *Anim. Prod.* **7**, 183
— — (1965c). *Anim. Prod.* **7**, 189
Walker, D. E. and McMeekan, C. P. (1944). *N.Z. Jl. Sci. Tech.* **26A**, 51
Yeates, N. T. M. (1952). *Aust. J. agric. Res.* **3**, 68

PROBLEMS IN THE EVALUATION
OF PIG CARCASSES

ZOFIA OSIŃSKA

Institute of Animal Physiology and Nutrition, Jablonna, nr. Warsaw

THE GOAL of pig production is shifting increasingly from just the production of 'a' market pig to the production of 'the' pig which, after slaughter, will yield carcass and cuts suitable for the demands of the modern meat industry and consumers. For this purpose the fullest application of the achievements of research is necessary. Therefore, problems in the evaluation of pig carcass quality are of interest to several groups of people, particularly to the meat industry and to research workers in such fields as genetics and nutrition. However, the approach of each of these groups differs widely. The meat industry aims at the evaluation of carcass quality in the broad or true sense of this expression, but, due to very large numbers of carcasses to be evaluated daily and limitations imposed by technical requirements, has to make use of the characteristics which are easiest to obtain, though rarely the most reliable, such as backfat thickness and subjective assessments. On the other hand, research workers can usually afford to devote more time and effort to this problem and so obtain more accurate information but on more limited material. Moreover, due to increasing specialization, sometimes only one group of carcass quality factors is investigated without reference to other characteristics. The dividing line seems to lie most often between carcass composition on the one hand and meat quality on the other.

Though investigations on the evaluation of pig carcass quality have been carried out for over a century and the number of papers on this subject appearing each year is increasing steadily (especially during the last 15 years), our knowledge is far from being complete. There is still confusion in the use of some expressions. In some instances, information on the techniques used is rather incomplete, which makes it difficult to replicate the respective procedure in investigations carried out at other institutions or in another country.

The use of the expressions 'carcass quality' or 'carcass evaluation' and 'carcass composition' as synonymous may be quoted as an example of the lack of precision in definition. As Harrington (1966) pointed out, 'carcass evaluation' has a wider meaning than just the morphological or chemical composition of the carcass, as it also includes characteristics of distribution (for example proportions of cuts) and those of meat quality. It should be supplemented further by giving different weighting (on an economic or other basis) to various characteristics, which would make the full evaluation of carcass quality still more complicated. This is probably the reason why much more work was and is being done on methods for estimating carcass composition than on carcass quality evaluation.

The first difficulty encountered in comparing different methods of estimating the composition (morphological or chemical) of the pig carcass is the lack of uniformity in the basic concept of what the pig carcass constitutes. Apart from taking the hot or cold carcass (or side) weights as denominators for computing the yields of cuts, 'the pig carcass' may mean 'head on, leaf and kidney fat, kidneys in' but also 'head with jowl and trotters off, leaf and kidney fat, kidneys and psoas muscles out, partially skinned'. For baconers (about 90 kg live weight) the difference in the 'carcass' weight in these extreme cases may amount to 11–12 kg. A comparison of carcass yields or yields of cuts, as well as carcass composition is in such cases impossible.

The inaccuracies in descriptions of the points of measurements are sometimes obvious, as in the case of belly thickness measurement stated as 'four fingers width from the sternum' (Clausen and Gerwig, 1958), but even such apparently accurate (and widely used) information as 'the loin eye area was assessed on the cross-section at the last rib' has to be regarded as insufficient. It may be used with equal justification for cross-section through the head of the last rib, i.e. between the last but one and the last thoracic vertebrae, but also tangentially to the tip of the last rib. These cross-sections may lie at least two vertebrae or approximately 6–7 cm apart from each other. Regardless of many investigations on relationships between the loin eye areas on different cross-sections, primarily in the thoracic region (Kliesch and Horst, 1965), very little is known on the variation of the loin eye area in the crucial region of the last rib (Pedersen, 1964).

Even when both 'the carcass' and the measurements or the method of cutting are defined in detail, comparisons of requirements or results are very difficult because of the scarcity of relevant data. The minimum requirement for the yield of ham, for example, in the Czechoslovakian progeny-testing scheme is 27 per cent; in the Polish scheme such a yield would be decidedly above average. In

Czechoslovakia, however, the ham is cut off between the last but one and the last lumbar vertebrae and the yield is computed in relation to the whole side, whereas in Poland the separating cut is between the second and third sacral vertebrae and the yield is expressed as a percentage of the weight of primal cuts. It needed a small-scale pilot test to reveal that the Czechoslovakian method yielded hams which were 2–2·5 kg heavier than the corresponding cuts obtained according to the Polish method (10·47 vs. 8·16 kg on average) and the respective yields were 29·3 and 26·6 per cent (Osińska, 1964). The preparation of conversion formulae or tables seems to be possible but would require data from more material to establish. Apparently there is insufficient interest so far in making such comparisons but with the increasing exchange of breeding stock (progeny- or sib-tested by different methods) it may become necessary.

There are at least as many methods of estimating carcass composition and/or its total value as there are countries in which pig production is carried out on a fairly large scale and, consequently, both the meat industries and research workers are interested in development in this field. The presentation of even only the most widely used methods would take too much time and would be hardly worth while, as it is extremely difficult to form an opinion on their relative worth and the possibilities of their wider applications. On the other hand, there are some problems common to all of these methods which seem to merit more attention than is apparently paid to them now.

There are two sources of variation which have to be taken into account whenever comparisons of results are made. These are (a) the technical errors and (b) the biological variation of the experimental material (Harrington, 1963) and their relative importance varies according to circumstances. The results for different groups in an experiment, especially a large-scale one, are wholly comparable as the unavoidable errors will be random and tend to cancel each other, provided that the technical data (for example measuring and cutting) were gathered by the same team. When results of experiments carried out at different places, or even in the same institution but over a number of years, are to be compared, it should be borne in mind that the technical errors may in such instances become systematic and bias the results. Unfortunately, there is no simple solution to these problems. Even periodic training sessions for the technical staff of some co-operating research centres, involving demonstrations of the standard technique, do not prevent some minor but systematic deviations from appearing between such sessions, as has been found out in the case of the Polish progeny-testing stations (Bielińska, 1960). As an example of the size of technical errors the following data selected from the records of the Institute of Animal Physiology and Nutrition can be used. Three groups of carcasses of Polish Large White pigs slaughtered at 90–95 kg live weight were measured, cut and dissected by three teams of skilled technicians following conscientiously the same, quite well-defined method. The average weights of such a readily accessible and easy to define and separate part as the psoas muscle and their relation to the lean content in primal cuts were:

Team (n)	Lean content in primal cuts kg	Psoas muscle weight kg	r
K (33)	15.88 ± 1.05	0.24 ± 0.03	0.17
KW (32)	15.93 ± 1.30	0.28 ± 0.04	-0.03
J (33)	16.04 ± 0.91	0.37 ± 0.03	$0.65*$

* Correlation significant at the 1 per cent level

Examples of the magnitude of the technical errors of measurements may be found in the review by Clausen and Gerwig (1958).

Some methods seem to be more liable to technical errors than others, as, for example, when the cutting of the carcass is done along curved or broken lines (separation of the 'picnic' shoulder or of the ham). Also the determination of specific gravity as an indicator of the composition of the carcass seems to be vulnerable in this respect, which makes it necessary to compute separate regression equations for each set of experimental conditions (Adam and Smith, 1964). A comparison of the original data of Adam and Smith (1964) and of Joblin (1966) with the results of estimating the fat content in the carcasses by means of the regression equations given by these authors and also by the formula of Claus (1957) gave the following results:

	Adam and Smith (1964)	Joblin (1966)
Original data	38.8 ± 2.3	43.5 ± 4.9
Fat content estimated by the equation of:		
Adam and Smith		42.0
Joblin	42.1	
Claus	31.5	34.4

419

The result of the application of Adam and Smith's equation to the data of Joblin seems to give confidence in this method of estimating carcass composition, but the reversal of the procedure, i.e. the application of Joblin's equation to the data of Adam and Smith, must raise some doubts, and the results of computations by the formula of Claus are completely out of line.

The selection of a method for estimating carcass composition depends quite often on traditional factors, including the practices of the meat industry. Such is the case in the selection of the cross-section for evaluating the area of the loin 'eye'. This is a very popular characteristic, as apart from its value (sometimes over-rated) as an indicator of carcass leanness it is of importance in itself as an often decisive factor in the consumer's acceptance of back rashers or loin chops. The most widely used cross-sections of the loin are: between the 10th and 11th ribs (U.S.A.), 13–14th ribs (Germany) and at the last rib. The latter region is now gaining popularity due to the increasing attention being paid to the so-called 'slight of lean', i.e. deformation of the longissimus dorsi muscle resulting from an excessive development of the third layer of backfat, which is most marked in the region of the last rib. The data of Nitzsche (1965), who evaluated subjectively the degree of loin eye deformation in German Improved Landrace pigs show this tendency quite clearly:

| | Deformation of 'eye' muscle at | | | | | | | |
| | 13–14th rib | | | | Last rib | | | |
	none	slight	moderate	strong	none	slight	moderate	strong
No. of Barrows	137	28	8	2	41	43	60	31
No. of Gilts	165	7	3	0	69	54	34	18

The repeatability of loin 'eye' area at different cross-sections along the vertebral column is low (Kliesch and Horst, 1965) and it is possible that the choice of a cross-section may have an influence on the interpretation of experimental results. Such was the case in a recent feeding experiment carried out jointly by the Institute of Animal Production (Instytut Zootechniki) and the Institute of Animal Physiology and Nutrition. There were two groups of four littermates (two barrows and two gilts) by each of 17 Polish Large White boars. Two levels of feeding were compared, the average daily rations over the live weight range 30–80 kg being 2·2 and 2·0 kg/pig of the same pelleted feed (individual feeding). The pigs were slaughtered at 85–87 kg live weight and three cross-sections were made through the right side: between the 10th and 11th thoracic vertebrae ($10\ T$), the 13th and 14th vertebrae ($13\ T$) and between the last thoracic and first lumbar vertebrae (LT). Tracings were made of cranial surfaces of the cross-sections of the longissimus dorsi muscle and their areas measured with a compensating polar planimeter. An analysis of variance gave the following results (mean squares):

Source of variance	d.f.	10 T	13 T	LT
Boars [B]	16	18·70*	14·95	16·78*
Level of feeding [L]	1	22·57	14·76	50·17*
Sex [S]	1	260·99†	374·90†	322·41†
B × L	16	8·85	6·59	8·02
B × S	16	8·49	6·01	7·62
L × S	1	0·04	2·02	16·47
B × L × S	16	4·00	2·18	3·94
Error	68	8·99	9·19	9·13

* Difference significant at the 5 per cent level
† Difference significant at the 1 per cent level

The influence of feeding level was significant only for areas of the loin eye at the last thoracic vertebra. On the cross-section between 13th and 14th thoracic vertebrae no significant differences were visible among boar progeny groups. If, in a similar experiment, the loin eye area was measured on one cross-section only, the choice of the cross-section would markedly influence the conclusions reached on the influence of feeding level and/or sire on the loin eye area. Moreover, there was a marked shift in the ranking of boars in respect to the loin eye area of their progeny when this was measured on different cross-sections; the rank correlation coefficient ($10\ T$) vs. LT) was only 0·5.

One of the biological factors which has a widely recognized influence on the composition of pig carcasses, and is therefore taken into account in the planning of experiments, is the sex, the carcasses of gilts being on average leaner than those of barrows. However, the responses of barrows and gilts to nutritional factors, such as the energy and/or protein level of the rations, may also be different.

Suchodolska-Rytel (1967) recently found highly significant differences in the variance of the fatty tissue content between barrows and gilts from feeding experiments, as well as in the regression equations relating such carcass characteristics as the lean and fatty tissue content in hams to the morphological composition of carcasses. Much less is known on the influence of breed of pig on carcass characteristics, as many of the comparisons published were made on comparatively small groups of animals; at least, one should be cautious in interpreting such differences. Kroeske (1966), for example, found a highly significant difference in the yield of the leaf and kidney fat between Dutch Landrace and Dutch Large White pigs slaughtered at about 100 kg live weight (3·06 vs 2·49 per cent). He commented that this might have been due to selection, as the yield of the leaf and kidney fat had been recorded for 30 years (and presumably taken into account by breeders) in the progeny testing scheme for the Large White but not for Landrace pigs. However, Bielińska (1960) found an even larger difference (3·5 vs 2·5 per cent) in this respect between the Polish Landrace and Polish Large White pigs, neither of which had been selected for this characteristic.

If our knowledge on the influence of various factors on carcass composition and yields of cuts is still incomplete, the situation in the field of meat quality is even worse. This is due to the fact that for a long time, the meat industry, which was carrying out most of the investigations in this field, commenced interest at the earliest on entry of pigs to the slaughterhouse and only from the carcass 'on the hook', disregarding such factors as breeding, feeding and management of pigs before delivery. Even now, when the importance of the agricultural factors affecting meat quality is recognized, it is sometimes impossible to obtain from the industry clear guidance in this respect for such characteristics as the colour of meat or its water-holding capacity.

Only two of the various systems of scoring pig carcasses gained wider application; the Smithfield system (Davidson, Hammond, Swain and Wright, 1936) and the Canadian ROP score. Even for these systems the accuracy of the quality estimation seems to leave some doubts (Osińska and Kielanowski, 1958; Fredeen, Berg, Bowland and Doornenbal, 1964). There were a few attempts at estimating the total value of carcasses on the basis of yields and average prices of different cuts (Zobrisky, Brady, Lasley and Weaver, 1959; Fredeen et al., 1964), but their usefulness seems to be limited by the fact that not only are the price relations different in various countries but they change within a country in course of time according to the consumers' demands. The following data on the relative prices of cuts (loin = 100) illustrate this point:

	U.S.A.		Switzerland		Hungary	Canada
	*1937–1947**	*1961†*	*1953‡*	*1962‡*	*1959§*	*1964 ‖*
Loin	100	100	100	100	100	100
Ham	90	91	91	84	83	115
Picnic shoulder	80	55	}77	}64	60	65
Butt	80	72			100	92
Belly	80	68	85	76	44	75

* Zobrisky *et al.*, 1959
† Sherrit and Ziegler, 1962
‡ Gerwig, 1966
§ Szigeti, 1959
‖ Fredeen *et al.*, 1964

Such changes make it necessary to use, in the course of time, different weighting factors which makes comparisons difficult.

The presentation of the shortcomings, misunderstandings and difficulties involved in attempts at comparisons of results obtained in different countries or at various times is not meant to discourage people working on the problems of carcass quality. Perhaps a clearer recognition of the present difficulties will make some progress towards the unification of working methods somewhat easier. If some small proportion of the time and effort devoted now to the search for new indicators of carcass composition and/or quality could be spared for comparative investigations or clarification of the current methods, further exchange of views and results might be greatly widened and made more comprehensive.

REFERENCES

Adam, J. L. and Smith, W. C. (1964). *Anim. Prod.* **6**, 97

Bielińska, Krystyna (1960). *Roczn. Nauk roln.* **B 75**, 583

Clàus, A. (1957). *Tierzüchter* **9**, 554

Clausen, H. and Gerwig, C. (1958). *Pig breeding, recording and progeny testing in European countries.* FAO Agricultural Studies No. 44

Davidson, H. R., Hammond, J., Swain, J. B. and Wright, N. L. (1936). *Pig Breed. A.* **16**, 49

Fredeen, H. T., Berg, R. T., Bowland, J. P. and Doornenbal, H. (1964). *Can. J. Anim. Sci.* **44**, 334

GROWTH AND DEVELOPMENT OF MAMMALS

Gerwig, C. (1966). *Untersuchungen über die Schlachtqualität von Schweinen bei verschiedenem Mastendgewicht*. Eidgenössische Technische Hochschule in Zürich. Prom. Nr. 3736

Harrington, G. (1963). *Ann. N.Y. Acad. Sci.* **110**, 642

— (1966). *Z. Tierzücht. ZüchtBiol.* **82**, 187

Joblin, A. D. H. (1966). *N.Z. Jl. agric. Res.* **9**, 227

Kliesch, J. and Horst, P. (1965). *Züchtungskunde* **37**, 97

Kroeske, D. (1966). 'Comparison of pig breeds in some fattening experiments.' EAAP Meeting, Edinburgh

Nitzsche, G. (1965). *Arch. Tierz.* **8**, 137

Osińska, Zofia (1964). *Zesz. probl. Postep. Nauk roln.* **43a**, 169

— and Kielanowski, J. (1958). *Roczn. Nauk roln.* **B 73**, 9

Pedersen, O. K. (1964). 'Methods for determination of carcass quality in pigs with special reference to fleshiness of back rasher.' EAAP Meeting, Lisbon

Sherritt, G. W. and Ziegler, J. H. (1962). 'Some comparisons of carcass evaluation methods.' *5th Nat. Pork Ind. Conf.*, Columbus, Ohio

Suchodolska-Rytel, Ewa (1967). Personal communication

Szigeti, J. (1959). *Allattenyesztes* **8**, 69

Zobrisky, S. E., Brady, D. E., Lasley, J. F. and Weaver, L. A. (1959). *J. Anim. Sci.* **18**, 594

DISCUSSION

Professor Butterfield (Sydney) asked why the expression of results in terms of percentage 'lean' was preferred to the more explicit terms of bone, muscle and fat. Mr. Carroll replied that one reason was the difficulty of achieving standardization of procedure in a dissection unit handling large numbers and this was helped by dissection into separable lean followed by chemical analysis; a second reason was the moisture loss which occurred if carcasses had to be stored and which could be attributed to the lean if chemical analysis was carried out. To a further question from Professor Butterfield on whether muscle/bone ratio could be influenced nutritionally, Mr. Carroll thought that it was possible to effect this by nutritional means.

Dr. Pálsson (Reykjavik) commented that in making breed comparisons it was necessary to allow for the early maturity, and therefore the lower optimum slaughter weight, of the traditional beef breeds, to which Mr. Carroll replied that serial slaughtering at two or three weights was necessary in valid genetic comparisons; the abdominal muscles, for example, were late maturing and 'this showed up in the data on the Friesian breed.

Professor Lucas (Bangor) commented that the data showed a correlation between carcass weight and total lean but this was only to be expected and that it might be more meaningful to attempt to correlate carcass weight and composition with the *proportions* of lean. Mr. Carroll thought that what really mattered was weight of total lean and the efficiency with which it could be produced, and that too much attention had been directed towards measuring things other than this.

424

Dr. Fowler (Aberdeen) thought that the data presented suggested that the point of inflection on the growth curve was determined not by age but by weight, and if this was so, was there any information on the heritability of weight at that point? Dr. Timon replied that his data was not the same as that of Brody, in that the latter was a regression of the logarithm of weight yet to be gained on time, whereas his own data was a regression of weight yet to be gained as a fraction of weight already gained at that point. There were preliminary data which indicated that there was little variation in weight at the point of inflection but considerable variation in age.

Mr. Harrington (London), referring to Mr. Carroll's paper, agreed that there was considerable need for experimentation, both nutritional and genetic, involving slaughter at a series of weights rather than rigid adherence to weights currently accepted commercially, because, (a) as animals became leaner the desirable slaughter weights would change, (b) information was needed on the optimum size of carcass to produce the maximum weight of consumer cuts, and (c) retail butchers did not reflect accurately changes in consumer preference. The need was to find the weight of animal at which optimum fatness was reached. He went on to compliment Dr. Timon on presenting his data in the form of the number of animals required to achieve the same degree of precision using different techniques.

Mr. Vial (Dublin) expressed doubts on the need for expensive serial slaughtering, in view of the indications from the data which had been presented that lean body mass was the most important characteristic and that this, as measured by muscle/bone ratio, was independent of fat.

Dr. Braude (Reading) thought that more attention should be paid to muscle composition in the assessment of carcasses from the genetic standpoint, to which Dr. Timon replied that there was little one could do to change the chemical composition of the fat-free mass; fatness and distribution of fat could be changed as they appeared to be highly heritable but muscle distribution showed little variation, as indicated by the generally good correlation obtained between the part and the whole for muscle.

Mr. Houston (Leeds) did not think that dissection of a carcass into butchers' joints would give any indication of the growth of the animal; he thought attention should be paid to the chemical rather than the physical aspects of growth and the parameter which should be used was that of water and dry fat-free residue.

Professor Butterfield asked whether better results might be obtained in evaluation of pig carcasses if anatomical dissection of individual muscles were used, but Professor Osińska doubted whether this would be so as it would reduce greatly the number of animals involved, since a full anatomical dissection required over 100 h to complete. Dr. Ingram (Bristol) commented that anatomical dissection of individual muscles could be risky as some were very variable and some ill-defined, for example the psoas.

Mr. Carroll commented that many people were dissecting carcasses and obtaining details of carcass composition but very few were studying the markets and attempting to assess what would be required in the future; he thought that there was a great need for economists at research centres.

Mr. O'Grady (Fermoy) asked for further information on the causes of variability in density determinations, to which Professor Osińska replied that there were three variables involved, not only muscle and fat but also bone, and the proportions between any two of these could vary. There was also the problem of variation in temperatures of carcass and water. Dr. Timon commented further that sp. gr. did not give independent estimates of fat and muscle but only an overall index; it was a measure of variation rather than of absolute values. DR. INGRAM (Cambridge) observed that it was necessary to distinguish between the fundamental difficulties, such as differences in sp. gr. between the tissues, and the technical difficulties, such as standardizing the temperatures of the animal and the water.

DR. MITCHELL (Reading) agreed with Professor Osińska that the great need was to obtain more exact data rather than to devise elaborate ways of manipulating figures which were basically inaccurate. He advocated an accepted code of practice for use by all workers. DR. FOWLER commented on variation in vertebral numbers which caused considerable difficulty in the defining of anatomical regions.

DR. INGRAM observed that there was considerable confusion of objectives, particularly between dissection and chemical methods of carcass assessment. The need was for some more fundamental measure to allow carcass data to be assessed in terms of growth, for example protoplasmic quantity. This did not include depot fat or intra-muscular fat but protoplasm did contain a proportion of fat and, therefore, some measure of this was necessary. It was necessary to know the quantities of intra-muscular fat, inter-muscular fat, the proportion of intra-muscular fat which was cellular, water and nitrogenous fractions.

VIII. PRACTICAL IMPLICATIONS OF FACTORS AFFECTING GROWTH

GENETIC AND ENVIRONMENTAL INFLUENCES ON GROWTH IN BEEF CATTLE

R. T. BERG

University of Alberta, Edmonton, Alberta, Canada

GROWTH can be expressed as live weight, carcass weight, or tissue (muscle, fat. or bone) gain per unit of time. Relative growth involves the changing relationships among and within the tissues relative to age, weight or stage of physiological development. The economic value of a beef carcass is influenced by the relative proportions of the contributing tissues—muscle, fat and bone—and the quantitative requirements in the carcass are best met when muscle is maximum, bone is minimum, and fat is at an optimum determined by local consumer preferences. A knowledge of normal growth patterns and the various genetic and environmental factors which can influence these patterns, should assist in formulating methods for the production of the most desirable product. The present discussion will deal with the quantitative aspects of growth, and the genetic and environmental factors which influence growth in an absolute or relative manner.

NORMAL PATTERNS OF GROWTH

Data from a research project of the Royal Smithfield Club, London (Anonymous, 1966) are plotted in *Figure 1* to illustrate normal growth patterns of the carcass and of tissues. Herefords, representing beef-type, and Friesians, representing dairy-type, were reared under the same management system from soon after birth to approximately 2 years of age. An attempt was made to slaughter 4 animals of each breed at six-monthly intervals but because of losses, there were only 3 Herefords slaughtered at 18 months and 2 Herefords and 3 Friesians at 24 months. Live weight showed a similar pattern to carcass weight and is not, therefore, included. Carcass weight showed an expected sigmoid curve with the point of inflection approximating to the stage of increased fat deposition, between 12 and 18 months for both breeds. At a given age, the

429

Friesian had greater size, more muscle, more bone, but essentially the same amount of fat as the Hereford. This figure will be referred to again when breed influences are discussed.

In *Figure 2* are plotted data from the same source as *Figure 1*, but this time the tissue growth is plotted against the total carcass weight.

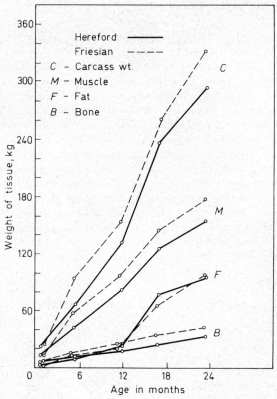

Figure 1. *Carcass and tissue weights from Hereford and Friesian steers slaughtered at* 6-monthly *intervals* (*Data from Anonymous, 1966*)

In the early stages, before fattening began, there were no marked differences in tissue growth between Hereford and Friesian steers. The fattening phase began at lighter carcass weights in the Herefords than in the Friesians. This is more pronounced in *Figure 3* where tissue weight is plotted against fat-free (muscle + bone) tissue.

430

Figures 1, 2 and *3* illustrate the relative growth of tissues, with bone shown as early developing, fat as late developing and muscle intermediate. The allometric relationships among the tissues have been described by Tulloh (1963) and Berg and Butterfield (1966). From

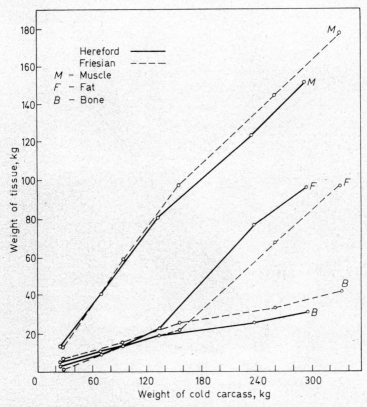

Figure 2. Growth of tissues relative to cold carcass weights from Hereford and Friesian steers (Data from Anonymous, 1966)

birth to maturity, muscle has a higher growth impetus than bone, i.e. its increase is proportionately greater. Under normal circumstances, after a certain stage of development is reached, the impetus for fat deposition is greater than that for muscle. These figures also illustrate the concepts involved in considering events of relative tissue growth as being age-dependent or weight-dependent. Bone growth is relatively slow and has been considered to have first demand on nutrients from diets inadequate for optimum growth. Thus, bone growth is often thought to be age-dependent. However, this conclusion may result from the fact that small changes in bone are difficult to detect and also that the higher impetus tissues show a

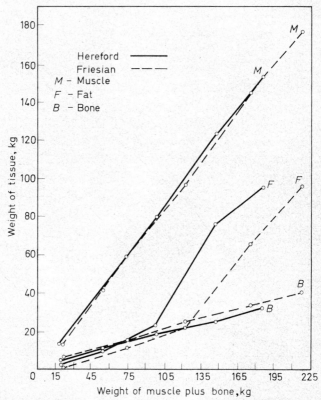

Figure 3. Growth of tissues relative to muscle-plus-bone growth (Data from Anonymous, 1966)

more marked retardation merely by slowing the normal patterns of all tissues along their expected growth curves. Thus, since muscle and fat show greatest growth under normal circumstances, it follows that growth retardation will be most clearly evident in these tissues. Certain stages of physiological development have been shown to be more weight-dependent than age-dependent, for example, pupation time in Drosophilla (Robertson, 1963) and sexual maturity in mice (Monteiro and Falconer, 1966). Relative tissue growth in cattle is probably basically weight-dependent (Berg and Butterfield, 1966; Butterfield and Berg, 1966b; Tulloh, 1963).

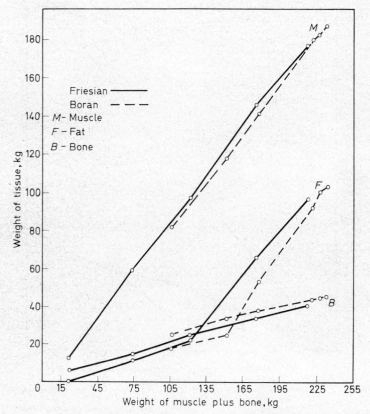

Figure 4. Comparison of tissue growth from Boran steers (From Ledger, 1965) and Friesian steers (Anonymous, 1966)
By courtesy of Cambridge University Press

However, there may be genetic and environmental factors which could alter this dependency.

In *Figure 4*, tissue growth data from Ledger (1965) are plotted on the same scale as the Friesian data of *Figure 3*. Ledger dissected East African Boran cattle from 1·5 to 5·5 years of age. Although these data extend over a much longer age span than the British

Friesians, their growth pattern is very similar when plotted against total muscle plus bone growth. *Figure 5* shows the data of Butterfield and Berg (1966a) plotted against the Friesian data. These data were from Australian cattle extending from day-old to more

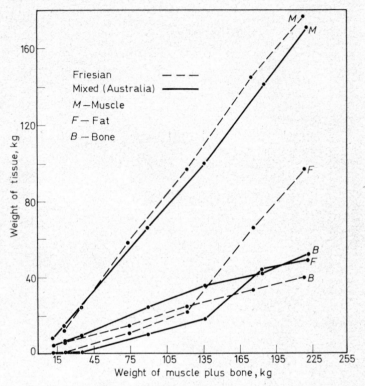

Figure 5. Comparison of tissue growth from mixed steers from Australia (Butterfield and Berg, 1966) and Friesian steers (Anonymous, 1966)

By courtesy of Blackwell Scientific Publications

than 5 years of age. Although the pattern of muscle and bone growth is similar to the British Friesians, there is more bone, less muscle, and less fat, at all stages of development. Many of these cattle were reared on a low level of nutrition and they did not deposit as much fat tissue relative to muscle plus bone development as did the British Friesians. Whether the markedly lower levels of fatness also affect muscle development will be discussed later.

Tentatively, we might conclude that under normal circumstances

434

muscle, bone and fat follow weight-dependent patterns of growth. These patterns seem to be clearer when considered relative to total muscle plus bone development rather than to carcass weight, which is influenced by fat, the most variable of the tissues.

INFLUENCE OF NUTRITION ON GROWTH AND RELATIVE GROWTH

1. Plane of nutrition

The influence of plane of nutrition on live or carcass gain per day need not be considered but rather the influence on relative growth of tissues. Meyer, Null, Weitkamp and Bonilla (1965) fed steers in three time periods on various combinations of high, medium and low planes of nutrition. They demonstrated that the high-level (a fattening ration throughout) resulted in a higher proportion of the net energy being used for growth than any other combination, decreasing for combinations which extended the maintenance period. The high levels of nutrition also seemed to increase the percentage of fat in the carcass and to decrease correspondingly the proportion of protein. Whether the more efficient rations, from the point of view of energy conversion, would be the best economically, would depend on the relative cost as well as the final carcass composition. It was not possible from the data presented to appraise carcass composition except with respect to the percentage of fat, from which one can conclude that level of nutrition can influence the proportion of fat at given carcass weights.

Data which illustrate more dramatically the effect of plane of nutrition on carcass composition were taken from Guenther, Bushman, Pope and Morrison (1965). They fed weaned (W) groups of half-sib Hereford steers on a high-plane (H) and a moderate-plane (M) of nutrition and estimated body composition at the start of the trial and as it progressed. *Figure 6* shows the results of their experiment plotted against age, while *Figure 7* shows the same data plotted against total muscle plus bone. Relative to age, in *Figure 6*, it is evident that the increased carcass weight on the high-plane was brought about by increased muscle and fat while bone seemed to be little affected. From *Figure 7*, it appears that the high-plane of nutrition eventually resulted in more fat in the carcasses relative to muscle plus bone. Callow (1961) slaughtered animals from four planes of nutrition (Table 1) and found a significant ($P < 0.05$) difference in the percentage of fat in the carcass; those on the high-plane in the final feeding period had fatter carcasses than those on the moderate-plane regimen. Hendrickson, Pope and Hendrickson (1965) performed a similar experiment with similar results (Table 1).

In order to examine more closely the relationship of muscle to bone, the muscle-bone ratios from the data of Guenther *et al.* (1965) are plotted in *Figure 8*. The high-plane of nutrition seemed to increase the muscle-bone ratio. The authors noted that, on a weight-

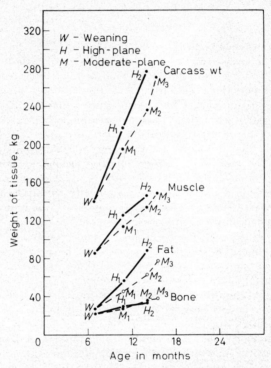

Figure 6. Carcass and tissue growth from half-sib Hereford steers on high and moderate planes of nutrition (Data from Guenther et al., *1965)*
By courtesy of The Animal Husbandry Research Division

constant basis, there was no significant difference in lean content but the moderate group averaged 2 kg (per side) more bone than did the high-plane calves. It would be interesting to know if the same conclusion would have resulted if bone weight had been adjusted to a constant muscle plus bone weight rather than carcass weight. As fat increases, the proportion of muscle and bone must decrease but the muscle-bone ratio may not be altered. Tayler

(1964) suggested that a minimum of 20 per cent of body fat is needed for maximum muscle-bone ratios. The percentages of fat in the carcasses in the groups of Guenther *et al.* (1965) was $H_1 - 25 \cdot 7$, $H_2 - 31 \cdot 8$, $M_1 - 23 \cdot 4$, $M_2 - 26 \cdot 5$ and $M_3 - 28 \cdot 6$, respectively and, therefore, fat level *per se* does not provide an adequate explanation

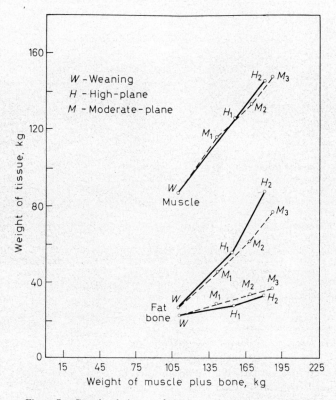

Figure 7. Growth of tissues relative to muscle-plus-bone growth from Hereford steers on two planes of nutrition (Data from Guenther et al., 1965)
By courtesy of The Animal Husbandry Research Division

for the apparent muscle-bone ratio differences from the two planes of nutrition. Callow (1961), whose groups had fat levels well over 20 per cent, found no difference in muscle-bone ratios from any of the four planes of nutrition used (Table 1). Although the data of Hendrickson *et al.* (1965) were not statistically analysed with respect to muscle-bone ratio, their results seem to be similar to Callow's (Table 1). Because of the conflicting results, it is not possible to conclude whether plane of nutrition has an effect on the relative growth of muscle and bone or if it is merely involved with slowing down or speeding up the whole process in a normal allometric manner. It is clear, however, that the fattening process can be

Table 1

Effect of plane of nutrition on carcass composition

Treatment	High-high	High-moderate	Moderate-high	Moderate-moderate	Sig. level†
Source					
I. Callow (1961)					
Wt. of carcass (kg)	347·3	329·2	308·7	338·7	
% Muscular tissue	55·2	59·2	56·5	58·7	†
% Fatty tissue	30·6	25·5	29·0	26·2	†
% Bone	11·8	12·8	12·3	12·7	
Muscle-bone ratio	4·7	4·6	4·6	4·6	
II. Hendrickson et al. (1965)					
Wt. of carcass (kg)	242·9	236·5	236·5	242·0	
% Muscle*	55·8ce	58·4c	57·3e	60·3	†
% Fat*	31·9e	27·4	29·5e	25·9	†
% Bone*	13·8c	14·9	14·3c	14·7	†
Muscle-bone ratio	4·0	3·9	4·0	4·1	n.a.

* Estimated from rib dissection
† $P < 0.05$, n.a.—not analysed;
  Means with the same superscript do not differ at $P < 0.05$

enhanced or retarded relative to muscle and bone development by altering the plane of nutrition.

2. Starvation and compensation

Monteiro and Falconer (1966) state that the phenomenon of compensatory growth assures that a certain final size is reached. Physiological factors associated with the growth curve (sexual maturity and inflection point of the growth curve) are weight-dependent rather than age-dependent, which results in compensatory growth. They cite Robertson (1963) who found that Drosophila larvae reared on sub-optimal diets had reduced rates of growth but

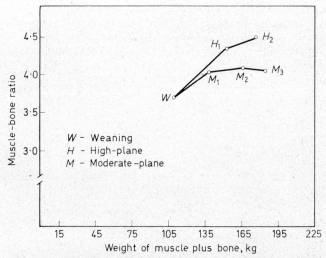

Figure 8. Effect of two planes of nutrition on muscle-bone ratio from Hereford steers (Data from Guenther et al., 1965)

By courtesy of The Animal Husbandry Research Division

Table 2

Effect of plane of nutrition and compensation on carcass composition in beef steers. (Adapted from Lawrence and Pearce, 1964)

Plane of nutrition	High	Medium	Low	Sig. level
No. of steers	12	12	12	
Carcass weight (kg)	274·2	271·3	271·5	
Composition (Predicted from 10th rib dissection)				
% Muscle	54·2	53·6	55·3	
% Dissectible fat	27·5	28·0	24·3	
% Bone	15·3	15·3	16·6	*
Muscle-bone ratio	3·55	3·49	3·33	n.a.

* Significant at $P < 0.05$
n.a.: not analysed

a lengthened larval period because pupation was weight-dependent. Compensation thus assured that the adult size remained essentially the same. Final size and perhaps composition in cattle may be subject to similar effects of compensation.

439

Periods of low-plane nutrition are said not to affect carcass composition when animals are allowed an adequate compensatory period of unrestricted recovery (Lawrence and Pearce, 1964; Tayler, 1964, citing earlier work). Lawrence and Pearce wintered

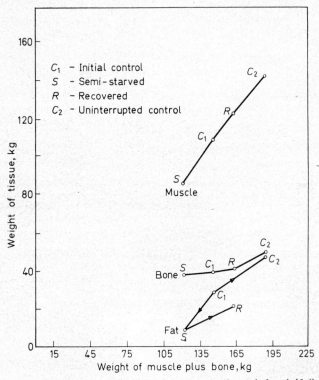

Figure 9. *Effect of starvation and recovery on tissue growth from half-sib Hereford steers (Data from Butterfield, 1966)*

By courtesy of Blackwell Scientific Publications

steers on three planes of nutrition for a 168-day period, following which they were allowed good pasture and were slaughtered at a fixed weight. The relevant data from their experiment are shown in Table 2. There is an indication that composition was affected in that percentage of bone was higher in the low-plane group and percentage fat slightly lower, though not significantly so. The muscle-bone ratio favours the uninterrupted group but this was not tested statistically. A similar effect with respect to percentage of fat was obtained by Meyer *et al.* (1965).

The relative growth of bone, muscle and fat under high-plane nutrition is reversed under semi-starvation. The depletion of fat is most rapid and the degree of involvement of muscle and bone depends on the severity and length of time on semi-starvation (Butterfield, 1966; Yeates, 1964). Butterfield (1966) performed an experiment which combined starvation and re-alimentation using 23 Poll Hereford half-sib steers. The data on tissue weight are plotted in *Figure 9*. At the beginning of the trial, a control group C_1 was slaughtered. S refers to a group slaughtered following semi-starvation, R—following semi-starvation and recovery, and C_2—an uninterrupted control. It would seem that muscle loss and gain follow a completely normal pattern. The absolute amount of bone under semi-starvation does not drop to a level corresponding to that found in a normal carcass of the same weight but it does seem to make near normal recovery on re-alimentation. Fat is depleted under semi-starvation and in the recovery period of the experiment did not reach the expected level relative to the controls.

Semi-starvation thus seems to deplete fat and muscle with a lesser effect on bone. Re-alimentation or compensation results in an increase in muscle toward a point of normal muscle-bone relationship, while the amount of fat tissue will only reach the same level as uninterrupted controls if a long enough period of compensation is involved. Presumably the nutritional level in the compensatory period would affect the speed of these recovery effects.

INFLUENCE OF BREED OR TYPE ON GROWTH AND RELATIVE GROWTH

Hankins, Knapp and Phillips (1943) proposed muscle-bone ratio as an index of merit in beef cattle and demonstrated a significant ($P < 0.01$) difference in this trait in favour of beef-type vs. dual-purpose Shorthorns slaughtered at similar live weights of approximately 410 kg. Berg and Butterfield (1966) extended this concept and suggested that muscle-bone ratio adjusted for carcass weight might be amenable to improvement by breeding methods. They suggested percentage of fat as a second criterion of carcass merit which might provide an index of acceptability with the optimum percentage probably differing for different markets. The acceptance of muscle-bone ratio and percentage fat as measures of carcass merit would hinge on the magnitude of genetic differences which might occur between breeds or types of cattle (*Figures 2* and *3* should be recalled). Because differential growth of tissues proceeds in a weight-dependent manner, comparisons should be made on an equal weight basis, preferably at equal muscle-plus-bone weights. Many comparisons between breeds and types have been made on a percentage basis with the result that carcasses which have a high proportion of fat automatically have a relatively low proportion of

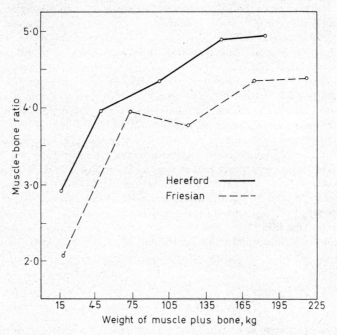

Figure 10. Muscle-bone ratios relative to total muscle-plus-bone weight from Hereford and Friesian steers (Adapted from Anonymous, 1966)

muscle and bone. Comparisons should be made, therefore, relative to a normal development pattern, or breeds should be compared with respect to their development patterns of muscle, bone and fat.

For example, in *Figure 3* the Hereford steers showed a marked tendency to begin the fattening phase at lower muscle-plus-bone weights than did the Friesians (Anonymous, 1966). There also was less bone relative to muscle-plus-bone in the Hereford steers. This is more clearly shown in *Figure 10* where the muscle-bone ratios for the two breeds are plotted against muscle-plus-bone weights. The Herefords maintained higher muscle-bone ratios throughout. Both muscle-bone ratio curves plateaued indicating that further increase in size would probably not be accompanied by further increases in muscle-bone ratio.

Most animals from breed comparison experiments have not been killed over a wide enough weight range to give a clear picture of the developmental patterns of muscle, bone and fat. Sometimes animals are slaughtered at a constant live weight leading to one set of conclusions, and at other times at a constant age often leading to quite different conclusions. However, in most comparisons between beef and dairy breeds, the results are generally similar to those in *Figures 3* and *10* when differences in carcass weights are taken into account. British beef breeds show clearly a tendency to fatten earlier than dairy breeds and they are usually superior in muscle-bone ratio at equal carcass or muscle plus bone weights. Carrol, Clegg and Kroger (1964) found Herefords at 325 kg carcass weight had an estimated 40 per cent of fat, significantly ($P < 0.05$) greater than the 35 per cent of 384 kg carcass-weight Holsteins. The muscle-bone ratio of the Herefords was 4·80, significantly ($P < 0.01$) greater than the 4·35 of the Holsteins. Callow (1961) compared Herefords, Dairy Shorthorns and Friesians at 304, 328 and 360 kg carcass weight, respectively. Fat percentage was significantly different ($P < 0.01$) between the breeds at 28·5, 30·4 and 24·5 per cent, respectively. Muscle-bone ratios did not differ in Callow's data but a difference might have been expected had the breeds been compared at equal carcass weights. Branaman, Pearson, Magee, Griswold and Brown (1962) compared beef and dairy cattle at nearly equal carcass weights and found the beef carcasses higher (though not significantly so) in percentage of fat and also higher in muscle-bone ratio.

Critical comparisons of relative tissue growth patterns between British beef breeds are lacking. Cole, Ramsey, Hobbs and Temple (1964) compared 133 steers of six breeds and one crossbred group, with the results shown in Table 3. Fat percentage differences were significant and differences in muscle-bone ratio appear to be quite large but were not statistically analysed.

Recently, at the University of Alberta, Charolais and Brown Swiss sires have been mated to British beef-breed dams (Hereford, Angus and Galloway) and the steer progeny compared with British-beef-breed crosses and with Canadian Holstein-Friesians. The resulting dissection data are given in Table 4. The Holstein steers were weaned at birth and castrated at approximately 1 month, but the other three groups were single suckled to approximately 6 months of age and castrated at that time. All animals were fed the same fattening ration from approximately 6 months of age to slaughter weight. Cross-bred progeny involving the large European breeds seem to fall between the British beef-breed crosses and the Holsteins in muscle-bone ratio and they tend to have less fat even though heavier in carcass weight. The tendency of the British beef-breed crosses to fatten at lighter weights and have higher

Table 3

Comparisons of carcass composition of breeds
(Adapted from Cole *et al.*, 1964)

Breed	Warm carcass wt. kg	Muscle + bone kg	Percentage separable fat	Muscle-bone ratio
Hereford	239	164	31·3	3·87
Angus	236	155	34·3	4·12
Brahman	228	172	24·4	3·93
Brahman Cross	248	182	26·9	3·89
Santa Gertrudis	242	176	27·1	3·64
Holstein	233	181	22·1	3·41
Jersey	198	144	26·2	3·52

muscle-bone ratios than the Holsteins agrees with other work cited earlier. In Canada, the Brown Swiss is used as a dairy breed and the Charolais is valued for beef production. It is interesting to note that at essentially equal weights and fatness, Brown Swiss crosses had lower muscle-bone ratios than Charolais crosses. The early fattening, slower growing British beef breeds generally have higher muscle-bone ratios at given muscle-plus-bone weights than do the later-fattening, faster-growing, larger European breeds. Whether these effects are coincidental or whether they are the result of correlated selection responses in the evolution of the British beef-breeds, is worthy of speculation. Early fattening and early muscle development have been considered indications of physiological maturity and may be genetically correlated. However, selection for blockiness in the development of the British beef breeds could have elicited a response in both early fattening and early muscle development even if these effects were essentially uncorrelated.

444

Some evidence for the latter view is found in the study of Martin, Walters and Whiteman (1966) who found carcasses of 'choice' (thick and full) conformation to have a significantly higher muscle-bone ratio than 'standard' carcasses which were characterized by being longer and narrower. They attempted to make their comparisons between carcass groups at equal weights and at equal fatness as indicated by fat cover at the 12th rib but, in spite of this precaution, there was nearly 3 per cent more fatty tissue in the 'choice' group.

Table 4

Comparison of carcass composition among British beef-breed crosses, European breed × British beef-breed crosses, and Canadian Holstein-Friesian steers

Breed	No.	Cold carcass wt. kg	Muscle + bone kg	Percentage separable fat	Muscle-bone ratio
Beef crosses	6	248·6	175·6	28·63	5·15
Charolais crosses	4	264·5	195·8	25·33	4·90
Brown Swiss crosses	6	269·4	199·1	24·89	4·60
Holstein-Friesian	3	248·2	179·4	26·11	4·02

SEX INFLUENCES ON RELATIVE GROWTH

Sex is known to have a marked influence on the onset of fattening, with heifers fattening at lighter weights than steers, and steers at lighter weights than intact males. Critical information is scarce on the muscle-bone ratios at various stages of development. Because of slower live-weight gains and earlier fattening of heifers, sex comparisons are often made at different weights. Bradley, Cundiff, Kemp and Greathouse (1966) compared 34 steer and 34 heifer carcasses at carcass weights of 298 and 267 kg, respectively. The fat percentages estimated from the 9–11th rib-cut were 40·7 and 44·6, and the muscle-bone ratios were 3·31 and 3·19, respectively. Breidenstein, Breidenstein, Gray, Garrigan and Norton (1963) selected, from a packing plant, 78 sides from steers and 93 sides from heifers. The side weights were similar for the sexes but heifers had a greater amount of waste indicating greater fatness. The heifers did have, however, 0·4 in^2 greater 'rib-eye' areas than did the steers $(P < 0·01)$ which would indicate that at equal weights heifers may be equal or superior to steers in muscling.

Comparisons involving steers and intact males have the same shortcomings as comparisons of heifers and steers. Intact males

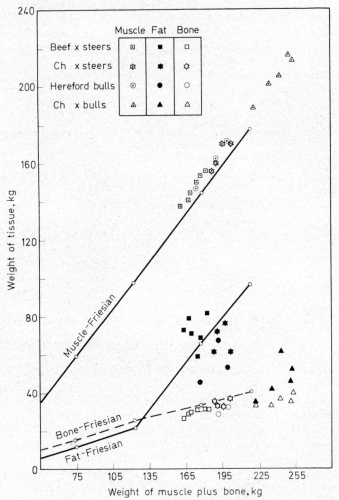

Figure 11. *Tissue growth from steers and bulls relative to muscle-plus-bone weight (Author's unpublished results. Friesian data from Anonymous, 1966)*

normally grow faster and fatten at heavier weights so that their carcasses are usually heavier than those from steers with which they are compared. Prescott and Lamming (1964) compared steers castrated at 7 months with intact males. The cold carcass weights were 218 and 229 kg for steers and bulls. The steers had 29·2 per cent fat in the 10th-rib cut compared to 16·8 per cent for the bulls and a muscle-bone ratio in the same cut of only 2·72 compared with 3·38 for bulls. Bailey, Probert and Bohman (1966) found similar results with steers and bulls at 254 and 259 kg carcass weight, respectively; fat percentages of the 9–11th rib were 40·1 and 30·8 and the muscle-bone ratios were 2·79 and 2·97, respectively.

Table 5

Comparison of carcass composition of steers and bulls of British beef breeds, and Charolais × British cross-breds. (University of Alberta data)

Sex and breed	No.	Cold carcass wt. kg	Muscle + bone kg	Percentage separable fat	Muscle-bone ratio
Steers					
British beef crosses	6	248·6	175·6	28·63	5·15
Charolais crosses	4	264·5	195·8	25·33	4·90
Bulls					
Hereford	3	249·7	190·8	22·09	5·27
Charolais crosses	5	294·3	241·3	16·03	5·81

Figure 11 shows tissue weights relative to muscle plus bone for two groups of steers and two groups of bulls, comprising British beef-breed cross-bred steers, Charolais × British beef-breed crosses, Hereford bulls and Charolais × British beef-breed cross-bred bulls (University of Alberta unpublished results). The British Friesian growth curves from *Figure 3* are included for comparison. The means for carcass composition from the same animals are given in Table 5. The carcasses from bulls had a lower proportion of fat than those from steers even though the bulls were carried to heavier weights. The muscle-bone ratios were higher from the bull groups, especially from the heavier Charolais cross-bred group. It should be recalled (*Figure 10*) that muscle-bone ratios in steer carcasses seemed to plateau at about 140–160 kg of muscle plus bone in the carcass. Therefore, the increased weight of the bull carcasses would probably not account for all of the apparent increase in muscle-bone ratio (Table 5). Perhaps intact males have a more prolonged impetus for muscle growth, and therefore muscle-bone ratios would rise higher

before levelling off. Clarifying this possibility and making more precise composition comparisons between bulls and steers would necessitate extending the slaughter range for both groups. *Figure 11* shows how breed and sex shift the onset of the fattening phase with steers fattening earlier than bulls and British beef breeds fattening earlier than Charolais cross-breds.

GENERAL DISCUSSION AND CONCLUSIONS

Differential growth of muscle, bone and fat in the carcasses of cattle results in changes in the proportions of these tissues as growth and development proceed. The immediate post-natal period is characterized by relatively rapid muscle growth and an increasing proportion of muscle in the carcass. At a later stage, fat deposition assumes major importance and as the percentage of fat increases, the percentage of muscle decreases. Bone growth proceeds at a slow rate relative to muscle and fat. Thus, under normal circumstances carcass composition will be determined by slaughter weight, with increase in muscle proportion characterizing the phase up to the inflection of the growth curve following which increase in fat proportion becomes more important.

Other environmental and genetic factors influence the normal differential growth patterns changing the expected tissue proportions at given slaughter weights. Plane of nutrition has a marked influence on relative tissue growth with high-plane nutrition causing the fattening phase to be earlier relative to muscle and bone development and a low-plane resulting in a delayed fattening phase. The effect of plane of nutrition on muscle-bone relationships is not clear. There is some indication that a high-plane of nutrition or an increase in carcass fatness increases muscle-bone ratio but other results show little influence of plane of nutrition on this trait. The importance of intra-muscular fat to muscle weight needs to be explored.

Genetic influences play a major part in carcass composition. Their effects are related in a large degree to differences which occur in the growth curve, in physiological maturity or in mature weight. The British beef breeds have been shown to be early fattening, that is, the inflection in their growth curve occurs at relatively light weights. These breeds are usually characterized by earlier, more pronounced muscular development as indicated by muscle-bone ratios at given muscle-plus-bone weights. On the other hand, the larger European breeds such as the Friesians, Charolais and Brown Swiss enter the fattening phase at heavier weights and their muscular development seems also to be somewhat delayed in comparison with the British beef breeds. The apparent relationships between early fattening and early muscle development may have resulted from physiological or genetic correlation or it may have been brought about by coincident selection for 'type' in the British beef breeds. Crosses between breeds which differ in growth patterns and tissue development seem to be intermediate and thus allow the possibility of producing the type of response desired by judicious use of breeds or breed combinations.

Sex has a major effect on carcass composition through its influence on the onset of the fattening phase. Heifers normally fatten at lighter weights than steers which, in turn, fatten at lighter weights than bulls. Muscle-bone ratios normally favour bulls over steers and steers over heifers, but this may be partly a reflection of normal muscle and bone growth since slaughter weight, being influenced by fatness, normally favours bulls over steers and steers over heifers.

It would seem that major influences on carcass composition can be brought about by altering slaughter weights or by shifting the onset of the fattening phase by either genetic or environmental means. It should thus be possible, by exercising some control over these factors, to produce carcasses to meet any specific composition requirements.

REFERENCES

Anonymous (1966). *Report of Major Beef Research Project*. London. The Royal Smithfield Club

Bailey, C. M., Probert, C. L. and Bohman, V. R. (1966). *J. Anim. Sci.* **25**, 132

Berg, R. T. and Butterfield, R. M. (1966). *Anim. Prod.* **8**, 1

Bradley, N. W., Cundiff, L. V., Kemp, J. D. and Greathouse, T. R. (1966). *J. Anim. Sci.* **25**, 783

Branaman, G. A., Pearson, A. M., Magee, W. T., Griswold, R. M. and Brown, G. A. (1962). *J. Anim. Sci.* **21**, 321

Breidenstein, B. C., Breidenstein, B. B., Gray, W. J., Garrigan, D. S. and Norton, H. W. (1963). *J. Anim. Sci.* **22**, 1113 (abstr.)

Butterfield, R. M. (1966). *Res. vet. Sci.* **7**, 168

— and Berg, R. T. (1966a). *Res. vet. Sci.* **7**, 326

— — (1966b). *Proc. Aust. Soc. Anim. Prod.* **6**, 298

Callow, E. H. (1961). *J. agric. Sci., Camb.* **56**, 265

Carroll, F. D., Clegg, M. T. and Kroger, D. (1964). *J. agric. Sci., Camb.* **62**, 1

Cole, J. W., Ramsey, C. B., Hobbs, C. S. and Temple, R. S. (1964). *J. Anim. Sci.* **23**, 71

Guenther, J. J., Bushman, D. H., Pope, L. S. and Morrison, R. D. (1965). *J. Anim. Sci.* **24**, 1184

Hankins, O. G., Knapp, B. Jr. and Phillips, R. W. (1943). *J. Anim. Sci.* **2**, 42

Hendrickson, R. L., Pope, L. S. and Hendrickson, R. F. (1965). *J. Anim. Sci.* **24**, 507

Lawrence, T. L. J. and Pearce, J. (1964). *J. agric. Sci., Camb.* **63**, 5

Ledger, H. P. (1965). *J. agric. Sci., Camb.* **65**, 261

Martin, E. L., Walters, L. E. and Whiteman, J. V. (1966). *J. Anim. Sci.* **25**, 682

Meyer, J. H., Null, J. L., Weitkamp, W. H. and Bonilla, S. (1965). *J. Anim. Sci.* **24**, 29

Monteiro, L. S. and Falconer, D. S. (1966). *Anim. Prod.* **8**, 179

Prescott, J. H. D. and Lamming, G. E. (1964). *J. agric. Sci., Camb.* **63**, 341

Robertson, F. W. (1963). *Genet. Res.* **4**, 74

Tayler, J. C. (1964). *Emp. J. exp. Agric.* **32**, 191

Tulloh, N. M. (1963). *Symp. Carcase Composition and Appraisal of Meat Animals*. Paper 5-1. (Ed. by D. E. Tribe.) Melbourne, C.S.I.R.O., 5-16

Yeates, N. T. M. (1964). *J. agric. Sci., Camb.* **62**, 267

PRACTICAL IMPLICATIONS OF GENETIC AND ENVIRONMENTAL INFLUENCES: SHEEP

C. R. W. SPEDDING

The Grassland Research Institute, Hurley, Berks.

GROWTH and development represent a major part of the meat and wool production process in sheep. Practice is concerned with the whole process, however, and the importance of changes in rate of growth and development must therefore be assessed, at some stage, in relation to the entire production system. The practical implications of genetic and environmental influences will thus depend upon both the importance of growth rate in the production process and the extent to which these genetic and environmental factors can be manipulated.

It is intended to deal chiefly with the former aspect in this paper but it should be emphasized that the conclusions so drawn must greatly affect the economic feasibility of the latter. There are few aspects of the environment that are beyond control and it must be largely considerations of cost that determine the degree of control that is used. Sheep farming is currently characterized by a high degree of acceptance of the natural environment; thus it is as common to choose a kind of sheep to fit the environment as it is to control the environment to suit the sheep. The extent of the choice is limited in both directions, often for different reasons, and genetic factors cannot be manipulated very greatly in the short term. Nevertheless, there is a wide range of genetic material to choose from at the present time and cross-breeding can effect very rapid changes in the sheep available.

In current practice, genetic limitations set a ceiling to performance, including rate of growth and development, but the environment, operating chiefly through climatic conditions, nutrition and disease, commonly restricts this to a level well below the genetic ceiling. The simplest illustration of this is to be seen on most sheep farms where twin and single lambs grow at very different rates although genetically similar. There is little doubt that the

451

majority of twins could grow as fast as singles if they received more milk (Brown, 1964), and it is extremely unlikely that even the growth rate of the singles is anywhere near their potential. Prolificacy is a potent factor in both productivity and profitability, yet where it is high it is most improbable that lamb growth rates are much more than 60 per cent of their potential.

Clearly, the importance of growth rate will vary with the reproductive rate and it is usually worth while having multiple lambs even if, within limits, this is associated with poorer individual growth rates. Similarly, the significance of a reduced growth rate must be different if it is caused by a disease in which food consumption is not decreased, compared with the same reduction caused by food shortage. Since production systems embody a great many different relationships which may all be affected differently by a change in growth rate, it is necessary to arrive at a limited number of expressions for the efficiency of the whole process in order to assess the influence of a change in one component, such as growth rate.

EFFICIENCY OF PRODUCTION

Efficiency is primarily a relationship between product output and the use of resources. Where profit is the motive, one measure must be the return on the total monetary investment. Even where biological relationships are of primary interest, it is difficult, if not impossible, to compare the efficiency of production of products differing as widely as, for example, meat and wool, in other than monetary terms.

The major input items are money, labour, time, food, land and fertilizer. There are others, of course, and where irrigation is practised, for example, a sensible calculation of efficiency could be based on water use. There is, in fact, a great need for assessments of the effect of varying any one component on the efficiency of whole systems.

No one expression will be satisfactory for all purposes but the one which is most useful to both farmer and biologist at the present time is the efficiency of food use, most usefully expressed in terms of digestible organic matter (D.O.M.). Practical implications can be considered most easily therefore by studying changes in the quantity of saleable product (carcass or clean wool) produced per unit of D.O.M. consumed. The main weakness of such an expression lies in the poor definition of the product; this is not always corrected simply by improving its description.

Where output from a system consists of more than one product, it is necessary at least to know what proportion of the total monetary output is derived from each (Table 1).

The proportions vary somewhat with stocking rate but major variations (as in column 2, Table 1) are due to changes in meat

Table 1

Proportion of income derived from wool. Results of an experiment with 30 Scottish Half-bred ewes at each stocking rate (1962, 1963, 1964, 1965)

	1962		1963		1964		1965		Mean
Stocking rate (ewes/acre/annum)	3	5	3	5	3·2	3·8	3·2	3·8	3·75
Total income (£/acre/annum)	33·8	37·9	33·4	24·4	35·2	39·9	35·4	36·5	34·6
Percentage of total income derived from wool	12·0	16·4	12·4	26·5	13·9	13·8	13·0	16·2	15·5

Table 2

Influence of litter size and fleece weight on the proportion of income derived from wool (as percentage of total income per ewe)

Fleece wt.* (kg/ewe)	Litter size				
	1	2	3	4	5
5·5 (e.g. Devon Longwool)	37·7	23·2	16·8	13·1	10·8
2·75 (e.g. Scottish Half-bred)	23·2	13·1	9·2	7·0	5·7

* Assuming no effect of litter size (with all lambs removed at birth), wool valued at 11s/kg, and a constant carcass value of £5

output rather than wool production. Large changes are to be expected with differences in breed, especially where fleece weight varies greatly, and with variation in mean litter size (Table 2).

The calculated values refer to income per ewe and the relative values per unit of land might be quite different. It is unlikely, however, that the percentage income from wool would amount to more than 40 per cent, unless due to very inefficient meat production or to deliberate concentration on wool-only systems.

It would be helpful if the resources used could be similarly allocated but this is rarely possible. Indeed, most sheep cannot be kept at all without producing some wool and the protective value of the fleece cannot be entirely ignored.

Finally, the efficiency of production can be assessed for individual lambs, for ewes and their lambs or for whole flocks over substantial periods. At some point, the last method has to be employed.

THE SIGNIFICANCE OF GROWTH RATE

Changes in growth rate will have different consequences in different kinds of sheep, related to carcass composition and other aspects of development, and at different phases of the animals' lives. Two of the most important of these aspects are the significance of birth weight and of final size.

Birth weight

The weight of a lamb at birth may be governed by several factors and the significance of birth weight will be expected to vary according to whether genetics, disease or nutrition is the major determinant. It is well known that birth weight is important in relation to survival and that lambs can be either too light or too heavy (Purser and Young, 1959; Gunn and Robinson, 1963). Although the optimum birth weight must therefore vary with the degree of stress immediately after birth, there must be a minimum birth weight for survival; for lowland breeds of the Clun-Hampshire type, Dawes and Parry (1965) have suggested minimal birth weights of 2·75 kg and 2·25 kg for male and female lambs respectively. These represent approximately 3·5–4·0 per cent of the ewe's mature weight.

Within a breed or cross, litter size is a major factor in influencing mean birth weight and Dickinson, Hancock, Hovell, Taylor and Wiener (1962) have suggested the following formula relating the major determinants:

$$W = 0 \cdot 181 \ D^{0 \cdot 83} \ [1 - 10^{-1 \cdot 1/M \ (R/D)^{0 \cdot 83}}]$$

where $W \equiv$ mean birth weight of the lamb (lb), $M \equiv$ multiplicity of birth, $D \equiv$ the live weight of the donor ewe (lb) and $R \equiv$ the live weight of the recipient ewe (lb). This formula was derived from studies of ovum transfer and it is of considerable interest that it expresses quite satisfactorily the relationships observed in our own work on superovulation by the injection of pregnant mare serum (P.M.S.). The actual values found in our experiments are compared with the calculated values in *Figure 1*. Since the calculation takes no account of the kind of ram used, values from three different ram breeds have been included.

The non-continuous line has been calculated from the formula using the actual ewe weights from our experiment; the continuous

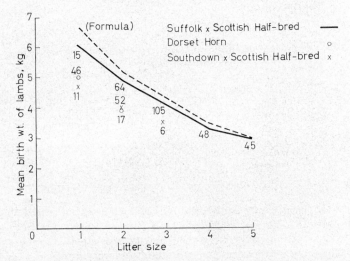

Figure 1. The relationship between lamb birth weight (kg) and litter size in experiments at Hurley and as predicted using the formula of Dickinson et al. (1962). The numbers of lambs involved are given at each point

line relates the actual birth weights of lambs from these same ewes to the size of the litter produced. Scottish half-bred ewes were crossed with a Suffolk ram and great care was taken to feed the ewes adequately, i.e. in relation to their foetal burden. The numbers of lambs on which the points are based are given in the figure. The additional points relate to pure Dorset Horns and to Southdown × Scottish Half-bred lambs; these data are more limited and less control was exercised over nutrition during pregnancy.

It is worth noting that if the minimal birth weights of Dawes and Parry (1965) are inserted into the formula of Dickinson *et al.* (1962), minimum ewe size can be calculated for each litter size. Table 3 shows the results, where minimal birth weight has been assumed to be 2·5 kg; the calculation is only valid, therefore, for situations where this figure is applicable.

There is less information on the influence of lamb birth weight on subsequent growth rate. In most situations where the lamb is suckled vigour and weight at birth may influence considerably the subsequent milk supply from the ewe. This particular interaction is avoided in artificial systems of rearing but there are few data for such situations where birth weights vary but ewes are known to have been fed in relation to their need or to the total foetal weight carried during pregnancy.

Table 3

Calculated minimum ewe weight (kg) related to mean litter size, where a minimal birth weight of 2·5 kg is assumed

	Mean litter size				
	1	2	3	4	5
Minimum live weight of a ewe (kg)	22	30	40	50	61

Observations on lambs produced by such ewes but reared on milk substitute suggest that growth rate is not consistently related to birth weight, although there is a tendency towards a positive relationship where low birth weights are the result of larger litters.

Within groups of singles and twins, reared artificially to avoid any effect of different milk supplies, we have been unable to detect a relationship between carcass gain from birth to slaughter and birth weight. Similarly, for the same lambs, no relationship could be established between birth weight and efficiency of food conversion.

Final size

The effect of lamb growth rate on the food conversion efficiency of the ewe/lamb unit varies with the weight of individual carcass produced. The latter is governed by such factors as breed and market demand, and the carcass weight at which the product is satisfactory to the market is also influenced by growth rate. Within the tolerance limits necessary for optimal carcass composition, it will improve overall food conversion efficiency to continue growth of the lamb until the daily food conversion ratio of the lamb itself approaches that of the whole system. Clearly, at low growth rates this point will tend to be reached earlier but where the growth rate is high, it will generally be more efficient to achieve maximum size. Although growth rate may influence mature size (Widdowson and Kennedy, 1962; McCance and Widdowson, 1962), it is by no means clear that the effect will be important in sheep (Dickinson, 1960); the maximum size referred to here, however, relates to the largest acceptable carcass. If this continued growth requires a higher cost feed then the greater conversion efficiency of the whole process may still be rendered uneconomic, as may also occur if the value per unit of product decreases markedly.

Accepting that the product can be defined as a weight of carcass (or a range of tolerable weights) and that growth rate can be measured as live weight increment per unit time, then it is possible to assess the effect of changes in growth rate on the biological efficiency of resource use.

THE EFFECT OF GROWTH RATE ON PRODUCTIVITY

Productivity may be expressed by various relationships but those considered will be only the major ones of carcass output to inputs of food and land. Monetary values may be assigned to either of these; labour is omitted because no relevant data are available.

Carcass production per unit of food

The relationship between carcass output and food input may be considered for an individual lamb, for a ewe and her progeny for one year, or for a sheep flock over several years.

The individual lamb—Recent work by Pattie and Williams (1966) illustrates one aspect of the relationship between efficiency and the growth rate of the individual lamb. Using weaned lambs from Merino flocks selected on the one hand for high weaning weight and on the other for low weaning weight, it was found that 'growth efficiency' measured as body weight change per unit of food intake was unchanged by the selection procedure, since food intake was greater when body weight was increased, but this was over a standard 51-day feeding period and not over the same body weight change.

With lambs reared artificially, a positive relationship can be demonstrated between food conversion efficiency and growth rate. With suckled lambs the situation is more complex. In Table 4, food conversion has been calculated as the carcass gain per unit of food (D.M.) consumed by the lamb itself and by the ewe in excess of her maintenance requirement calculated by subtracting the quantity of food required by similar ewes maintained at constant body weight.

Twins, growing more slowly than singles, show lower food conversion efficiency values, although the combined growth rate of a pair of twins was greater. Within groups of both singles and twins, however, considerable variation existed in conversion efficiency but relatively little in growth rate.

Table 4

Food conversion efficiency (carcass gain per unit of food D.M.) in relation to the rate of carcass gain in single and twin lambs (food consumption is shown for the 10 ewes and their lambs separately)

	Food conversion efficiency	Carcass growth rate (kg/day)	Food consumption	
			ewes (kg DM)	lambs (kg DM)
Singles	0·245	0·23	40	40
	0·176	0·23	74	46
	0·159	0·21	72	40
	0·147	0·22	76	54
	0·128	0·20	57	79
Twins	0·144	0·33	112	122
	0·132	0·40	152	119
	0·121	0·38	168	97
	0·112	0·37	199	104
	0·090	0·36	228	126

The most probable explanation of this is that similar growth rates resulted from diets differing markedly in their proportions of milk and solid food. If a unit of lamb growth required more total solid food when a double conversion through milk was involved, then lower conversion efficiencies would be associated with relatively higher solid intakes by the ewes. The relevant solid food intakes are shown in Table 4 as kilogrammes of D.M. consumed by ewes above maintenance and by lambs. These results have considerable implications for the effect of weaning on food conversion efficiency of the ewe/lamb unit.

The ewe/lamb unit—When the carcass gains referred to in the previous section are expressed per unit of total food intake, including maintenance requirements, by ewes and lambs for the period from birth to slaughter of each lamb, food conversion efficiency did not differ greatly between ewes with singles and ewes with twins (Table 5). If the food consumed by the dry ewe is taken into account either for the whole year (Large, 1966) or for the 6 months after parturition (Spedding, 1966), food conversion efficiency is markedly higher for ewes with twins, although the difference is less for the shorter period.

Table 5

Food conversion efficiency (carcass gain per unit of food DM) and rate of carcass gain, for the period from birth to slaughter of each single lamb and of each pair of twins

	Food conversion efficiency	Carcass growth rate (kg/day)
Singles	0·088	0·23
	0·082	0·23
	0·079	0·22
	0·058	0·20
	0·048	0·21
Twins	0·087	0·40
	0·073	0·37
	0·059	0·36
	0·057	0·38
	0·049	0·33

It is possible, therefore, that if early weaning can be carried out without a great reduction in lamb growth rate, the overall food conversion efficiency will be improved.

The calculations in Table 6 are based on the assumptions that the carcass output from a pair of twins is 34·8 kg, that the carcass output from a single lamb is 18·2 kg (which are actual values but held constant throughout the calculation); and that growth rates are unaffected by weaning. The D.O.M. consumed was calculated from that actually consumed by the ewes when dry, pregnant and lactating, added to the amount required by a ewe of similar body weight for maintenance at constant weight after weaning and the amount consumed by a lamb from weaning to slaughter (the latter was calculated from the quantity eaten by artificially reared lambs over the appropriate period). The D.O.M. involved was therefore sometimes represented by dried lucerne and sometimes by dried milk substitute. Clearly, efficiency of food conversion changed very little for weaning at birth or at 1–4 months of age. The difference between the efficiency of ewes with twins and those with singles is very marked and has been slightly exaggerated by the method of calculation.

Table 6

Food conversion efficiency (carcass gain per unit DOM consumed) in relation to time of weaning

Food conversion efficiency	Time of weaning (months after birth)				
	0	1	2	3	4
Ewes with twins	0·074	0·073	0·072	0·072	0·072
Ewes with singles	0·045	0·044	0·044	0·043	0·043

The sheep flock—The chief difference between assessment on the basis of the ewe/lamb unit and the whole flock lies in the need to maintain the breeding population. This can be expressed as the extra financial cost of buying young ewes, or, alternatively, the additional food required to rear them can be incorporated into the calculations. In any event, much depends upon the necessary replacement rate and, if replacements are retained within a flock, the percentage to be retained is greatly influenced by the reproductive rate.

Granger (1944) (quoted by Mattner and Moule, 1965) developed a formula for determining the percentage of young ewes to be selected as replacements (S):

$$S = \frac{2,000,000}{MN(100-d)}$$

where $M \equiv$ mean percentage of lambs marked to ewes mustered at mating; $N \equiv$ mean number of times that a ewe is mated in her lifetime; $d =$ mean percentage death rate in the young ewes prior to mating age.

Table 7 has been compiled using this formula, where d is assumed to remain constant at 5 per cent, for a range of values for M and N; this illustrates the importance of both longevity and mean litter size, assuming no great increase in mortality.

Carcass production per unit of land

If the D.O.M. output per unit of land and the proportion of what is grown that is harvested remain constant throughout, the calculations based on feed can be translated directly into terms of land use.

Table 7

Variation in replacement rate (S) with number of times mated (N) and lambing percentage (M)

N	M				
	100	150	200	250	300
1	210·5*	140·4*	105·2*	84·2	70·2
2	105·2*	70·2	52·6	42·1	35·1
3	70·2	46·8	35·1	28·1	23·4
4	52·6	35·1	26·3	21·1	17·5
5	42·1	28·1	21·1	16·8	14·0

* Values in excess of 100 are clearly impossible, i.e. the population cannot be maintained

Neither of these propositions is likely to be true under grazing conditions, however, and most sheep are kept in this way. Furthermore, since the pattern of distribution of grass production differs from the pattern of food requirement for a sheep population, the efficiency of conservation must also enter the calculation. This begins to approach the complexity of whole system analysis and only two examples can be given here.

Growth rate and weaned lambs—In many grazing circumstances, growth rate of weaned lambs is negatively related to grazing pressure. Thus, if the same pasture D.O.M. is produced and grazing pressure is approximately proportional to stocking density, the growth rate of the individual will be reduced as the efficiency of harvesting is increased.

An example may clarify the problem. In an experiment involving weaned Suffolk × Scottish Half-bred lambs, it was found (Spedding, 1966) that lamb growth rate (g/day: y) was related to stocking rate (lambs/ha: x) in the following way:

461

$$y = 303 - 2 \cdot 14\,x$$

Table 8 illustrates the probable relationship between stocking rate, the quantity of D.O.M. harvested per ha and the efficiency of food conversion. The latter has been calculated using the D.M. requirements for lambs of 40 kg (A.R.C., 1965) and the growth rates actually recorded in the experiment referred to above. The quantities harvested have been calculated from the appropriate D.M. requirement figures for each group of lambs multiplied by the actual stocking rate.

Table 8

The relationship between food conversion efficiency and the quantity harvested by grazing

Stocking rate (lambs/ha)	45·0	56·25	67·5	78·75	90
DM harvested (kg/ha/day)	90·0	98·4	112·7	117·3	115·2
Food conversion efficiency $\left(\dfrac{\text{live weight gain (g/day)}}{\text{DM consumed (g/day)}} \right)$	0·105	0·101	0·098	0·091	0·084

As expected, the efficiency of food conversion falls as the quantity harvested rises. This is primarily, however, a problem associated with grazing animals in which the required level of production can only be met under conditions of lenient grazing, i.e. at a low grazing pressure.

One way of combining a low grazing pressure and efficient harvesting is to integrate grazing and cutting for conservation.

The influence of conservation efficiency—Since the proportion of food devoted to production, chiefly for lamb growth and lactation, is increased with increase in litter size, it follows that a smaller proportion of the total is conserved. Losses in conservation are caused by wastage during the process (of ensilage or haymaking) and wastage during feeding; these total losses may amount to more than 30 per cent but commonly vary between 5 per cent and 25 per cent for silage making and feeding.

It has been suggested (Spedding, 1966) that the percentage of the total D.O.M. that is consumed during the 6 months from lambing, varies with mean litter size as follows:

	Litter size		
	1	2	3
DOM (%)	60	67	74

Thus, 40 per cent, 33 per cent and 26 per cent of the respective totals would be consumed during the other half of the year, approximately half of which would be in a conserved form. Table 9 illustrates the effect of conservation efficiency for these situations.

Table 9

The effect of conservation efficiency on the percentage reduction of herbage utilized per unit of land

Litter size	Percentage loss in conservation				
	5	10	15	20	25
1	1·0	2·0	3·0	4·0	5·0
2	0·8	1·6	2·5	3·3	4·1
3	0·6	1·3	2·0	2·6	3·3

Clearly, conservation losses become less important as litter size increases but it should be remembered that the quality of the conserved product becomes more important.

The influence of fodder conservation in Australian pastoral systems where wool is the main product, has been reported recently in a note by Hutchinson (1967).

WOOL PRODUCTION

Finally, in assessing the practical implications of the foregoing it is necessary to refer briefly to efficiency of wool production.

Ferguson (1962) concluded that sheep on a seasonally fluctuating intake would show the same efficiency of wool production as sheep on the same average intake fed at a constant rate throughout the year. Furthermore, he concluded that sheep maintained in different states of body condition, which require different intakes, would show similar efficiencies.

More recently, Wodzicka-Tomaszewska (1966) has reported investigations on New Zealand Romneys, in which sheep selected for high (H) and low (L) wool yields were compared on both restricted and *ad lib.* intakes.

The H sheep produced over 30 per cent more scoured wool per unit body weight and 29 per cent more wool per unit of D.M. intake, when on restricted intake, than the originally lower yielders (L). On *ad lib.* feeding, at a different season, H outyielded L by 35 per cent and 23 per cent, respectively. Thus, higher yielders were more efficient wool producers under both conditions; which is consistent with the findings of Schinckel (1960) and Dolling and Moore (1960) that efficiency was higher where more wool was grown, even if food intakes were also higher.

There appears to be general agreement that sheep with a higher genetic capacity for wool growth are more efficient at converting food into wool and that this is not associated with significant differences in digestive efficiency (Dunlop, Dolling and Carpenter, 1966; Clark, Keshary, Coop and Henderson, 1965; Weston, 1959). There is considerable evidence, however, that the gross efficiency of conversion of feed to wool, i.e.

$$\frac{\text{wool growth (g/day)}}{\text{feed intake (g/day)}}$$

decreases for an individual sheep as its level of intake increases (Ahmed, Dun and Winston, 1963; Williams, 1966; Ferguson, 1962). There is thus a sound basis for the findings that wool output per acre increases with increasing stocking rate, while wool yield per head falls relatively little (Sharkey, Davis and Kenney, 1962; Spedding, 1966).

This represents a genetic/environment interaction which is and has been of immense practical significance. A high stocking rate ensures that a high proportion of the herbage grown is eaten and, provided that pasture production is not reduced, more wool per acre must result if efficiency of conversion does not fall. If it increases with reduced intakes and if genetically superior wool producing sheep maintain their advantage under these conditions, then wool production is an ideal process to be associated with the grazing situation.

It should be noted, however, that fleece growth may be markedly reduced by lactation and, to a lesser extent, by pregnancy (Corbett, 1964; Doney, 1964; Ray and Sidwell, 1964), although the magnitude of these effects may be influenced by breed (Doney, 1966).

ACKNOWLEDGEMENTS

I wish to thank my colleagues Mr. R. V. Large, Miss Alison Stanton, Miss Diana Seal for their help in the preparation of this paper, and Mr. L. C. Chapas for checking the calculations.

REFERENCES

Ahmed, W., Dunn, R. B. and Winston, R. J. (1963). *Aust. J. exp. Agric. Anim. Husb.* **3**, 269

A.R.C. (Agricultural Research Council) (1965). *The Nutrient Requirements of Farm Livestock*. No. 2. Ruminants. Summaries of estimated requirements. London

Brown, T. H. (1964). *J. agric. Sci., Camb.* **63**, 191

Clark, V. R., Keshary, K. R., Coop, I. E. and Henderson, A. E. (1965). *N.Z. Jl. agric. Res.* **8**, 511

Corbett, J. L. (1964). *Proc. Aust. Soc. Anim. Prod.* **V**, 138

Dawes, G. S. and Parry, H. B. (1965). *Nature, Lond.* **207** (4994), 330

Dickinson, A. G. (1960). *J. agric. Sci., Camb.* **54**, 378

— Hancock, J. L., Hovell, G. J. R., Taylor, St. C. S. and Wiener, G. (1962). *Anim. Prod.* **4**, 64

Dolling, C. H. S. and Moore, R. W. (1960). *Aust. J. agric. Res.* **11**, 836

Doney, J. M. (1964). *J. agric. Sci., Camb.* **62**, 59

— (1966). *J. agric. Sci., Camb.* **67**, 25

Dunlop, A. A., Dolling, C. H. S. and Carpenter, M. T. (1966). *Aust. J. agric. Res.* **17**, 81

Ferguson, K. A. (1962). In: *The Simple Fleece*. (Ed. by A. Barnard.) Ch. 11. Melbourne Univ. Press

Granger, W. (1944). *Aust. vet. J.* **20**, 253

Gunn, R. G. and Robinson, J. F. (1963). *Anim. Prod.* **5** (1), 67

Hutchinson, K. J. (1967). *J. Br. Grassld. Soc.* **21**, 303

Large, R. V. (1966). *Proc. 9th Int. Congr. Anim. Prod.*, Edinburgh

Mattner, P. E. and Moule, G. R. (1965). *Field Investigations with Sheep: a Manual of Techniques*. Melbourne, C.S.I.R.O.

McCance, R. A. and Widdowson, E. M. (1962). *Proc. R. Soc.*, B. **156**, 326

Pattie, W. A. and Williams, A. J. (1966). *Proc. Aust. Soc. Anim. Prod.* **6**, 305

Purser, A. F. and Young, G. B. (1959). *Anim. Prod.* **1**, 85

Ray, E. E. and Sidwell, G. M. (1964). *J. Anim. Sci.* **23**, 989

Schinckel, P. G. (1960). *Aust. J. agric. Res.* **11**, 585

Sharkey, M. J., Davis, I. F. and Kenney, P. A. (1962). *Aust. J. exp. Agric. Anim. Husb.* **2**, 160

Spedding, C. R. W. (1966). *Proc. 9th Int. Congr. Anim. Prod.*, Edinburgh

Weston, R. H. (1959). *Aust. J. agric. Res.* **10**, 865

Widdowson, E. M. and Kennedy, G. C. (1962). *Proc. R. Soc.*, B. **156**, 96

Williams, A. J. (1966). *Aust. J. exp. Agric. Anim. Husb.* **6**, 90

Wodzicka-Tomaszewska, M. (1966). *N.Z. Jl. agric. Res.* **9**, 909

PRACTICAL IMPLICATIONS OF SOME GENETIC AND ENVIRONMENTAL INFLUENCES ON GROWTH AND DEVELOPMENT IN PIGS

I. A. M. LUCAS

University College of North Wales, Bangor

THE ASPECTS of growth of primary importance to producers of meat pigs, excluding breeding stock, are rate of live weight gain, carcass composition and feed conversion efficiency.

Rapid rate of gain allows maximum rate of throughput in piggeries and hence best capital utilization and cash turnover.

The carcass composition demanded by both the meat trade and consumer frequently commands a financial premium, but even when it does not, it may be argued that producers can only retain their market by offering the most desired article. Definitions of optimum carcass composition have frequently foundered in excessive detail. Nowadays, a common generalization is that maximum lean content is the best single criterion, although there are other aspects of commercial importance. Absolute size and shape of the 'eye' muscle (transverse section of longissimus dorsi), colour and wetness of muscle and the presence of undesirable flavours influence saleability. The proportion of offal discarded, reflected by killing-out percentage, affects profit when pigs are sold by carcass weight. The so-called 'balance' of side, indicated by proportions of fore and ham and length, are important to the meat trade but, provided variability is not great, they now appear to be less so than formerly because of new techniques of cutting and marketing which make good use of the shoulder joint.

Feed eaten per unit weight gain is partly a function of the diet and to this extent is not an aspect of growth. For example, digestibility of dietary energy, which can markedly affect feed conversion efficiency, decreases with increase in dietary fibre. However, efficiency varies with stage of growth when diets of equal Digestible Energy value are given and is influenced by the growth processes and metabolic transformations of nutrients which are taking place.

Efficiency also varies between pigs, but the relative contributions to this of variations in growth processes, in nutrient requirements or in metabolic efficiency are not clear.

The most meaningful way of comparing the practical importance of different measures of growth is to consider their financial implications, the disadvantage being that these are subject to change. Estimates made recently by the Pig Industry Development Authority (1966) for use in devising a selection index are shown in Table 1, along with the variability in each trait and the average financial advantage accruing per pig over the mean in the best 25 per cent of the population. It must be noted, however, that the variabilities, and hence the financial advantages calculated from them, are for pigs kept in standard conditions of progeny test stations and do not take account of the effects of variations in nutrition, housing and disease. Nevertheless, considered against a potential profit of say 50s per pig, the table highlights the importance of feed conversion efficiency and carcass lean, each of which contributes a potential average financial advantage of about 25 per cent in one pig in four. On the same basis there is a 19 per cent improvement associated with killing-out percentage but only 5–14 per cent improvement with rate of weight gain, 6 per cent with dressing percentage and only 3 per cent with the component of eye-muscle area not already accounted for by the associated change in carcass lean.

It is clear, however, that these figures require qualification. Rate of gain may be changed at will by restricting or increasing feed allowances, a circumstance which does not operate by intention in progeny test stations. It has been suggested (A.R.C., 1967) that a 15 per cent reduction in daily feed intake below that of test station pigs would lead to little change in feed conversion efficiency and hence to a 15 per cent reduction in rate of weight gain, which quantitatively is about 0·25 lb/day, implying a reduced profit of 3s. 9d. or 10s./pig, depending on whether or not increased throughput is taken into account. Such feed restriction would be imposed to reduce carcass fatness and so improve grading. If carcass lean were increased by 1·8 per cent (A.R.C., 1967), and if an accurate method of paying for lean were used, which is not so, there would be a gain of 8s. to offset the loss on growth rate. Thus changes in growth rate can be of greater financial importance than is indicated in Table 1. Moreover, under some circumstances growth rate is sufficiently important that a conscious reduction to improve carcass leanness may not be of immediate financial advantage, even in bacon pig production.

Effects of changes in feed conversion efficiency are of direct benefit to producers, but dressing percentage* is neither usually measured in individual pigs nor is the financial gain passed to producers, although an increase would be of advantage to curers. Also, in practice, leanness is measured indirectly through fat measurements,

Table 1

Financial gains (1966) associated with improvements in performance*

Change in	Financial gain per pig from 50–200 lb	Standard deviation of trait[†]	Average financial advantage per pig over the mean in the best 25% of pigs[‡]
			s. *d.*
Daily gain	18*d*/0·1 lb/day§	0·14	2 8
	48*d*/0·1 lb/day ‖		7 1
lb feed/lb live weight gain	48*d*/0·1 lb	0·25	12 9
Killing-out percentage¶	54*d*/1 per cent	1·65	9 5
Dressing percentage**	38*d*/1 per cent	1·0	4 0
Carcass lean percentage	54*d*/1 per cent††	2·4	13 9
Back and ham as percentage of carcass	25*d*/1 per cent		
Proportion of carcass lean in the back and ham	18*d*/1 per cent	1·4	2 8
'Eye' muscle area	6*d*/cm^2‡‡	2·7	1 9

* Columns 1–3, P.I.D.A. (1966)
† The variability of traits (for example daily gain) in the national pig progeny test stations. *See also* Table 3
‡ Calculated from 1·27 × (SD in column 3). Hence the pigs referred to are those kept under uniform conditions of diet and environment
§ Reduction in overheads per pig
‖ Reduction in overheads per pig plus increase in output of pigs with faster throughput in piggeries
¶ Carcass as percentage of live weight
** Proportion of carcass, after trimming, put into cure. May be termed 'trimming percentage'
†† Assuming that payment is made for each 1 per cent lean
‡‡ In addition to financial gain associated with the related increase in carcass lean

most of which are not well correlated with lean content. Fat thicknesses are then grouped to give a series of grades insufficiently narrow to allow payment to be made for an estimated 1 per cent change in lean. Thus, although improved lean content benefits producers, calculation of increased return per unit increase in percentage lean has, at present, little practical meaning. This

* Defined in footnote ** to Table 1.

argument, however, is likely to be less true when applied to a large number of pigs than to a small number.

After identifying the aspects of growth which are financially important, the problem for agriculturalists is to improve them by applying knowledge of genetics, nutrition, disease control and physical environment. Of particular practical significance are the *amount* of improvement which can be achieved, say by variation in nutrition, the *rate* of improvement through breeding and selection, and the *financial economies* of these changes.

SELECTION AND GROWTH

Because of space limitation, discussion of genetic influences will be confined to selection and cannot include other important aspects

Table 2

Estimates of heritabilities of growth traits in pigs

	Large Whites*	Landrace†	P.I.D.A. data‡	Average from eight studies§
Daily gain	0·41	0·41	0·4	0·42
lb feed/lb live weight gain	0·50		0·45	0·48
lb feed/lb carcass gain	0·58	0·48		
Killing-out percentage	0·40	0·26	0·40	0·32
Dressing percentage			0·50	
Percentage lean in carcass			0·65	
Proportion of lean in back and ham			0·50	
'Eye' muscle area	0·35	0·49	0·40	0·42
Carcass length (to first rib)	0·60	0·87		0·62
Fat depth at 'C'	0·65	0·62		

* Smith, King and Gilbert, 1962
† Smith and Ross, 1965
‡ P.I.D.A., 1966. Data derived partly from * and †
§ Smith and Ross, 1965. Data derived from * and † and six other studies

such as the influence of breed, inbreeding, cross-breeding and the production of new breeds.

In general, heritabilities of reproductive characters are low but those relating to growth are higher. Estimates for Large White and Landrace pigs derived from data from British progeny test stations are shown in Table 2, along with average parameters from eight studies. Heritabilities of growth rate and feed conversion efficiency are of the order of 0·4–0·5, while those of the important carcass characteristics are about 0·6–0·7. Heritability of muscle colour in Danish pigs is estimated as 0·06 in castrates and 0·32 in females (Jonsson, 1963), but data from the British test stations (Pease and Smith, 1965) suggest moderate heritabilities in Large White gilts and castrates and Landrace gilts and castrates of 0·17, 0·34, 0·55 and 0·41, respectively. Pale, watery muscle and 'two-tone' muscle is financially important to meat factories.

Thus it should be possible to improve growth traits by selection but the rate of improvement depends upon the genetic variabilities

Table 3

Means and standard deviations of some traits in pigs in progeny test stations

| | Large Whites* | | | | Landrace† | |
| | Castrates | | Gilts | | Combined sexes | |
	Mean	SD	Mean	SD	Mean	SD
Daily gain (lb)	1·52	0·14	1·51	0·12	1·45	0·11
lb feed/lb live weight gain	3·41	0·24	3·31	0·23		
lb feed/lb carcass weight gain	4·37	0·29	4·22	0·28	4·34	0·26
Killing-out percentage	73·5	1·69	73·9	1·62	74·2	1·60
'Eye' muscle area (cm²)	24·7	3·13	28·1	3·16	27·8	2·88
Carcass length (mm) (to first rib)	802·2	19·6	809·7	20·0	811·7	19·5
Fat depth at 'C' (mm)	23·1	4·41	18·1	3·42	22·5	3·76

* Smith, King and Gilbert, 1962. Pigs 50–200 lb
† Smith and Ross, 1965. Pigs 50–200 lb

of the traits, their measurement in potential breeding stock, the proportions of tested animals selected for breeding on the basis of superior performance, and upon the generation interval.

Some variabilities are given in Table 1, others are shown in Table 3, but not all the sets of data are independent. As all are for pigs in a constant environment and given standard diets it is assumed that variability is genetic, but this is not strictly true since some may be due to disease level, which cannot be measured, and some to treatment before the pigs enter the test stations. Smith, King and Gilbert (1962) comment that, on the basis of overseas experience, the latter might not be insignificant but tends to be contemporary at a test station. In Britain (Smith, 1965) the percentages of total variation in test performance accounted for by pretest environment for Large Whites and Landrace were 2 per cent and 1 per cent for daily gain, 0 per cent and 1 per cent for feed conversion efficiency, 8 per cent and 0 per cent for average backfat thickness and 5 per cent and 0 per cent for carcass length. Of these, the percentages for fat thickness and length in Large Whites were significantly different from zero and could inflate estimates of heritability of these traits.

With these reservations, the figures in Tables 1 and 3 can be considered in quantitative biological terms in much the same way as they have been considered in quantitative financial terms. One pig in four, out of a large tested population, should have average genetic advantages over the means of 0·16 lb/day in rate of weight gain, 0·3 in feed conversion efficiency and 3·0 in percentage carcass lean. Alternatively, on average one in four have growth rates better than the mean by 11 per cent, feed conversion efficiencies better by 9 per cent, carcass lean better by 6 per cent, length and killing-out percentage better by 3 per cent and area of 'eye' muscle and fat depth at 'C' better by 15 per cent and 23 per cent respectively. Clearly, with the associated high heritability, the opportunity exists to make particularly rapid proportionate changes in fat thickness, but the proportionate change in carcass lean, with selection for lean, would be considerably less rapid.

Problems of accuracy in the measurement of traits, upon which the effectiveness of selection must to some extent depend, will not be considered in detail here. Some, such as rate of weight gain and feed conversion efficiency are easy to measure, and some are more difficult, either technically or because of cost. These are percentage lean, 'eye' muscle area and colour, which involve non-commercial cutting of the carcass. Thus, less accurate measurements (Buck, Harrington and Johnson, 1962), such as mid-line backfat thicknesses on the normal commercial side, have frequently been taken as inverse criteria of leanness. On the other hand, the composition of the back joint correlates well with the composition of the whole side (Cuthbertson and Pease, 1965) and may be used as a compromise in progeny and sib tests, muscle colour is associated with pH measured 45 min after slaughter (Pig Industry Development Authority, 1966) and fat thickness at 'C' is well correlated with lean content and can be measured either with an optical probe on the intact carcass or with ultrasonics on the live pig. This use of ultrasonics has made possible the performance testing of boars for leanness. Nevertheless, indirect measurements inevitably introduce inaccuracies when applied to individuals, for example, fat at 'C' left unexplained 36 per cent of the variance within sexes in percentage carcass lean (Buck, Harrington and Johnson, 1962). Such inaccuracies must reduce the rate of genetic improvement to less than that theoretically possible.

The quantitative effects of long term testing and selection are illustrated by changes with time in Danish Landrace (*Figure 1*;

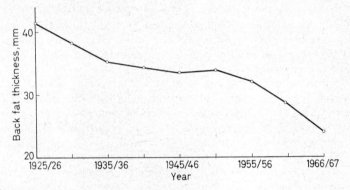

Figure 1. Changes with time in fat thickness of pigs on progeny tests in Denmark (From Clausen and Nørtoft Thomsen, 1967)

By courtesy of *Forsøgslaboratoriet*

Clausen and Nørtoft Thomsen, 1967), but testing with little selection is, of course, ineffective; over 800 boars were progeny tested in Britain between 1959 and 1963, and little genetic progress was made. One reason for this was that the boars were from 300 herds and three quarters of breeders each tested less than 4 boars over 4 years. Selection differentials were low, at only 0·05–0·30 standard deviation units, for daily gain, feed conversion efficiency, average backfat thickness and carcass length (Smith, 1965).

However, with a selection differential for single traits of 1 in 4, the theoretical rates of improvement may be calculated (Table 4; Smith, King and Gilbert, 1962). In each generation of Large Whites, percentage improvements (from data in Table 4) in daily gain when selecting for gain would be about 2·5, in feed conversion efficiency when selecting for efficiency would be about 2·8 and in average backfat when selecting against that trait would be about 5·7.

With time the rate of improvement would decline, partly because there must be biological limits to change and partly because of the possibility of reduced variability. Smith, King and Gilbert (1962) speculated on whether lower variability in Danish Landrace than in British Large Whites was due to the effects of long term selection in the former. They commented that the difference may also be due partly to the different feeding systems employed for testing in the two countries. Unfortunately there are no data on the effect of diet upon variability in pig testing.

Table 4

Expected response from one generation of selection for individual traits in Large Whites when the best 25 per cent of boars for each trait are used (Smith, King and Gilbert, 1962, by courtesy of Oliver and Boyd)

	Change in			
Selection for improvement in	Daily gain lb/day	lb feed/ lb carcass gain	Average back fat thickness (mm)	'Eye' muscle area (cm^2)
Daily gain	0·039	−0·06	−0·05	−0·34
lb feed/lb carcass gain	0·025	−0·12	−0·41	0·33
Average backfat thickness	0·001	−0·03	−1·84	0·46
Area of 'eye' muscle	−0·014	−0·014	−0·72	0·88

In addition to showing the response of a trait to selection for that trait, Table 4 shows the changes which may be expected to occur in other growth characters due to genetic correlations (Table 5*). Fortunately, the genetic correlations are such that as one important trait is improved there are usually improvements in other important traits. The main incompatibility is in the improvement in killing-out percentage, which is associated with decreases in carcass length and rate of gain and increased fatness. Also, increased rate of gain is associated with a small reduction in 'eye' muscle area.

The previous discussion centres on the quantitative effects of selection for single traits but in practice it is frequently desired to improve several facets of growth simultaneously. This would lead

* Genetic correlations for Large Whites only were used by Smith, King and Gilbert (1962) to obtain the figures in Table 4. Averages from several studies are shown in Table 5 as these can be 'treated with more confidence than estimates from only one analysis' (Smith and Ross, 1965).

to the rate of progress for any single one being reduced to a rate of about

$$\frac{1}{\sqrt{\text{number of traits selected}}}$$

if the traits were independent. Also, it is easy when selecting for several traits to over-emphasize some which are relatively un-important and to reduce progress by marked changes in emphasis from generation to generation. Hence, to achieve a correct and

Table 5

Average parameters from eight studies; phenotypic correlations above the diagonal, genetic correlations below the diagonal (Smith and Ross, 1965, by courtesy Oliver and Boyd)

	Daily gain	Feed conversion efficiency*	Killing out percentage	Backfat thickness	'Eye' muscle area	Carcass length
Daily gain		−0·73	−0·17	−0·07	−0·03	0·07
Feed conversion efficiency*	−0·76		−0·05	0·19	−0·16	−0·04
Killing-out percentage	−0·19	0·01		0·19	0·15	−0·19
Backfat thickness	−0·15	0·21	0·28		−0·13	−0·22
'Eye' muscle area	−0·11	−0·34	0·36	−0·28		−0·05
Carcass length	0·14	−0·08	−0·40	−0·30	−0·08	

* A higher figure is a poorer efficiency. Hence a negative sign implies an improvement

consistent balance a weighted index may be compiled, taking account of the relative economic importance of the traits, their variabilities, heritabilities and genetic correlations. Such an index, based on data from a performance tested boar and his sibs has been devised by the Pig Industry Development Authority (1966) for use in the new British test system, which emphasizes performance rather than progeny testing. With performance testing more boars can be evaluated and the facilities are used primarily by up to 70 élite herds. Thus a reasonably high selection pressure can be applied to improving a limited pig population, rather than dissipating im-provement potential by spreading it over a larger population.

After making assumptions on the intensity of selection practised, the extent of home performance testing of gilts and the frequency with which boars and gilts are replaced, the Pig Industry Development Authority (1966) made a 'conservative estimate' that an average élite herd might improve at the rate of 1 index point per month. Hence the average superiority corresponding to a 10 point superiority in index (Table 6) would be achieved in 10 months. The percentage changes in particular illustrate that, under practical

Table 6

Average superiority corresponding to a 10 point superiority in the P.I.D.A. selection index

(By courtesy of Pig Industry Development Authority, 1966)

Character	Improvement		
	Absolute*	Percentage†	Pence per pig*
Daily weight gain	0·0076 lb/day	0·5	1·4
Feed per lb weight gain	0·03 lb	0·9	14·6
Lean in side	0·31 per cent	[0·6]	14·4
Killing-out percentage	−0·008	−0·01	−0·4
Dressing percentage	0·016		0·6
Lean distribution	0·0	0·0	0·0
'Eye' muscle area	0·1 cm²	0·5	0·4

* P.I.D.A. figures
† Calculated using mean performance shown in Table 3

conditions, the rates of change of individual facets of growth which can be achieved through selection based on a national testing scheme are not high but the sum of their financial contributions is significant, at 31d per pig in 10 months, or 37d in 12 months; equivalent to about a 6 per cent increase in profit in 12 months, if this were initially 50s. per pig.

Finally, in this section, it may be noted that the measurements of growth which have been discussed are those conventionally used in pig production but with poultry greater refinements have been introduced. For example, in populations of turkeys the shape of the growth curve has been altered by selecting either for high 8-week weight, high 24-week weight, or for high 8-week weight without increase in 24-week weight (Abplanalp, Ogasawara and Asmundson,

1963). It would be interesting to know whether selection for high average growth rate in pigs to 200 lb (say to 50 per cent of mature weight) influences either mature weight, age at mature weight or the overall shape of growth curve.

NUTRITION AND GROWTH*

Pre-natal nutrition

In considering practical implications of nutritional influences upon growth, a valid question is whether these are important as early as the foetal stage. Moustgaard (1962) commented 'it is generally accepted that the foetus has first priority with regard to certain nutrients such as amino acids, calcium, phosphorus and B-vitamins, which are all, as far as is known, transferred actively against a concentration gradient'. Thus growth and development of the foetus should only be affected by severe deficiencies of nutrients in the sow's diet. Nutrients which have been implicated in the birth of dead, weak or deformed piglets include calcium, manganese, iodine, vitamin A, riboflavine and pantothenic acid (National Research Council, 1964). Such effects are financially important when they occur but they appear to be uncommon in practice because of the relative ease and cheapness with which mineral and vitamin supplements are added to the diet. Greater interest at present centres on protein and energy intakes, because it is by reducing these that the greatest economies in feeding the sow herd can be achieved.

In Danish experiments (Jespersen and Olsen, 1940), quoted by Moustgaard (1962), pregnant gilts and sows given 320 g *digestible* protein daily had one more pig per litter than those given 275 g daily but average birth weights were slightly lower. In a British experiment, however, birth weight and number born were not reduced by lowering daily protein intake to 280 g (Boaz, 1962). Also, in the United States of America, neither litter size, birth weight, nor survival of piglets was influenced by protein intake during pregnancy, where the comparison was between 136 and 545 g soya protein daily (Clawson, Richards, Matrone and Barrick, 1963). In an even more extreme example, a daily intake during pregnancy of 90 g protein from maize and soya bean, or sesame or gelatin, did

---

* Throughout this review dietary nutrient concentrations are expressed on an air dry basis, unless otherwise stated.

not affect the number of pigs farrowed, birth weight or survival to 2 weeks of age (Rippel, Rasmussen, Jensen, Norton and Becker, 1965).

Lack of effect on birth weight, however, does not necessarily mean a lack of change in physiological development or nutritional status. Low protein (176 g daily) intakes during pregnancy may reduce the nitrogen contents of piglets at birth (Elsley and MacPherson, 1964, Elsley, McDonald and MacPherson, 1966) and piglets from sows given 200–250 g protein daily had reduced muscle fibre diameters at birth. The effect on musculature did not persist, however, when the piglets were given a diet high in energy and protein (Livingston, 1962). In another experiment, no measurements were made on body composition at birth but a low protein level of 215 g cereal protein daily during gestation did not affect growth of the offspring to slaughter at 200 lb, or carcass conformation at that weight (Livingstone, MacPherson, Elsley, Lucas and Lodge, 1966).

Although the majority of the limited evidence suggests that low protein gestation diets have no effects of practical significance on the young, there are reports that they reduced survival rate (Stevenson and Ellis, 1957) and lowered resistance to an enteric infection (Hanson, 1958). These, and the Danish experiment (Jespersen and Olsen, 1940), introduce an element of doubt but it requires confirmation that the effects were specific to protein in the diet. Also, the pregnant sow's requirements for individual essential amino acids have yet to be determined and until that is done apparent inconsistencies in results of 'protein' experiments are to be expected. In some of the above experiments, however, where 200–250 g protein daily had little or no long term effect, the protein was from cereals and so would have a low proportion of lysine and methionine + cystine. It is not surprising, therefore, that with a gestation intake of about 350 g protein daily, derived mostly from fish meal *or* groundnut meal, the number and weights of piglets were not affected by protein source (Salmon-Legagneur, 1964).

It is well known that low feed intakes during pregnancy can reduce birth weight in cattle and sheep but until recently there was no evidence of a similar effect with pigs. However, recent trials have shown that sows given only 3 lb meal daily throughout pregnancy produced piglets about 0·2–0·5 lb lighter on average than sows given 5 or 6 lb meal daily (Clawson, Richards, Matrone and Barrick, 1963; Henson, Eason and Clawson, 1964; Lodge, Elsley and MacPherson, 1966; Elsley *et al.* 1967; O'Grady, 1967); increasing feed intake from 3 to 6 lb for the last 5½ weeks of pregnancy increased birth weight by almost 0·2 lb (Lodge, Elsley and MacPherson, 1966). There is evidence that there is little advantage to birth weight in raising the meal intake during gestation above 4·5 lb/day (Salmon-Legagneur and Jacquot, 1961; Salmon-Legagneur, 1962) but this requires confirmation.

From a practical standpoint it must be asked whether a reduction of up to 0·5 lb in birth weight significantly affects later productivity. Logically, low birth weight would increase susceptibility to cold, although this might be less than supposed due to the reduced heat losses which occur with huddling (Mount, 1960). Also, a piglet's large glycogen reserves at birth (Brooks, Fontenot, Vipperman, Thomas and Graham, 1964; Jones and Elsley, 1966) might be reduced disproportionately with reduction in birth weight, although there is no evidence on this point. The influence of birth weight upon post-natal growth is the subject of some controversy. There is no *prima facie* evidence from the experiments quoted above that reduced weight, within the limits due to treatments, affects survival or subsequent performance. Some statistical studies have shown a weak correlation between birth weight and 8 week weight, the former accounting for no more than 25 per cent of the variability in the latter, and have stressed that 8 week weight is affected mostly by milk and creep-feed intake (Lucas and Lodge, 1961, review; Lodge, McDonald and MacPherson, 1961). Wood (1964) concluded from two sets of data on Poland, China and Yorkshire pigs (Weaver and Bogart, 1943; Waldern and Wood, 1954) that an increase of 0·5 lb in average birth weight leads to an average increase of 4–5 lb in 8 week weight. Such an association between averages is not inconsistent with a low correlation coefficient, which may reflect inaccuracy of the relationship as it applies to individuals.

Variations in birth weight within litters are more likely to cause variations in growth than are differences between litters (Lodge and McDonald, 1959; Lodge, McDonald and MacPherson, 1961), perhaps because small pigs have less ability to compete for higher yielding teats (McBride, James and Wyeth, 1965), but the protein and energy values of gestation diets have not yet been shown to influence variability within litters.

Effects of birth weight upon survival may be of more practical importance than those on growth, and both the data of Lodge, McDonald and MacPherson (1961) and others quoted by these authors have shown that mortality was greatest in the lightest pigs, although this may have been confounded with the number of pigs in a litter. There may tend to be a critical birth weight below which the chance of survival is small (Pomeroy, 1960); in the experiments of Lodge *et al.* (1961) this was about 2 lb.

The general conclusions from this section are that a protein intake as low as or lower than 200 g daily, in a diet given at 3 lb daily, might affect the compositions and weights of piglets at birth, although there is little evidence that these effects are sufficiently large to influence subsequent biological or economic performances. However, such influences are difficult to detect in small experiments and reservations exist. An accurate evaluation in terms of input and output is not possible, but an increase from the amounts above to 4·5 lb daily of a 14·5 per cent crude protein diet, so giving about 300 g protein daily, would cost about 40–45s. per pregnancy; equivalent in value to less than one half of a weaner saved, or about 2 lb extra live weight per weaner, or to £400 extra cost per annum for a 100 sow herd.

Nutrition from birth to 8 weeks

Increase in weight gain in young pigs is very rapid, and during the first week approaches 10 per cent of body weight daily. A change to a slower relative and absolute rate of gain has been observed between the 15th and 20th day of life (Wood, 1964). The reason for this is not known but it may reflect reduction in energy intake from milk per unit body weight. Eight week weight is frequently limited by restricted total feed intake, including both milk and meal, but where intakes are high it may be as much as 60 lb.

Growth at this time is characterized by a very rapid increase in percentage body lipid from a low level at birth, and a smaller initial increase in percentage protein, with decreases in percentage ash and water. The chemical compositions of individual tissues also change, for example, water contents of both fat and muscle become lower. These effects are illustrated by data on changes in chemical composition of the whole body of a single, rapidly growing pig determined by an *in vivo* method (*Figure 2*; Wood, 1964; Groves and Wood, 1965; for more data *see also* Wood and Groves, 1965), by data on average chemical compositions of skinned carcasses, less head and feet (Table 7) and on the compositions of fat and muscle tissues (Table 8). There is at present no indication that these changes in the young pig are affected by sex (Wood and Groves, 1965; Whitelaw, Elsley, Jones and Boyne, 1966).

Energy of lipid deposited daily exceeds that of protein from birth, usually by a factor of at least two; weight of lipid deposited daily also usually exceeds that of protein from birth, although exceptions may be found (Wood, 1926) but because of the greater amount of water associated with protein than with lipid in body tissues (for example, Table 8), weight of fat tissue deposited daily may not exceed that of muscle until pigs weigh 100–200 lb.

In experiments on the protein requirements of young weaned pigs, growth rate and feed conversion efficiency have been used as criteria of adequacy. A review of results suggested that 52 and 43 g digestible crude protein per Mcal Digestible Energy are needed by

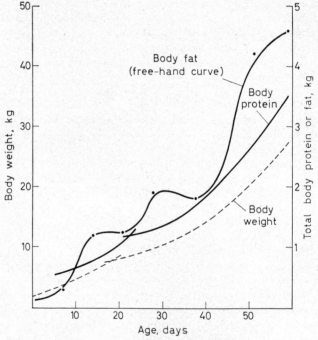

Figure 2. Change in body composition with age (From Wood, 1964)
By courtesy of E. and S. Livingstone

pigs of 10–20 and 20–45 lb. These are equivalent to about 23 per cent and 17 per cent crude protein in diets with 1·6 and 1·5 Mcal DE/lb, respectively (Lucas and Lodge, 1961).

Suckled pigs obtain their protein for the first 3 weeks of life from sow's milk, which contains about 25 per cent crude protein, rising to 33 per cent crude protein in the dry matter at week 8 (Lucas and Lodge, 1961). From 3 weeks onwards supplementary meal makes up an increasing proportion of the daily ration and it has been calculated that there should be about 15 per cent crude protein in the air-dry meal to meet the young pigs requirements. Feeding

Table 7

Average compositions of skinless carcasses of Hampshire pigs, with heads and feet removed (From Brooks, Fontenot, Vipperman, Thomas and Graham, 1964, by courtesy of The Editor, *J. Anim. Sci.*)

Age days	Live weight lb	Protein percentage	Lipid percentage	Water percentage	Ash percentage
Birth	2·4	11·7	1·4	74·5	5·1
6	4·4	14·2	9·7	71·1	3·5
12	7·5	15·0	15·3	66·2	3·3
18	8·5	15·5	13·7	66·2	3·2
24	17·3	15·0	17·3	63·5	3·2
30	26·6	13·8	18·4	63·8	3·1

Table 8

Age changes in the chemical compositions of fat and muscle tissue (From McMeekan, 1940, by courtesy of Cambridge University Press)

Age weeks	Outer backfat			m. longissimus dorsi		
	Lipid percentage	Tissue* percentage	Water percentage	Lipid percentage	Residue† percentage	Water percentage
Birth	6·2	8·9	84·9	1·9	16·6	81·5
4	75·4	5·1	19·5	4·3	19·9	75·7
8	76·7	6·6	16·7	4·7	19·0	76·2
16	84·4	6·4	9·1	3·4	20·9	75·7
20	85·9	3·9	10·2	4·0	21·6	74·4
28	92·4	3·6	4·9	5·6	22·6	71·8

* Presumably mostly protein
† Dry fat-free muscle

trials, which are few, give general support to this figure (Terrill, Meade, Nelson and Becker, 1952; Aumaître, Jouandet and Salmon-Legagneur, 1964). A recent British report (Whitelaw, Elsley, Jones and Boyne, 1966) describes a trial in which the supplementary feeds contained 13·8, 17·8 or 21·0 per cent crude protein, with calculated total lysine levels of 4·4, 4·8 and 4·9 per cent of the protein. Protein level did not affect growth rate and feed conversion efficiency but each increase was associated with a slight increase in the proportion of protein in body gain between 21 and 56 days of age. A rough calculation indicates that the 13·8 per cent protein supplementary feed, plus milk, was equivalent to an all-meal diet with about 17 per cent crude protein for pigs of 30–35 lb, which is the estimated requirement of weaned pigs.

Both the data of Whitelaw *et al.* (1966) and that for older pigs referred to later, indicate that an increase above 17 per cent dietary crude protein might lead to some increase in body protein in 8-week old animals, but even if this is so there is no evidence that it is of practical importance. Whitelaw *et al.* (1966) took pigs from their three treatments to slaughter at about 200 lb, and there were no differences in growth from weaning to 200 lb or in final carcass composition. Smith and Lucas (1957) gave weaned pigs of 25–50 lb diets with either 18, 21, 24 or 27 per cent crude protein, and these treatments also had no effect upon growth to 200 lb or upon carcass measurements.

Rate of weight gain in young pigs is influenced primarily by intake of dietary energy. Over recent years there has been considerable resurgence of interest in the effects of early rate of gain on later development and somewhat different results have occurred between pigs weaned at 8 weeks and those weaned earlier.

In an analysis of data from 1,345 pigs used in nutrition experiments between weaning and 205 lb live weight (Table 9), an increase in 8 week weight was associated with increased carcass length and, apparently, with change in distribution of carcass fat (confused by a significant interaction with sex), but the proportions of variation accounted for were very small, stressing the overriding importance of treatment from 8 weeks onwards. The effect on length, however, is interesting when considered in relation to an analysis of litter recording data from over 106,000 pigs, in which greater 8 week weight within the range from under 25 lb to over 45 lb was associated with an increased proportion in the premium bacon grades after slaughter at 200 lb (Table 10). This improvement was particularly in the AA+ grade, where an additional premium was given for length.

In survey data there is no indication of the cause of variation in early growth, whereas in experiments, variations may be imposed through nutrition. In one trial, access to supplementary feed was controlled so that weights at 8 weeks were 30, 40 or 50 lb, after which the pigs were kept to a single scale of feeding to slaughter at 200 lb. Neither growth rate nor feed conversion efficiency from 50–200 lb, nor any carcass measurement was affected (Boaz and Elsley, 1962). A similar result was obtained from a second trial at the same centre (Fowler and Boaz, 1963), but in other experiments

Table 9

Data from an analysis of the effects of 8-week weight on subsequent growth
(From Bowland, Braude and Rowell, 1965, by courtesy of Oliver and Boyd)

	Differences associated with a 10 lb increase in 8-week weight				
	Killing out percentage	Carcass length mm	Shoulder backfat mm	Loin backfat mm	lb carcass weight gain per day 8 week–205 lb
Between pigs within litters	−0·35	5·7*	−0·4	0·5	0·013*
Percentage of variation accounted for	0·3	0·8	0·1	0·1	0·3
Between litters	0·25	5·3*	−0·8*	0·6	0·006
Percentage of variation accounted for	1·0	3·4	1·6	0·6	0·5

* Statistically significant at or above the 5 per cent level

(Duckworth, 1965; Vanschoubroek, de Wilde and Van Spaendonck, 1965) high 8-week weights (51 cf. 44 lb and 38 cf. 23 lb, respectively) were associated with reductions in post weaning growth rate of 4 or 8 per cent, with increases in lb feed per lb gain of 4 or 7 per cent and with increases, not statistically significant, of about 0·5–1·5 mm in backfat thickness.

With early weaning, two questions arise. First, do early weaned and suckled pigs of the same weight at 8 weeks perform equally well from then on and, second, what are the effects of variations imposed upon the growth of early weaned pigs?

In one comparison (Lucas, Calder and Smith, 1959) both suckled pigs and others weaned at 10 days of age weighed 51 lb at 59 days, but from 51–205 lb the early weaned pigs grew more slowly (1·37 cf. 1·47 lb/day) and required more feed per lb gain (3·37 cf. 3·48 lb).

After slaughter at 200 lb their fat measurements were greater (+3 mm in fat over 'eye' muscle), indicating an increase of about 2·5 in percentage of carcass fat (A.R.C., 1967). In another similar comparison, however, the early weaned pigs contained more fat and less lean at 49 lb live weight but did not have significantly poorer growth or yield significantly fatter carcasses when slaughtered at 200 lb (Boaz and Elsley, 1962; Elsley, 1963a, 1963b).

Table 10

Percentage of pigs in the bacon carcass grades related to 8 week weight*
(By courtesy of the Pig Industry Development Authority, 1962)

Grade†	Average weight per pig at 8 weeks (lb)					
	under 25	25–29	30–34	35–39	40–44	45 and over
AA+	52·0	55·2	57·7	60·3	63·9	66·4
AA and A	13·2	13·8	14·7	13·9	13·1	13·4
B+	25·4	22·2	20·7	19·4	17·7	15·9
B	1·3	1·5	1·3	1·1	1·0	0·7
C	6·5	6·2	4·5	4·3	3·4	2·7
F	1·1	0·7	0·6	0·5	0·3	0·3
L	0·5	0·4	0·5	0·5	0·6	0·6

* For 106,377 pigs
† Increasing order of fatness from AA, A, B, C to F. L down graded because of insufficient fat;
+ indicates attainment of a standard for carcass length

The possibility that cessation of weight gain for 7–10 days which usually occurs after weaning at 10 days might be responsible for the difference in performance between suckled and early weaned pigs, when these occur, was not supported by observations on the effects of more prolonged growth checks imposed at that stage (Lucas, Livingstone and McDonald, 1964).

There is general agreement that reduction in growth rate of early weaned pigs before about 50 lb live weight leads to improved later performance (Lucas, Calder and Smith, 1959; Boaz and Elsley, 1962; Elsley, 1963a, 1963b; Nielsen, 1964). Tables 11 and 12 give examples from two experiments. In one (Table 11) rate of gain in pigs weaned at 10 days was controlled by adjusting the feed allowance, so that the pigs weighed 51 lb at 60, 73 or 90 days. From then to slaughter at 200 lb all pigs were kept to a single high scale of intake. It may be calculated from the two extreme treatments that for each extra day taken to reach 51 lb, there were subsequent improvements of about 0·005 lb in daily gain, 0·006 lb in feed required per lb gain, 0·07 mm in thickness of fat over the 'eye' muscle and 0·06 cm² in area of eye muscle. These changes are in remarkable agreement with regressions (Table 12) calculated by Nielsen (1964). In his experiment, groups of pigs fed to reach 44 lb at either 59, 68,

Table 11

Effect of variation in early rate of growth of pigs weaned at 10 days on subsequent performance (From Lucas, Calder and Smith, 1959, by courtesy of Cambridge University Press)

	Age in days at 51 lb *live weight*		
	60	73	90
Daily weight gain (51–205 lb) (lb)	1·37	1·43	1·47
Feed/lb gain (51–205 lb) (lb)	3·74	3·63	3·55
Killing-out percentage	75·5	75·1	75·8
Shoulder fat (mm)	47	45	44
Fat over 'eye' muscle (mm)	24	21	22
Area of 'eye' muscle (cm²)	25·4	27·1	27·3

Table 12

Regression of various characters on age at 44 lb live weight (From Nielsen, 1964, by courtesy of Oliver and Boyd)

Regression on days at 44 lb	Level of feeding 44–198 lb	
	Low	*High*
Daily weight gain (44–198 lb) (lb)	0·0018	0·0057
Daily feed intake (44–198 lb) (SFU)	—	0·006
(SFU)/kg gain*	−0·004	−0·009
Fat over 'eye' muscle (mm)	−0·04	−0·004
Area of 'eye' muscle (cm²)	0·07	0·06

* As one SFU is the equivalent in feed value of 1 kg barley, SFU/kg gain is similar to lb feed/lb gain

80 or 91 days were each divided and given, from 44 to 198 lb, separated milk and a protein supplement, with barley either to a fixed scale (low) or to appetite (high). With high scale feeding compensatory growth from 44–198 lb was more marked and, perhaps in consequence, the effects of early growth rate on the final carcasses were less.

A summary of the practical importance of variations in energy intake and rate of gain in young pigs is made difficult by apparent contradictions in the evidence but may be attempted as follows:

For pigs weaned at 8 weeks, survey data suggest a positive association between 8-week weight and carcass quality at 200 lb, although it is possible that the improvement is mostly in carcass length. Also, the regression analysis of Bowland *et al.* (1965) indicates that 8-week weight only accounts for small proportions of the variations in carcass measurements. A weakness of surveys is that they do not establish cause and effect. For example, farmers having the skill to produce heavy weaners might also have the skill to promote good carcass quality, or disease might reduce both weaning weight and carcass quality by affecting nitrogen balance. Variations in early growth rate imposed by controlling feed intake either have had no effect on growth to 200 lb, or high early rate of growth has been followed by poorer performance. Perhaps it is fortunate that early rate of growth in suckled pigs is difficult to control closely in practice, as it depends upon the sows milk yield and the intake of creep feed, which is usually given to appetite. There is at present no evidence to support a change in this practice. *With pigs weaned at 10–21 days* reduction in rate of gain before 40 or 50 lb due to reduced feed allowance has improved later performance. Tables 11 and 12 suggest that to improve feed conversion efficiency from 50–200 lb by 0·1 (≡48*d.* per pig; Table 1), weight gain by 0·07 lb/day (≡13*d.*) and percentage carcass lean by, say, 0·5 (≡27*d.*), growth in young pigs would have to be reduced so that they would stay in their pens for an extra 17 days. The practical problem is then to assess the capital and labour charges for these 17 days against a potential extra profit (if lean were paid for accurately) of about 7*s.* per pig reared to 200 lb. Food costs up to 8 weeks are not affected by level of intake within wide limits, but they would be expected to increase if restriction were severe enough to cause a delay of 30 days in reaching 44 lb live weight (Lucas, Calder and Smith, 1959; Nielsen, 1964).

Nutrition from 50 lb to slaughter

Growth from 50 lb to slaughter is characterized by increase in the proportion of fatty tissue in the body, with smaller declines in the proportions of muscle, bone, blood and intestines. With the pigs described in Table 13 the percentage of fat increased by 16 units between 44 and 264 lb live weight, but the percentage of meat decreased by only 3 units, a ratio of 5:1. Emphasis on fat content has been used in arguments against high slaughter weights; conversely those who advocate high slaughter weights can properly point out that the extra fat is not deposited solely at the expense of lean, hence fat is not necessarily a disadvantage in itself when butchering involves fat trimming, although it would clearly be a disadvantage if it were shown to be financially uneconomical to produce.

Table 13

Composition of Danish Landrace pigs
(From Clausen, 1953, by courtesy of Queen's University, Belfast)

Live weight lb	Net weight* lb	Proportion of net weight					
		Fat† per-centage	Meat per-centage	Bones per-centage	Blood per-centage	Gristle and hoofs per-centage	Total offal per-centage
44	39	19	45	15	5·5	0·5	15
110	98	25	45	12	4·6	0·4	13
198	180	31	43	10	4·6	0·4	11
264	241	35	42	9	3·7	0·3	10

* The pig without contents of stomach and intestines, and other losses such as evaporation
† Including leaf and intestinal fat and skin

Although the weight of fat tissue in the carcass may not exceed that of lean until live weight is over 260 lb (for example, Cuthbertson and Pomeroy, 1962; Clausen, 1953), the proportion of fatty tissue in gain may exceed that of lean when pigs are 150–200 lb, or even lighter (A.R.C., 1967).

The pattern of growth and development varies between pigs and with environmental influences. The main nutritional effects, as with younger pigs, are through protein and energy intakes. These were reviewed recently in detail by the A.R.C. (1967) who agreed with others, mostly economists, that for practical purposes the concept of single figure 'requirements' are of less value than knowledge of the quantitative changes in growth with variations in level of protein and energy supply. But so far there are insufficient data to estimate the latter accurately.

The levels of dietary protein below which growth rate and feed conversion efficiency are reduced are about 16–17·5 per cent for pigs of 40–110 lb and 13–14·5 per cent for pigs of 110–210 lb.   Corresponding figures for L-lysine, which is usually the first limiting amino acid in British pig diets, are about 0·8 and 0·63 per cent respectively (A.R.C., 1967).   There is evidence that optimal levels of protein and amino acids are related to dietary energy, but the A.R.C. (1967) considered that a precise relationship cannot yet be estimated.   The effects on carcass composition of reducing protein intake have varied but there is frequently reduced lean content. With increases in protein level to above 17·5–20 per cent, experiments are more consistent in showing increased lean.   Some examples are shown in *Figure 3*.   Although in each case the diets were given from weaning to slaughter, other experiments could be quoted (for example Chamberlain and Lucas, 1966) where lean content rose with dietary protein when this was given to the more conventional pattern of a higher dietary concentration up to 110 lb and a lower concentration from then to slaughter.   In this context, some nutritionists are questioning the virtue of reducing dietary protein concentration for the heavier pigs (for example Lewis, 1965) and more data are required on the effects upon growth and development of changes in the distribution of protein supply from birth to slaughter.

No significance should be attached to comparisons of lean content between experiments in *Figure 3*, as dissection procedures were not necessarily standard, but the within-experiment comparison of pigs of different slaughter weights (Blair, Raeburn and Ward, 1966) illustrates the decline in percentage lean with increased weight.

Within experiments regressions may be calculated to show the average quantitative change in lean output with change in protein or amino acid input.   From these the cost of extra lean may be obtained.   For example, Chamberlain (1966), from his work on diets of separated milk and supplemented barley, calculated that 1 lb extra carcass lean cost either 3s. or 3s. 4d. to produce, depending upon the particular feeding regime.   This type of costing would be made more complex if there were also changes in rate of weight gain and feed conversion efficiency.   It also raises again the problem of whether the methods of assessing extra lean in meat factories are sufficiently accurate to enable additional payment to be made.   It is difficult to advise a farmer to increase cash input if there is not a reasonable chance of increased financial return.

Changes in the energy, or daily feed, intakes of pigs between 50 lb and slaughter have long been known to have marked effects upon rate of weight gain and body development.   With reduction in intake, fat deposition is reduced and rate of weight gain declines, but the evidence on the important practical aspect of feed conversion efficiency is conflicting (Lucas and Calder, 1956; Lucas, McDonald and Calder, 1960; A.R.C., 1967; Vanschoubroek, de Wilde and Lampo, 1967).

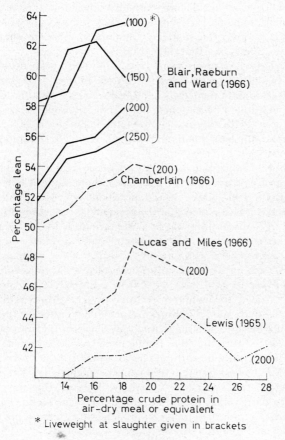

Figure 3. Increase in carcass lean with increase in dietary protein intake

The A.R.C. (1967) used data from progeny test pigs fed twice daily 'to appetite' as a standard 'base line' of energy input, and suggested that the error would not be large if it were accepted for the present that feed conversion efficiency is not affected by varying intakes within the range $+5$ per cent to -15 per cent of this 'base line'. A known percentage increase or decrease in daily energy intake would then lead to the same percentage change in rate of weight gain.

But Vanschoubroek, de Wilde and Lampo (1967), from a statistical analysis of published data, derived equations indicating improved feed conversion efficiency with feed restriction; the maximum improvement being 6·6 per cent when daily intake was reduced by 25 per cent. From this it follows that the percentage change in growth rate is less than that of feed intake.

It is perhaps dangerous to make assumptions in an attempt to reconcile two views, nevertheless this is done in *Figure 4*. It may be calculated from data reported by the A.R.C. (1967) that in terms of dietary Digestible Energy, their 'base line' is about 2·2 × Maintenance for a 44 lb pig and 3·2 × Maintenance for a 200 lb pig. Thus, overall, the base line is, on average, about 2·5 × Maintenance. Vanschoubroek *et al.* (1967) indicate that they consider *ad lib.* feeding to be 4 × Maintenance, but this appears high. If intakes of self-fed pigs are on average 117 per cent of those of the A.R.C. 'base line' (A.R.C., 1967), they would be about the 3 × Maintenance level. Hence it is assumed that the percentage reductions in intake described by Vanschoubroek *et al.* (1967) are from 3 × Maintenance. Even if this is incorrect, the two sets of data are now considered using the same general assumptions. For *Figure 4* it is also assumed that a pig with an average daily intake of 2·5 × Maintenance eats 3·3 lb feed per lb gain from 44–200 lb. Lines relating intake to efficiency, drawn through this point, represent the hypotheses of the A.R.C. and Vanschoubroek *et al.* They indicate that the A.R.C. may be correct in suggesting that the error may not be large (it would be 1–2 per cent) in assuming linearity over a limited range, and that it may only be with intakes at, say, 15–20 per cent higher than the A.R.C. base line that feed conversion efficiency deteriorates significantly. It is reasonable to assume, however, that the general response to daily intake should be curvilinear, as lb feed per lb gain must approach infinity as intake approaches maintenance, as shown by the sketched line in *Figure 4*.

Clearly the validity of the argument in the preceding paragraph rests partly upon the choice of the levels of feed intake against which percentage changes are considered.

Level of feed intake also influences fatness. Data from the A.R.C. (1967) and Vanschoubroek *et al.* (1967) relating reductions in level of intake to decrease in backfat thickness are shown in *Figure 5*, again using 2·5 × Maintenance and 3·0 × Maintenance as the points of origin. The regressions are different, with the A.R.C. suggesting the greater rate of change in fatness, but some of this difference is associated with the effect, reported by Vanschoebroek *et al.* and indicated by the broken line in *Figure 5*, that as feed restriction becomes greater, the decrease in backfat thickness becomes relatively smaller. Also, some may be due to the particular fat measurements recorded. For example, the A.R.C. suggest that

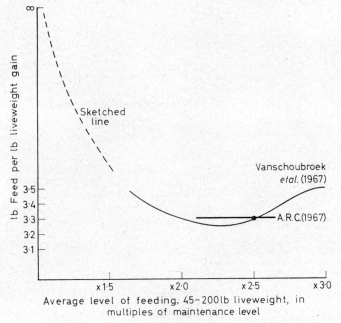

Figure 4. Effect of level of feeding on feed conversion efficiency

thickness of shoulder fat declines more rapidly (in absolute terms but not necessarily proportionately) than mid backfat with reduced energy intake. The A.R.C. (1967) suggest the approximation that a reduction of 10 per cent in daily energy intake from 44–200 lb would reduce carcass fat by about 1·2 per cent. The increase in percentage lean would be less than this. Unfortunately, however, in most experiments relating stated energy or feed intakes to growth and development, linear measurements have been the criteria of carcass composition.

491

The Econometric Approach

Although quantitative relationships between dietary proteins and amino acids on the one hand and growth and development on the other have yet to be established with certainty, it is clear that such data would be of great value as a basis for recommendations on nutrition and growth to give maximum financial returns. As this is a matter of great practical importance a start has been made, using such data as are available.

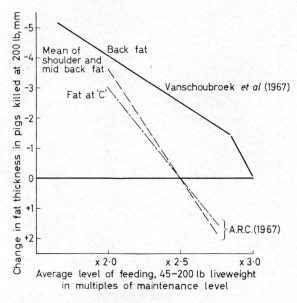

Figure 5. Effect of level of feeding on carcass fat measurements

There are several complementary lines of approach to this problem, but they fall within two general categories. These are (1) the estimation of inputs of nutrients which give a known output in terms of growth and (2), knowing the nutrient inputs, to devise the ration which will supply them at least cost.

Dent (1964, 1966) used the data of Robinson, Morgan and Lewis (1964) and Robinson and Lewis (1964) to calculate those combinations of dietary energy and protein which will each result in weight gains of 1·0 or 1·25 lb/day for pigs of 60–110 lb and 1·3 or 1·5 lb/day from 110–200 lb live weight. As an example, one curve (Dent,

1966) is shown in *Figure 6*. This indicates that a growth rate of
1·3 lb daily can be achieved by giving rations ranging from a low
level of protein combined with a high level of energy, to a higher
level of protein combined with a lower level of energy. Dent points
out, however, that there are fundamental inadequacies in the data
because, on the input side, amino acid balance, particularly for

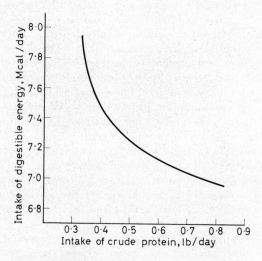

*Figure 6. Iso-output curve for 1·3 lb live weight gain daily
in a finishing pig (From Dent, 1966)*

By courtesy of Pig. Ind. Dev. Authy

lysine, would be a valuable addition, and a 'complete' model could
include many variables, such as physical environment. Although
production surfaces could then no longer be presented in three-
dimensional diagram form, they could be described algebraically.
On the output side, daily gain is an inadequate description as it
takes no account of changes in body development, which can occur
in pigs gaining weight at the same rate. Nevertheless, in future it
may be possible to express output in a term meaningful in practice;
perhaps an index of the type developed by geneticists could be used.

Having estimated iso-output curves Dent (1964), using a linear
programming technique and introducing certain constraints such as
amino acid content of protein, fibre level and maxima for various
feeds, calculated least cost rations to give chosen rates of gain (*see also*
Dent and English, 1966). Essentially, this involved computing the
least cost solution at each combination of protein and energy along

the chosen iso-output curves and then accepting the cheapest of these as the overall least cost rations. These rations were tested (Dent and English, 1966) and although there were variations both in rate of weight gain and in carcass composition, performances were close to those predicted.

Godsell (1966) also used linear programming to devise least cost rations, but instead of being based upon iso-output curves depending on two nutritional variables, these were based upon more conventional estimates of 'minimum requirements for nutrients', with more nutritional variables and constraints. The aim was to compute diets, one of which would be the least cost solution, giving equal growth rates and carcass quality, with absolute growth rates and carcass quality depending upon the level of feed allowance. However, if protein and amino acid levels are stated only as minima, it appears possible that one computed diet might have higher levels than another and give carcasses with greater lean content. It may also be argued that growth rate and carcass composition cannot be predicted precisely as these vary with genetic make up, disease and environment as well as with nutrition. Godsell's (1966) diets, in common with those of Dent and English (1966), are not conventional in their proportions of feedingstuffs and are thus of particular interest in applied pig feeding.

Several assumptions are made with linear programming which are not necessarily true, or which at present disregard biological variability (Dent, 1964, 1966; Godsell, 1966). Some are summarized by the statement (Dent and English, 1966) that 'in using linear programming to meet any set of nutritional requirements it is assumed that no matter how the diet is formulated it will produce the same amount and quality of growth'. It follows that any comparison of different rations which should give equivalent growth is relevant to the evaluation of linear programming. This is essentially a practical test of nutritional theory.

Computed diets based upon tables of nutrient requirements were found in Canadian experiments (Bell, 1961; Bowland, 1962) to give poorer performance than when 15 per cent margins were added, but it was not possible to determine which estimate of requirement was defective in the original tables. Many tables of requirements may indeed be insufficiently precise to allow prediction of performance. For example, lysine levels are sometimes quoted but there are differences in availability of lysine to the pig and, for precision, inputs expressed as available lysine would be better. When different sources of protein were given to pigs from weaning to 200 lb, so that the diets contained the same concentrations of available lysine but different concentrations of total lysine and crude protein, growth rates, feed conversion efficiencies and linear carcass measurements were the same (Jones, Livingstone and Cadenhead, 1965). In another test of nutritional equivalence (Chamberlain and Clent, 1967) a comparison was made between six diets compounded from a variety of ingredients, each diet having the same ratios of protein and lysine to energy. When given on an equi-caloric basis mean growth rate ranged from 1·18 to 1·31 lb/day (100:111) but there were only small differences in energy intake per lb gain and in percentage lean in the back joint.

Clearly the econometric approach will contribute to the practical application of knowledge of nutrition, environment and growth. But more basic data on quantitative input-output relationships and on variability have to be obtained to allow its full exploitation. It is tempting to decry the present inadequacy of these data but nutritional science is itself young; for example, input-output data on available lysine could hardly have preceded the comparatively recent increase in knowledge of amino acid availability as an important determinant of non-ruminant growth. Even now, the essential techniques of assessing availability have not been fully developed.

REFERENCES

Abplanalp, H., Ogasawara, F. X. and Asmundson, V. S. (1963). *Br. Poult. Sci.* **4**, 71

A.R.C. (Agricultural Research Council) (1967). *The Nutrient Requirements of Farm Livestock.* Part 3, Pigs. London

Aumaître, A., Jouandet, C. and Salmon-Legagneur, E. (1964). *Annls Zootech.* **13**, 241

Bell, J. M. (1961). *J. Anim. Sci.* **20**, 925 (abstr.)

Blair, R., Raeburn, J. R. and Ward, J. (1966). *Proc. Pig Industry Development Authority Second Conf. Agric. Res. Workers and Agric. Economists.* p. 34. London; P.I.D.A.

Boaz, T. G. (1962). *Vet. Rec.* **74**, 1482

— and Elsley, F. W. H. (1962). *Anim. Prod.* **4**, 13

Bowland, J. P. (1962). *Can. J. Anim. Sci.* **42**, 191

— Braude, R. and Rowell, J. G. (1965). *Anim. Prod.* **7**, 389

Brooks, C. C., Fontenot, J. P., Vipperman, P. E., Thomas, H. R. and Graham, P. P. (1964). *J. Anim. Sci.* **23**, 1022

Buck, S. F., Harrington, G. and Johnson, R. F. (1962). *Anim. Prod.* **4**, 25

Chamberlain, A. G. (1966). Unpublished data, University College of North Wales, Bangor

Chamberlain, A. G. and Clent, E. G. (1967). *Anim. Prod.* **9**, 273
— and Lucas, I. A. M. (1966). *Anim. Prod.* **8**, 361 (abstr.)
Clausen, H. (1953). *George Scott Watson Memorial Lecture.* Belfast; Queens University
— and Nørtoft Thomsen, R. (1967). *327 Beretn. Forsøgslab.*
Clawson, A. J., Richards, H. L., Matrone, G. and Barrick, E. R. (1963). *J. Anim. Sci.* **22**, 662
Cuthbertson, A. and Pease, A. H. R. (1965). *Anim. Prod.* **7**, 284 (abstr.)
— and Pomeroy, R. W. (1962). *J. agric. Sci., Camb.* **59**, 251
Dent, J. B. (1964). *J. agric. Econ.* **16**, 68
— (1966). *Proc. Pig Industry Development Authority Second Ann. Conf. Agric. Res. Workers and Agric. Economists.* p. 58. London; P.I.D.A.
— and English, P. R. (1966). *Anim. Prod.* **8**, 213
Duckworth, J. E. (1965). *Anim. Prod.* **7**, 165
Elsley, F. W. H. (1963a). *J. agric. Sci., Camb.* **61**, 233
— (1963b). *J. agric. Sci., Camb.* **61**, 243
Elsley, F. W. H., Bannerman, Mary, Bathurst, E. V. J., Bracewell, A. G., Cunningham, J. M. M., Dodsworth, T. L., England, G. T., Forbes, T. J. and Laird, R. (1967), *Anim. Prod.* **9**, 270
— McDonald, I. and MacPherson, R. M. (1966). *Anim. Prod.* **8**, 353 (abstr.)
— and MacPherson, R. M. (1964). *Anim. Prod.* **6**, 295 (abstr.)
Fowler, V. R. and Boaz, T. G. (1963). *Anim. Prod.* **5**, 222 (abstr.)
Godsell, T. E. (1966). *Proc. Pig Industry Development Authority Second Ann. Conf. Agric. Res. Workers and Agric. Economists.* p. 25. London; P.I.D.A.
Groves, T. D. D. and Wood, A. J. (1965). *Can. J. Anim. Sci.* **45**, 14
Hanson, L. E. (1958). *Proc. Cornell. Nutr. Conf. Feed Manufacturers,* p. 10
Henson, D. B., Eason, D. W. and Clawson, A. J. (1964). *J. Anim. Sci.* **23**, 878 (abstr.)
Jespersen, J. and Olsen, H. M. (1940). *192 Beretn. K. Vet.-og Landbohøjsk. Lab. landøkon Forsøg.*
Jones, A. S. and Elsley, F. W. H. (1966). *Rowett Research Institute Rep.* **22**, 97
— Livingstone, R. M. and Cadenhead, A. (1965). *Anim. Prod.* **7**, 286 (abstr.)
Jonsson, P. (1963). *Z. Tierzücht. ZüchtBiol.* **78**, 205
Lewis, D. (1965). *Feedstuffs, Minneapolis* **37** (21), 18
Livingston, D. M. S. (1962). *Anim. Prod.* **4**, 296 (abstr.)
Livingstone, R. M., MacPherson, R. M., Elsley, F. W. H., Lucas, I. A. M. and Lodge, G. A. (1966). *Anim. Prod.* **8**, 337
Lodge, G. A., Elsley, F. W. H. and MacPherson, R. M. (1966). *Anim. Prod.* **8**, 29
— and McDonald, I. (1959). *Anim. Prod.* **1**, 139
— McDonald, I. and MacPherson, R. M. (1961). *Anim. Prod.* **3**, 261
Lucas, I. A. M. and Calder, A. F. C. (1956). *J. agric. Sci., Camb.* **47**, 287
— — and Smith, H. (1959). *J. agric. Sci., Camb.* **53**, 136
— Livingstone, R. M. and McDonald, I. (1964). *Proc. 6th Int. Congr. Nutr.,* Edinburgh, p. 568 (abstr.)

Lucas, I. A. M. and Lodge, G. A. (1961). *Tech. Comm. 22., Commonw. Bur. Anim. Nutr.*, Aberdeen
— McDonald, I. and Calder, A. F. C. (1960). *J. agric. Sci., Camb.* **54**, 81
— and Miles, K. (1966). Unpublished data, University College of North Wales, Bangor
McBride, G., James, J. W. and Wyeth, G. S. F. (1965). *Anim. Prod.* **7**, 67
McMeekan, C. P. (1940). *J. agric. Sci., Camb.* **30**, 276
Mount, L. E. (1960). *J. agric. Sci., Camb.* **55**, 101
Moustgaard, J. (1962). In: *Nutrition of Pigs and Poultry.* (Ed. by J. T. Morgan and D. Lewis.) p. 189. London; Butterworths
National Research Council (1964). *Nutrient Requirements of Swine,* 5th ed. Washington; *National Academy of Sciences,* Publ. 1192
Nielsen, H. E. (1964). *Anim. Prod.* **6**, 301
O'Grady, J. F. (1963). *Ir. J. agric. Res.* **6**, 57
Pease, A. H. R. and Smith, C. (1965). *Anim. Prod.* **7**, 273
Pig Industry Development Authority (1962). *Litter recording: a report on two six-month periods, April–Sept. 1961–March 1962.* London
— (1966). Mimeo Document DA 188. London
Pomeroy, R. W. (1960). *J. agric. Sci., Camb.* **54**, 31
Rippel, R. H., Rasmussen, O. G., Jensen, A. H., Norton, H. W. and Becker, D. E. (1965). *J. Anim. Sci.* **24**, 203
Robinson, D. W. and Lewis, D. (1964). *J. agric. Sci., Camb.* **63**, 185
— Morgan, J. T. and Lewis, D. (1964). *J. agric. Sci., Camb.* **62**, 369
Salmon-Legagneur, E. (1962). *Annls. Zootech.* **11**, 173
— (1964). *Annls. Zootech.* **13**, 51
— and Jacquot, R. (1961). *C.r. hebd. Séanc. Acad. Sci., Paris* **253**, 1497
Smith, C. (1965). *Anim. Prod.* **7**, 133
— King, J. W. B. and Gilbert, N. (1962). *Anim. Prod.* **4**, 128
— and Ross, G. J. S. (1965). *Anim. Prod.* **7**, 291
Smith, H. and Lucas, I. A. M. (1957). *J. agric. Sci., Camb.* **49**, 409
Stevenson, J. W. and Ellis, N. R. (1957). *J. Anim. Sci.* **16**, 877
Terrill, S. W., Meade, R. J., Nelson, T. S. and Becker, D. E. (1952). *J. Anim. Sci.* **11**, 777 (abstr.)
Vanschoubroek, F., de Wilde, R. and Lampo, Ph. (1967). *Anim. Prod.* **9**, 67
Vanschoubroek, F. X., de Wilde, R. O. and Van Spaendonck, R. L. (1965). *Anim. Prod.* **7**, 111
Waldern, D. E. and Wood, A. J. (1954). *Proc. Can. Soc. Anim. Prod.,* Western Section, June 1952, p. 96
Weaver, L. A. and Bogart, R. (1943). *Bull. Mo. agric. Exp. Stn.* 461
Whitelaw, A. W. W., Elsley, F. W. H., Jones, A. S. and Boyne, A. W. (1966). *J. agric. Sci., Camb.* **66**, 203
Wood, A. J. (1964). *Proc. 6th Int. Congr. Nutr.,* Edinburgh, p. 89
— and Groves, T. D. D. (1965). *Can. J. Anim. Sci.* **45**, 8
Wood, T. B. (1926). *J. agric. Sci., Camb.* **16**, 425

DISCUSSION

DR. PRESCOTT (Newcastle upon Tyne), on the question of effect of nutrition on muscle/bone ratio, commented that Missouri data showed that 'high-plane' animals had a much higher muscle/bone ratio than 'medium' and 'low-plane' groups; in comparison between the 1-year fat-free weight of the HP group, the 2-year fat-free weight of the MP group and 3-year fat-free weight of the LP group, the LP group had the highest fat-free weight but considerably lower muscle/bone ratio. PROFESSOR BERG replied that a certain fat level appeared to be necessary for maximum muscle/bone ratio. DR. TAYLOR (Hurley) commented that muscle/bone ratio was impaired by low-plane feeding but asked whether a high weight-for-age might worsen efficiency of feed conversion by increasing the maintenance requirement. PROFESSOR BERG thought that the important measure was feed conversion into muscle rather than live weight and that the high impetus of muscle relative to bone would improve the efficiency in these terms.

DR. ARMSTRONG (Newcastle upon Tyne) commented that if an animal was on a calorie intake inadequate for full growth potential, but receiving adequate protein and calcium, then muscle/bone ratio would be depressed. Consideration should also be given to the effect of intake of mineral elements on muscle/bone ratio, in that calcium intake, for example, could markedly affect bone density. PROFESSOR BERG agreed with this and observed that protein intake should also be considered in view of the findings that boars responded to higher protein levels than castrates and wether lambs to higher protein levels than ewe lambs. DR. HARTE (Dublin) sought further comment on the ages of cattle from which data on nutritional effects on muscle/bone ratio had been obtained, to which PROFESSOR BERG replied that Gunter's were about 6 months old for the obtaining of 'W' values and, overall, were much younger than Callow's; there was a point at which low-plane nutrition would inhibit muscle relative to bone but information was needed on the location of this point.

DR. EVERITT (Ruakura) stressed the importance of time of application of nutritional treatment, particularly whether or not it occurred before 6 months of age. PROFESSOR BERG agreed and observed that most of the data presented had been derived from animals older than this when fat deposition was assuming greater importance.

DR. INGRAM (Bristol) stated that the authorship of the data comparing Friesian and Hereford breeds presented by Professor Berg as 'anonymous' was, in fact, Williams and Pomeroy and the experiments had been designed by a committee.

DR. ARMSTRONG referred to the graph relating efficiency of feed conversion to rate of carcass gain and asked whether the more the food consumed by the lambs the greater was the efficiency for both singles and twins. DR. SPEDDING replied that for the twins it was quite clear that at the same growth rate the higher efficiency was associated with greater food consumption by the lambs. For singles the situation was less clear but again for equal growth rates high direct food consumption by the lambs was associated with high efficiency; in the case where the proportion of food consumed by the lambs was increased but there was a fall in growth rate, then efficiency was not improved. DR. ARMSTRONG then asked whether this indicated a desirability of rapid rumen development and DR. SPEDDING replied that it did but the question of cost had to be considered; if the effect was to reduce growth rate, as it often had been, then the calculation was not simple. DR. TAYLOR (Edinburgh) asked whether it could be equally efficient to have a 22 kg ewe producing one lamb as a much larger ewe producing two. DR. SPEDDING replied that we exploit smaller sheep more fully than larger sheep by taking them closer to their mature size; maximum efficiency lay in having the smallest possible ewe for the desired litter size but by using a larger sire to obtain the maximum sized product from it.

DR. MITCHELL (Reading) asked whether there was any experimental evidence to support the suggestion that the most efficient way to feed the pregnant ewe was to achieve the same weight after lambing as at conception. DR. SPEDDING replied that there was no reason to suppose that fluctuations in ewe weight would improve efficiency; maintenance of constant weight should be most efficient nutritionally, although the situation would be different if the intention was to exploit an environment with an irregular food supply. During pregnancy itself there were two main factors to consider; first, a satisfactory lamb at birth and, second, the subsequent lactation. If lambs were to be removed at birth, as an adjunct of increased litter size, then lactation did not matter; in present circumstances a significant reduction in ewe weight during pregnancy might result in a lowered milk yield. It was quite possible to feed unnecessarily heavily during pregnancy a ewe with a single lamb.

DR. JONES (Aberdeen) asked at what stage the number of foetuses could be detected and whether this was when the varying feed levels were introduced. DR. SPEDDING replied that x-ray examination was used at 90–100 days of gestation, which was in time to vary planes of feeding at least for numbers up to triplets. MR. TREACHER (Hurley) commented here that although in the earlier experiments feed allowances had been based on weight gains and losses, more recently the allowances had been based on FFA levels in blood; considerable gains were needed in late pregnancy to prevent utilization of body reserves, with twins it was about 24 per cent of the body weight at 15 weeks.

DR. EVERITT commented on Finnish Landrace sheep with litters of 5 or 6 and the finding that when crossed with sires of other breeds abortion tended to occur in late pregnancy, and enquired whether this was also a feature of the pure breed. DR. SPEDDING thought that it was very difficult to obtain valid data on this as the general nutrition of Finnish sheep was very poor and associated with an average mean litter size of about 2·4. DR. JONES asked whether plasma FFA levels would be the same if ewes were in different states of body condition at the point when they started to utilize body reserves, to which MR. TREACHER replied that data on this were lacking.

MR. LITTLE (B.O.C.M. Ltd.) wondered whether some of the discrepancies between level of feeding and carcass quality in pigs could be due to inaccurate estimates of nutrient input through lack of precise information on dietary composition. PROFESSOR LUCAS thought that estimates of energy in terms of DE or ME were sufficiently accurate for all practical purposes to give the desired input–output relationships, but with protein it was much less certain. It was known that there were responses to protein in terms of growth and carcass lean but there was a technical failure in assessment of available amino acid levels; input data which would allow prediction of performance was therefore lacking.

DR. JONES asked whether some of the differences which had been found in the effect of 56-day weight on subsequent performance might be due to genetic differences between animals, to which PROFESSOR LUCAS replied that there was every indication that this must be so.

DR. BRAUDE (Reading) commented that the feeding regime of the suckled pigs in Nielsen's experiment was substandard and this might have accounted for the subsequent effects of low weaning weight. PROFESSOR LUCAS thought, however, that even if lean content at 8 weeks was slightly reduced through low quality creep feed, the Rowett evidence indicated that this would be compensated for subsequently.

LIST OF MEMBERS

ADAMSON, A. H.	N.A.A.S., Government Buildings, Lawnswood, Leeds 8
ADLER, Dr. J.	The Hebrew University, Jerusalem, Israel
AMIR, Dr. S.	Volcani Institute of Agricultural Research, Rehovoth, Israel
ANDREWS, Dr. R. J.	Research and Advisory Services (Agric.) Ltd., Poole, Dorset
ANNISON, Dr. E. F.	Unilever Research Laboratory, Colworth House, Bedford
ARMSTRONG, Dr. D. G.	Dept. of Agricultural Biochemistry, University of Newcastle upon Tyne
AUMAITRE, A. L.	C.N.R.Z., Jouy-en-Josas, France
BARKES, J. N.	Dept. of Agriculture, University of Newcastle upon Tyne
BEAUMONT, W. H.	Cooper Nutrition Products Ltd., Berkhamsted Hill, Herts.
BERG, Prof. R. T.	University of Alberta, Edmonton, Canada
BERGSTRÖM, Dr. P. L.	'Schoonoord' Research Institute for Animal Husbandry, Zeist, The Netherlands
BICHARD, Dr. M.	School of Agriculture, University of Newcastle upon Tyne
BINES, Dr. J. A.	N.I.R.D., Shinfield, Reading, Berks
BLAXTER, Dr. K. L.	Rowett Research Institute, Aberdeen
BOAZ, T. G.	Dept. of Agriculture, University of Leeds
BOUCHARD, K. A.	Cooper Nutrition Products Ltd., Ilford, Essex
BOWIE, Dr. R. A.	I.C.I. Ltd., Alderley Park, Cheshire
BOWMAN, Prof. J. C.	Faculty of Agriculture, University of Reading, Berks.
BRADFIELD, P. G. E.	G.R.I., Hurley, Berks.
BRAMBELL, Prof. F.W. ROGERS	U.C.N.W., Bangor, Caernarvonshire
BRAUDE, Dr. R.	N.I.R.D., Shinfield, Reading
BROOME, Dr. A. W. J.	I.C.I. Ltd., Alderley Park, Cheshire
BROWNLIE, W. M.	Boots Pure Drug Co. Ltd., Thurgarton, Notts.
BURT, Dr. A. W. A.	Unilever Research Laboratory, Colworth House, Bedford
BUSH, T. J.	Unilever Ltd., Leatherhead, Surrey
BUTTERFIELD, Prof. R. M.	University of Sydney, Australia
CAMPBELL, C.	U.S. Feed Grains Council, London

501

CAMPLING, Dr. R. C.	Dept. of Agriculture, Wye College, Kent
CARROLL, M. A.	The Agricultural Institute, Dublin
CLARKE, J. N.	Ruakura Agricultural Research Centre, Hamilton, New Zealand
CONIFFE, D.	The Agricultural Institute, Dublin
COOPER, Prof. M. M.	School of Agriculture, Newcastle upon Tyne
CRANWELL, P. D.	Unilever Research Laboratory, Colworth House, Bedford
CROWLEY, Dr. J. P.	The Agricultural Institute, Dublin
CYRZEK, Dr. P.	University of Wroclaw, Poland
DE BOER, H.	'Schoonoord' Research Institute for Animal Husbandry, Zeist, The Netherlands
DOW, J. K. D.	Unilever Research Laboratory, Colworth House, Bedford
DRAPER, Dr. S. A.	School of Agriculture, University of Nottingham
ELSLEY, Dr. F. W. H.	Rowett Research Institute, Aberdeen
EVANS, A. J.	P.R.C., Kings Buildings, Edinburgh
EVANS, Miss E.	Queen Elizabeth College, University of London
EVERETT, W. H.	T. Wall & Sons Ltd., Willesden, London
EVERITT, Dr. G. C.	Ruakura Agricultural Research Centre, Hamilton, New Zealand
FALCONER, Dr. I.	School of Agriculture, University of Nottingham
FEVRIER, C. A.	C.N.R.Z., Jouy-en-Josas, France
FOWLER, Dr. V. R.	Rowett Research Institute, Aberdeen
FREDERICK, Dr. G. L.	Canada Dept. of Agriculture, Nappan, Nova Scotia
FRIEND, Dr. D. W.	Canada Dept. of Agriculture, Nappan, Nova Scotia
GRAHAM, G.	G. W. Plowman Ltd., Spalding, Lincs.
GROVES, T. W.	I.C.I. Ltd., Alderley Park, Cheshire
GREENWOOD, Dr. F. C.	Imperial Cancer Research Institute, London
GOSLING, Miss J. M.	Unilever Research Laboratory, Colworth House, Bedford
HALLEY, R. J.	Seale-Hayne Agricultural College, Newton Abbot, Devon
HARRINGTON, G.	Produce Studies Ltd., London, W.C.2
HARTE, Dr. F. J.	The Agricultural Institute, Dunsany, Co. Meath
HAYNES, Dr. N. B.	School of Agriculture, University of Nottingham
HEGARTY, Dr. P. V. J.	The Agricultural Institute, Dublin
HOLME, Dr. D. W.	Unilever Research Laboratory, Colworth House, Bedford
HOUSTON, T. W.	Procter Dept. of Food & Leather Science, University of Leeds
HOWELL, Prof. and Mrs. W. E.	Dept. of Animal Science, University of Saskatchewan

HUNTER, Dr. W. M.	M.R.C. Clinical Endocrinology Research Unit, Edinburgh
INGRAM, Dr. M.	Meat Research Institute, Bristol
JAGGERS, Dr. S. E.	I.C.I. Ltd., Alderley Park, Cheshire
JONES, D. R.	N.A.A.S., Wolverhampton
JONES, DR. A. S.	Rowett Research Institute, Aberdeen
LAMMING, Prof. G. E.	School of Agriculture, University of Nottingham
LERNER, J. T.	Estacion Experimental, Buenos Aires, Argentine
LEWIS, Prof. D.	School of Agriculture, University of Nottingham
LITTLE, R. B.	B.O.C.M., Hull, Yorks.
LOANE, Dr. D. J.	Unilever Research Laboratory, Colworth House, Bedford
LODGE, Dr. G. A.	School of Agriculture, University of Nottingham
LUCAS, Prof. I. A. M.	Dept. of Agriculture, U.C.N.W., Bangor
LYNCH, G.	The Agricultural Institute, Dublin
MARRABLE, Dr. A. W.	Dept. of Veterinary Anatomy, University of Bristol
MCCARTHY, Dr. J.	Faculty of Agriculture, University College, Dublin
MESSAGE, Dr. M. A.	Anatomy School, University of Cambridge
MILLS, Dr. C. F.	Rowett Research Institute, Aberdeen
MITCHELL, Dr. K. G.	N.I.R.D., Shinfield, Reading
O'GRADY, J. F.	The Agricultural Institute, Fermoy
OSIŃSKA, Prof. Z.	Institute of Animal Physiology and Nutrition, Jablonna, Warsaw
OWEN, Dr. E.	Unilever Research Laboratory, Colworth House, Bedford
PÁLSSON, Dr. and Mrs. H.	Agricultural Society of Iceland
PICKARD, D. W.	School of Agriculture, University of Nottingham
PIKE, I. H.	Dept. of Agriculture, University of Leeds
PRATT, P. D.	E.C.F. Ltd., Ware, Herts.
PORTER-SMITH, Miss	Unilever Research Laboratory, Colworth House, Bedford
POMMIER, R. E.	University of Reading, Berks.
PRESCOTT, Dr. J. H. D.	School of Agriculture, University of Newcastle upon Tyne
PRYOR, Miss M. J.	Institute of Animal Physiology, Babraham, Cambridge
RAY, Mrs. P. M.	P.I.D.A., London
REED, Dr. R.	Procter Dept. of Food & Leather Science, University of Leeds
RERAT, Dr. A. A.	C.N.R.Z., Jouy-en-Josas, France
ROY, Dr. J. H. B.	N.I.R.D., Shinfield, Reading
RUNCIE, K. V.	University of Edinburgh, School of Agriculture
SALMON-LEGAGNEUR, Dr. E.	C.N.R.Z., Jouy-en-Josas, France

SAMUEL, Miss P. D.	Queen Elizabeth College, University of London
SHAW, Dr. J. C.	Merck Sharpe & Dohm (Europe) Inc., Brussels
SHRIMPTON, Dr. D. H.	Unilever, Sharnbrook, Beds.
SISSONS, Dr. H. A.	The Institute of Orthopaedics, London
SMITH, R. J.	P.I.D.A., Stotfold, Herts.
SMITH, W. C.	School of Agriculture, University of Newcastle upon Tyne
SPEAKE, Dr. R. N.	I.C.I. Ltd., Alderley Park, Cheshire
SPEDDING, Dr. C. W. R.	G.R.I., Hurley, Berks.
SPEIGHT, D.	Nitrovit Ltd., Dalton, Yorks.
STOBO, Dr. I. J. F.	N.I.R.D., Shinfield, Reading
SWAN, H.	School of Agriculture, University of Nottingham
TALLACK, R. C. M.	Beef Recording Association Ltd., Reading, Berks.
TANNER, Prof. J. M.	Institute of Child Health, University of London
TAYLER, Dr. J. C.	G.R.I., Hurley, Berks.
TAYLOR, Dr. ST. C. S.	A.B.R.O., Edinburgh
THOMAS, J. G.	M.A.F.F., Leeds
THOMSON, D. J.	G.R.I., Hurley, Berks.
THURLEY, D. C.	Ministry of Agriculture, Central Veterinary Laboratory, Weybridge
TIMON, Dr. V. M.	The Agricultural Institute, Dublin
TODD, P. M.	Procter Dept. of Food & Leather Science, University of Leeds
TOPPS, Dr. J. H.	School of Agriculture, University of Aberdeen
TREACHER, T. T.	G.R.I., Hurley, Berks.
VIAL, V. E.	The Agricultural Institute, Dublin
VANSCHOUBROEK, Prof. F.	Veeartsonijschool, Ghent, Belgium
VANSPAENDONCK, R. L.	Veeartsonijschool, Ghent, Belgium
WALKER, Dr. N.	Agricultural Research Institute, Hillsborough, N. Ireland
WARNER, Dr. R. G.	Dept. of Animal Science, Cornell University, New York
WHITWORTH, Mrs. V. A.	Procter Dept. of Food & Leather Science, University of Leeds
WIDDOWSON, Dr. E. M.	Dunn Nutritional Laboratory, Cambridge
WILLIAMS, D. R.	Meat Research Institute, Cambridge
WILSON, Dr. P. N.	Unilever Research Laboratory, Colworth House, Bedford
WRIGHT, D.	N. Ireland Pig Testing Station, Antrim
WOOD, J. D.	University of Newcastle upon Tyne

AUTHOR INDEX

AUTHOR INDEX

507

SUBJECT INDEX